AF803138

Analytical Techniques in Biochemistry and Molecular Biology

Rajan Katoch

Analytical Techniques in Biochemistry and Molecular Biology

Springer

Rajan Katoch
Biochemistry Laboratory
Department of Crop Improvement
CSKHPKV
Palampur, HP
India
rajankatoch@yahoo.com

ISBN 978-1-4419-9784-5 e-ISBN 978-1-4419-9785-2
DOI 10.1007/978-1-4419-9785-2
Springer New York Dordrecht Heidelberg London

Library of Congress Control Number: 2011931968

© Springer Science+Business Media, LLC 2011
All rights reserved. This work may not be translated or copied in whole or in part without the written permission of the publisher (Springer Science+Business Media, LLC, 233 Spring Street, New York, NY 10013, USA), except for brief excerpts in connection with reviews or scholarly analysis. Use in connection with any form of information storage and retrieval, electronic adaptation, computer software, or by similar or dissimilar methodology now known or hereafter developed is forbidden.
The use in this publication of trade names, trademarks, service marks, and similar terms, even if they are not identified as such, is not to be taken as an expression of opinion as to whether or not they are subject to proprietary rights.

Printed on acid-free paper

Springer is part of Springer Science+Business Media (www.springer.com)

Foreword

Biochemistry is one of the fundamental subjects in life sciences. Biological systems of virtually all sorts can be controlled in ways not thought possible as recently as a decade ago. Different disciplines are now being translated through common language of biochemistry. The twenty-first century is going to be an era of Plant Biology with emphasis on Biochemistry and Biotechnology. These disciplines are still expected to answer some of the most urgent questions in the discipline of life sciences. Basic knowledge has been linked, wherever appropriate, to the applicability of that knowledge for understanding various physiological processes. Inter-relatedness and regulations of biological processes are emphasized all through. In fact, this is a work of trans-disciplinary synthesis where excellence and relevance are sought to be combined. For successful researchers, it is necessary to acquire new skills and knowledge in the fields of biochemistry. Therefore, there was a dire need for such a book which would provide various biochemical and biotechnological procedures which are in frequent use in modern research.

Dr. Rajan Katoch must be congratulated for producing what I consider a truly monumental and unique work, thereby rendering a most valuable service to plant sciences and in particular to biochemistry. The book is targeted to all students of plant sciences. Scientists and researchers are to be benefitted by this book for intelligent and enlightened teaching and research work. I hope that the present book "Techniques in biochemistry and Molecular Biology" will equip students and teachers alike with the present day concept of understanding of biochemistry.

Dr. S.K. Sharma
Vice-Chancellor

Preface

In the recent years we have seen a remarkable increase of interest in biochemical and molecular methods for the elucidation of structural and functional relationship among different physiological processes. Science and its application to biochemistry today are facing the greatest opportunities. Detailed mechanistic knowledge resulting from the application of biochemical methods combined with numerous interdisciplinary techniques has aided the understanding of biological processes. Now a variety of new, faster and sensitive methods have enabled us for the examination of vital processes at biochemical and molecular level. The present day developments in the field of biochemistry and biotechnology have been made possible by isolation and purification of numerous enzymes, understanding nucleic acid metabolism, by refinement of existing techniques and the development of new ones. Biochemistry can and does contribute to the understanding and solutions of problems involved in many of the more specialized aspects of plant biology.

The aim of bringing out this book was not to produce a comprehensive text, but a general and wherever necessary, simplified methods for postgraduate students and researchers who have recourse to use various techniques during their research programmes. The book is mainly written for postgraduate students, researchers, lecturers and scientists in biochemistry and biotechnology. The methods described are also important components of courses in microbiology, genetics, plant physiology, etc. The text is designed, therefore, to bring researchers of different disciplines to a level of competence in biochemistry and molecular biology. I believe that this book can help students and research workers in these diverse fields by providing them with ready source of biochemical information directly applicable to plant sciences.

The book has been divided into 15 different chapters covering different aspects. The first three chapters of the book deal with fundamental aspects which are necessary for conducting any biological experimentation. The later part of the book deals with various advanced techniques in biochemistry. Chapter 1 introduces the readers to the concept of preparation of different solutions which are indispensible part of any experiment. Chapter 2 presents important fundamentals of expression of concentration of solutions. Buffers are important for any enzymatic work and the concept of buffers and their preparation have been described in the Chap. 3. Chapter 4 introduces the readers to different techniques used during biochemical analysis. Carbohydrates and their estimations have been described in Chap. 5. Chapters 6 and 7 deal with the estimation of lipids, amino acids and proteins. Protein purification techniques, cell disruption techniques, estimation of enzymes and isozyme analysis have been described in Chaps. 8–11. Chapter 12 concentrates on various chromatographic separation procedures. Chapter 13 deals exclusively with the nutritional and anti-nutritional evaluation of different food stuffs, whereas the methodology for the nutritional evaluation of forages has been described in Chap. 14. Different biotechnological procedures which are in vogue in modern research have been described in Chap. 15. Thus, every effort has been made to cover the different topics which are used in biochemistry and related disciplines. I hope that the reader, after having studied the concepts and methodology will find himself quantified to go to the laboratory and start investigating process applicable to his field.

I am indebted to my family for untiring support that helped me in bringing out this manuscript.

Palampur, HP, India Rajan Katoch

Contents

Chapter 1
Preparation of Solutions

The techniques in biochemistry generally involve the use of solutions. These solutions have different uses, e.g. extraction of biomolecules from tissues and their separation, quantitative estimation, purification and characterization of different compounds. Different kinds of solutions are integral part of molecular biology techniques. Therefore, it is important that one must be fully familiar with the proper preparation of solutions. A solution is homogeneous mixture of two or more non-reacting substances and has uniform properties such as chemical composition, density, refractive index, etc. However, its composition can be varied within certain fixed limits. A solution which is made up of two components is called binary solution. The dissolved substance in solution is called the solute and the medium in which it is dissolved is known as the solvent. The solute in a solution is always present in a smaller quantity than the solvent. The most commonly used solutions in biochemical work are of solid–liquid and liquid–liquid type.

Composition of a solution can be expressed in two ways: quantity and concentration. Quantity is the amount of any substance (solute) present in a solution/solvent irrespective of the amount/volume of the solvent/solution. In contrast, concentration refers to the quantity of the solute present in an exact or a specific amount of solvent or that of solution. For example, a solution containing 4 g NaOH in a volume of 100 mL of a solution has a total quantity of 4 g NaOH. When this amount is expressed per unit volume such as 4 g/100 mL of the solution, then it is termed as 4% (concentration) solution.

1.1 Types of Solutions

1.1.1 Standard Solutions

A standard solution is one that contains a precisely known concentration of solute. We express the concentration in terms of molality, molarity and normality. Sometimes the concentrations are also expressed as % solution and in ppm (parts per million)

R. Katoch, *Analytical Techniques in Biochemistry and Molecular Biology*,
DOI 10.1007/978-1-4419-9785-2_1, © Springer Science+Business Media, LLC 2011

Molality of standard solution (m): The molality of solution is defined as the number of moles of solute dissolved in 1,000 g of the solvent. We can make 1 molal solution by weighing out precisely 1 mole of the substance and then dissolving it in 1,000 g solvent. A 0.5 molal solution contains one half mole of solute per 1,000 g of solvent. A 2 M solution has 2 moles of solute in 1,000 g of solvent. This term has no relevance with the volume.

Molarity of standard solution (M): The molarity of a solution is defined as the number of moles of solute present per litre of the solution. The standard solution of 1 M concentration contains 1 mole of solute per litre of solution. The gram-formula weight of NaCl is 58.5 g. Weigh out this quantity of NaCI and dissolve it in water and make its volume to 1 L in a volumetric flask of 1,000 mL capacity. This will be called 1 M solution of NaCI. If we use only one half mole i.e. 29.25 g in 1 L of solution, our solution would be 0.5 M. Using 2 moles or 117 g in 1 L of solution makes a 2 M solution.

Normality of standard solution (N): The normality of solution is defined as the number of gram equivalent of substance present/litre of the solution. 1 N solution contains 1 GEW of substance per litre of solution. In the case of acids it is the number of gram equivalent of H^+ present in 1 L of the solution. Thus, 1 N acid solution contains 1 GEW of H^+ per litre of solution. To prepare a litre of 0.1 N HCl (the mol. wt. of HCl is 36.5 g) only 0.1 gEq of HCl per litre of solution is required. Hence, use 3.65 g of HCl diluted to 1 L. But this is 3.65 g of the anhydrous hydrogen chloride in one 1 L of solution, not 3.65 g of the concentrated hydrochloric acid in bottle of the laboratory. How may we determine the volume of concentrated hydrochloric acid to be measured out which contain 3.64 g of hydrogen chloride? This may be found very easily as the assay information is printed on the manufacture's label on the bottle of concentration HCl.

The concentrated acid is generally 37.23% hydrogen chloride by weight and has a specific gravity of 1.19; it means 1 mL of conc. HCl weighs 1.19 g of which 37.23% is hydrogen chloride.

1 mL conc. HCl contains: 0.3723×1.19 g $= 0.443$ g of hydrogen chloride.

Thus, the volume of solution needed to provide 3.65 g of HCl is 3.65/0.443 g/mL $=$ 8.43 mL of conc. HCl.

If 8.43 mL of conc. HCl is made to 1 L with distilled water with earlier mentioned information, it will make 0.1 N HCl.

We know that 1 mole (98 g) of H_2SO_4 contains 2 gEq of H^+ and is a dibasic acid. Therefore, 1 N H_2SO_4 contains (98/2) or 49 g of H_2SO_4 per litre of solution. A 5 N solution contains 245 g (49×5) of H_2SO_4 per litre, and 0.01 N H_2SO_4 contains 0.49 g (49/100) H_2SO_4 per litre of solution. Concentrated sulphuric acid is usually 95–98% and has a specific gravity of about 1.84. (for details see appendix. Reagent 1).

Percentage of standard solution (%): Parts of solute dissolved to form hundred parts of solution is termed as % solution. One gram of any compound dissolved in to form 100 mL of solution is termed as 1% solution and 2 g of NaCI dissolved in water to make to a final volume of 100 mL is a 2% solution.

Standard solution of known concentration in parts per million (ppm): Parts of solute dissolved to form million parts of solution is termed as a ppm solution. In a simple way 1 g dissolved in. 10,00,000 g or in other words 1 mg in 1 L makes 1 ppm solution.

Standard solution of hydrated salts: Crystalline salts containing water of hydration must be given special consideration in making up standard solutions. For example, crystalline cupric sulphate has the empirical formula $CuSO_4{\cdot}5H_2O$, a 1 M $CuSO_4$ solution. However, the formula weight of this hydrated salt is 249.5. This fact has to be taken into consideration when weighing out moles or gram equivalent of such hydrated salts i.e. 249.9 g of salt is to be weighed for making 1 M solution 1 L instead of 159.5 g.

$$\text{One millimolar or } 1\,\text{mM} = 1/1,000\,\text{M} = 0.001\,\text{M} = 1 \times 10^{-3}\,\text{M}$$

$$\text{One micromolar or } 1\,\mu\text{M} = 1/10,00,000\,\text{M} = 1 \times 10^{-6}\,\text{M}$$

Where N_1 and V_1 are normality of a solution and N_2 and V_2 are normality and volume of the solution to be prepared.

1.1.2 *Stock Solution*

Stock solution of a substance is the one having a concentration many folds higher than that actually required in the experiment. Stock solutions are prepared for the substances that are to be used frequently and are stable at higher concentration for several days and can be used after appropriate dilution just before use. It is sometimes convenient to weigh out a relatively large amount of the compound and prepare a stock solution of that compound from which small amounts can be withdrawn at convenience and added to solution of other components. The use of a stock solution thus cuts down on the amount of pipetting and at the same time reduces variability between a number of similar incubation mixtures, assay mixture etc. A particular volume of solution containing a desired concentration of a substance has to be prepared by using its stock solution. A simple mathematical equation can be used to calculate the volume of the stock solution needed to prepare a solution of required volume containing desired concentration of the compound:

$$N_1V_1 = N_2V_2$$

Where

N_1 = concentration of the solution to b prepared
V_1 = volume of the solution to be prepared
N_2 = concentration of the stock solution
V_2 = volume of the stock solution.

For example, for preparing 200 mL of a solution containing 15 mM Tris from a stock of 80 mM Tris, the volume of the stock solution which should be used can be calculated as:

$$V_2 = \frac{N_1 V1}{N2} = \frac{10 \times 200}{80} = 25\,\text{mL}$$

Therefore, by taking 25 mL of 80 mM stock solution of Tris and making its final volume to 200 mL (by adding water or other solutions) a solution containing 10 mM Tris is obtained.

1.1.3 Saturated Solution

Solubility of any substance in a particular medium varies with temperature. When a solution contains the solute in an amount in excess of that which can completely be dissolved at a given temperature and the solute in solution is in equilibrium with the excess of undissolved solute, the solution is said to be saturated.

1.1.4 Solution of Acids

The commonly used acids like hydrochloric acid and sulphuric acid are not pure and their effective strength is low which is to be taken into account while making their solutions of desired concentrations. The following formula can be applied to compute the volume of desired concentrations. The following formula can be applied to calculate the volume of the commercially available concentrated acid required for preparing a solution of required normality:

$$V_1 = \frac{\text{Eq. wt. of acid} \times V_2 \times \text{normality} \times 100}{1,000 \times \text{specific gravity} \times \text{purity } (\%)}$$

Where

V_1 = required volume of the concentrated acid
V_2 = total volume of the acid solution of desired normality to be prepared.

The use of above formula could be explained by considering the following example.

Prepare 1,000 mL of 1 N sulphuric acid from concentrated acid (purity, 96%; specific gravity, 1.84; Mol. wt., 98).

Eq. wt. of sulphuric acid = 49 (since sulphuric acid is disprotic).

Putting various values in the above equation, we get

$$V_1 = \frac{49 \times 1,000 \times 1 \times 100}{100 \times 1.84 \times 96} = 27.7\,\text{mL}$$

Thus, take 27.7 mL of concentrated sulphuric acid and dilute it to 1,000 mL with water to get 1 N solution of sulphuric acid.

1.1.5 General Precautions

1.1.5.1 Cleaning of Glassware

1. Ordinary glassware should be thoroughly cleaned with washing soda or any detergent followed by washing with ordinary tap water and rinsing with distilled water. Care should be taken to remove previous markings on the glassware, if any while cleaning.
2. Cleaning of oil flasks, used while estimating ether extract, should be done by slight boiling with dilute alkali (NaOH) followed by same procedure adopted for cleaning ordinary glassware. Care, however, should be taken not to use any brush for cleaning inside the flask to avoid scratch formations.
3. Graduated glassware may be cleaned by initially keeping in chromic acid solution (dissolve about 60 g potassium dichromate in 300 mL tap water by thorough stirring and boiling, to which around 450 mL of commercial sulphuric acid is added slowly after cooling) kept in cylindrical jar for reasonable time followed by washing and cleaning as per ordinary glassware. Discard chromic solution when it develops green colour.

1.1.5.2 Drying of Glassware

1. Ordinary glassware can be dried by keeping in hot air oven at low temperatures.
2. Graduated glassware (pipettes, burettes, measuring cylinders, volumetric flasks etc.) should never be dried in hot air oven as high temperature would change the volume for which they are graduated for.

1.1.5.3 Other Laboratory Precautions

1. Laboratory floor, working tables and water sinks should be kept neat and clean and it should be well ventilated and provided with an exhaust fan to remove unwanted gases, fumes and smoke.
2. One should work fully protected in laboratory by wearing white drill aprons and shoes.
3. Store chemicals and glassware in alphabetical order in well protected cupboards.
4. Reagents should be properly labelled with date of preparation before placing them on shelf.
5. Systematic breakage record should be maintained.

6. Always use acid and alkali gloves while handling strong acids and alkalies.
7. While working in Kjeldahl digestion room, use fume protecting face mask to avoid inhalation of highly irritating sulphur dioxide fumes.
8. Distilled water bottles should be kept tightly corked to avoid absorption of atmospheric gases.
9. Always add acid to water slowly from the sides of the container near the sink.
10. While opening liquor ammonia bottles, especially during summer season, cool it for some time in a freezer to avoid sudden spurt of ammonia gas accumulated in the bottles.
11. Set the balance and check the oscillation of pans before using analytical balances. After use the pan and platform of the balance should be cleaned with a camel hair brush, in case of any spillage of chemical, sample etc. during weighing.
12. Proper record of usage of special equipment should be made in log book meant for it showing date, time and condition of the equipment.
13. Fire extinguishers should be provided in each laboratory.
14. Always work in company during odd hours for the fear of fire and electrical accidents.
15. All used filter papers and other materials should be deposited in water baskets.
16. Smoking in laboratory premises should be restricted to avoid catching accidental firing by highly inflammable chemicals.
17. All observations should be recorded at least in duplicate.
18. Never pipette strong acids and alkalies with mouth. Always use adopter or rubber bulb or bulb pipette.
19. Never blow the solution left at the tip of the pipette and delivery of the reagent drawn into pipette should be uniform giving appropriate time, varying from 10 to 30 s. for quantities of 2–50 mL.
20. Mouth should be washed quickly with water or weak solution of washing soda during accidental sucking of acids.
21. Acid and alkali spillage on working tables, floor and clothes should be thoroughly washed with water after suitable neutralising with either weak alkali in case of acid and weak acid in case of alkali.
22. Use always glass distilled water while analysing minerals.
23. Commercial sulphuric acid should only be used for digestion of samples for nitrogen/protein estimation.
24. No need to use distilled water while making up the volume of digested sample for protein estimation and during the analysis of crude fibre.
25. Always use self prepared reagents and indicators.
26. Consider lower meniscus for clear colourless and upper meniscus for coloured solutions while recording observations with the help of measuring glassware.
27. During cooling samples in a desiccator, the lid should be displaced to leave a small space, which can be closed after complete cooling.

28. Do not turn on fans during decarbonization of a sample for ashing. Turn on the exhaust during decarbonization and while handling fuming acids and other chemicals.
29. Always keep decarbonized samples in a closed container like desiccator while carrying to muffle, otherwise due to light weight the material in silica basin may be displaced due to external air movement.

Chapter 2
Expression of Concentration

Standard solution: A standard solution is one that contains a precisely known concentration of solute. We express the concentration in terms of molality, molarity and normality. Sometimes the concentrations are also expressed as % solution and in ppm (parts per million). This section explains the common ways of expressing the concentration of solutions that are required for conducting different experiments.

2.1 Molarity (M)

This is the most common method for expressing the concentration of a solution in biochemical studies. The molarity of a solution is the number of moles of the solute dissolved per litre of the solution. A solution which contains 1 mole of the solute in 1 L of the solution is called a molar solution. Molarity of a solution can be calculated as follows:

$$\text{Molarity} = \frac{\text{Wt. of a solute in g/L of solution}}{\text{Mol. wt. of solute}}$$

It may be noted that in case of molar solutions, the combined total volume of the solute and solvent is 1 L. Therefore, for preparing 0.1 M NaOH, one may proceed as follows:

$$\text{MW of NaOH} = 40$$
$$\text{Required molarity of solution} = 0.1\ \text{M}$$

$$\begin{aligned}\text{Amount of NaOH (g) per litre of solution} &= \text{MW of NaOH} \times \text{molarity}\\ &= 40 \times 0.1 = 4\,\text{g}\end{aligned}$$

R. Katoch, *Analytical Techniques in Biochemistry and Molecular Biology*,
DOI 10.1007/978-1-4419-9785-2_2, © Springer Science+Business Media, LLC 2011

Weigh 4 g of NaOH, dissolve it in a small volume of solvent (water) and make the final volume to 1 L with the solvent. The number of moles of a substance in a reaction mixture can be calculated using a simple relationship:

$$\begin{aligned} 1\,\text{M solution} &= 1\,\text{mole of the substance/L of solution} \\ &= 1\,\text{mmole/mL of solution} \\ &= 1\,\mu\text{mole}/\mu\text{L of solution} \\ 1\,\text{mM solution} &= 1\,\text{mmole/L of solution} \\ &= 1\,\mu\text{mole/mL of solution} \end{aligned}$$

2.2 Molality (m)

A solution which contains 1 mole of the solute dissolved in 1 kg of the solvent is called a molal solution. Hence,

$$\text{Molality} = \frac{\text{Weight of a solute in g/kg of solvent}}{\text{Mol. wt. of solute}}$$

It is important to remember that in a molal solution, the amount of solvent is 1,000 g. Thus, in case of aqueous solutions, 1 molal solution is obtained by dissolving 1 mole of the solute in 1,000 mL (since Sp. gravity = 1) of water. For example, for preparing 1 m Na_2CO_3 solution, dissolve 106 g Na_2CO_3 (Mol. Wt. of Na_2CO_3 = 106) in 1 kg of water.

2.3 Normality (N)

The normality of a solution is the number of gram equivalents of the solute per litre of the solution.

A solution having 1 gEq of the solute per litre of solution is called 1 N solution. Therefore,

$$\text{Normality} = \frac{\text{Amount of a substance in g/L of solvent}}{\text{Eq. wt. of solute}}$$

For preparing 0.1 N Na_2CO_3 (Eq. wt. of Na_2CO_3 = 53) solution, dissolve 5.3 g Na_2CO_3 in a final volume of 1 L of solution.

Therefore, the normality of solution is defined as the number of gram equivalent of substance present per litre of the solution. 1N solution contains 1 GEW of substance per litre of solution. In the case of acids, it is the number of gram equivalent of H^+ present in 1 L of the solution. Thus, 1 N acid solution contains 1 GEW of H^+ per litre of solution. To prepare a litre of 0.1 N HCl, we know that the mol. wt. of HCl is 36.5 g. We want to have only 0.1 gEq of HCl per litre of solution. Therefore, we need to use 3.65 g of HCl diluted to 1 L. But this is 3.65 g of the anhydrous hydrogen chloride in one 1 L of solution, not 3.65 g of the concentrated hydrochloric acid in bottle in the laboratory.

If the given concentrated acid is 37.23% hydrogen chloride by weight and has a specific gravity of 1.19, it means 1 mL of conc. HCl weight 1.19 g of which 37.23% is hydrogen chloride. Or 1 mL conc. HCl contains:

$$0.3723 \times 1.19\,\text{g} = 0.443\,\text{g of hydrogen chloride}$$

Thus, the volume of solution needed to provide 3.65 g of HCl is

$$3.65/0.443\,\text{g/mL} = 8.43\,\text{mL of conc. HCl}$$

If 8.43 mL of conc. HCl is made to 1 L with distilled water with above-mentioned informations, it will make 0.1 N HCl.

Therefore, 1 mole (98 g) of H_2SO_4 contains 2 gEq of H^+ and is a dibasic acid. So, 1 N H_2SO_4 contains (98/2) or 49 g of H_2SO_4 per litre of solution. A 5N solution contains 245 g (49 × 5) of H_2SO_4 per litre, and 0.01 N H_2SO_4 contains 0.49 g (49/100) H_2SO_4 per litre of solution. Concentrated sulphuric acid is usually 95–98% and has a specific gravity of about 1.84.

2.4 Mass Concentration

Generally the substances like proteins, nucleic acids, etc. which do not have a uniformly defined composition, their concentration is expressed in terms of weight per unit volume rather than moles per unit volumes. The unit of volume is litre, so all concentrations should be expressed with L (g/L, mg/L, μg/L, etc.). The term percent (%) is also quite commonly used. To avoid any ambiguity it is necessary to properly define the basis of % solution as illustrated by the following example. A 2% solution of acetic acid could mean:

2 g of acetic acid per 100 g of solution (w/w)
2 g of acetic acid per 100 mL of solution (w/v)
2 mL of acetic acid per 100 mL of solution (v/v)

Thus, 1% (w/v) solution of trypsin would imply 1 g of trypsin dissolved in solvent to give a final volume of 100 mL of the solution.

Percentage of standard solution (%): Parts of solute dissolved to form hundred parts of solution is termed as % solution. One gram of any compound dissolved in to form 100 mL of solution is termed as 1% solution and 2 g of NaCl dissolved in water to make to a final volume of 100 mL is a 2% solution.

Standard solution of known concentration in parts per million (ppm): Parts of solute dissolved to form million parts of solution is termed as a ppm solution. In a simple way 1 g dissolved in. 10,00,000 g or in other words 1 mg in 1 L makes 1 ppm solution.

Standard solution of hydrated salts: Crystalline salts containing water of hydration must be given special consideration in making up standard solutions. For example, crystalline cupric sulphate has the empirical formula $CuSO_4{\cdot}5H_2O$. A 1 M $CuSO_4$ solution. However, the formula weight of this hydrated salt is 249.5. This fact has to be taken into consideration when weighing out moles or gram-equivalent of such hydrated salts, i.e. 249.9 g of salt is to be weighed for making 1 M solution of 1 L instead of 159.5 g.

2.5 Mass Fraction

The mass fraction of a component in solution is the mass of that particular component per unit mass of the solution. If W_A and W_B represent the masses of components A and B in a solution, then

$$\text{Mass fraction of A } (W_A) = \frac{W_A}{W_A + W_B}$$

$$\text{Mass fraction of B } (W_B) = \frac{W_B}{W_A + W_B}$$

2.6 Mass Percent % (w/w)

It is defined as the weight of the component present in 100 parts by weight of the solution. In a solution containing 10 g of sugar in 40 g of water, then

$$\text{Mass \% of sugar} = \frac{10 \times 100}{(10 + 40)} = 20\%$$

2.7 Percentage by Volume or % (v/v)

It is the volume of component in 100 parts by volume of the solution. In a solution containing 20 mL alcohol in 80 mL of water, the % volume of alcohol will be

$$\frac{20 \times 100}{(20 + 80)} = 20\%$$

2.8 Parts per Million (ppm)

This is generally employed for those solutions in which a substance is present in a very small quantity. It represents gram of a solute per million grams of solution of the gram of a solute per million millilitre of the solution.

$$\text{ppm} = \frac{\text{mass of the component}}{\text{total mass of solution}} \times 10^6$$

or

$$\text{ppm} = \frac{\text{g or mL of solute or substance}}{\text{g or mL of solution}} \times 10^6$$

thus, 1 ppm of solution KCl in water represents

$$1\,\text{ppm} = 1\,\text{mg KCl/L of solution}$$

or $$= 1\,\text{mg KCl}/1,000\,\text{mL of solution}$$

or $$= 1\,\mu\text{g KCl/mL of solution}$$

It can be elaborated for better understanding as follows:

A million is 10^6. So one part of 10^6 is 1 ppm.
For example:

$$1\,\text{kg} = 1,000\,\text{g}$$
$$1\,\text{g} = 1,000\,\text{mg}$$
$$\text{Therefore, } 1\,\text{kg} = 1,000 \times 1,000 \text{ or } 10^6\,\text{mg}$$

1 mg of a substance dissolved in 1 kg or 1,000 mL water (taking density of water is 1 g/mL) then the concentration of the resulting solution in 1 ppm. Therefore, ppm is 10^{-6}th part of the amount of water used to prepare a solution.

2.8.1 Preparation of 1,000 ppm Solutions

(a) From pure metal
Weigh out accurately 1,000 g of pure metal. Dissolve it in concentrated nitric acid or hydrochloric acid (concentrated acid: one part water + one part acid), and make up to the mark in 1 L measuring flask with deionised/double distilled water.

From a salt

(b) To prepare a 1,000 ppm solution say standard for sodium (Na) using the sodium chloride (NaCl)

Formula weight (FW) of NaCl; (atomic weight of sodium = 23 and atomic weight of Cl = 35.44); hence, the formula weight is 58.44 g. It means that to have 23 g of sodium we need to have 58.44 g NaCl. Therefore, to have 1,000 mg or 1 g Na weight 58.44/23 = 2.542 g NaCl and dissolve in 1 L volume of water to make a 1,000 ppm Na standard.

From an acidic radical of salt

(c) To prepare 1,000 ppm phosphate standard using salt KH_2PO_4
FW of KH_2PO_4 = 136.09; FW of radical PO_4^{3-} w.r.t. FW KH_2PO_4 salt 136.09/95 = 1.432 g.

Weigh 1.432 g KH_2PO_4 and dissolve in 1 L volume of water to make a 1,000 ppm PO_4 standard.

2.9 Parts per Billion (ppb)

Parts per billion (ppb) = ng/mL,

Thereby, it is 10^{-9}th part of the amount of water used to prepare a solution.

2.10 Preparation of 10^{-1}–10^{-10} M Solutions from 1 M Stock Solution by Serial Dilutions

Requirement: Ten measuring flasks of 100 mL capacity (Table 2.1).

Table 2.1 Preparation of solutions by serial dilutions

10^{-1} M	Pipette out 10 mL of the 1 M stock solution into the flask and make the volume up to the mark with water
10^{-2} M	Pipette out 10 mL of the 10^{-1} M solution into the flask and make the volume up to the mark with water
10^{-3} M	Pipette out 10 mL of the 10^{-2} M solution into the flask and make the volume up to the mark with water
10^{-4} M	Pipette out 10 mL of the 10^{-3} M solution into the flask and make the volume up to the mark with water
10^{-5} M	Pipette out 10 mL of the 10^{-4} M solution into the flask and make the volume up to the mark with water
10^{-6} M	Pipette out 10 mL of the 10^{-5} M solution into the flask and make the volume up to the mark with water.
10^{-7} M	Pipette out 10 mL of the 10^{-6} M solution into the flask and make the volume up to the mark with water
10^{-8} M	Pipette out 10 mL of the 10^{-7} M solution into the flask and make the volume up to the mark with water
10^{-9} M	Pipette out 10 mL of the 10^{-8} M solution into the flask and make the volume up to the mark with water
10^{-10} M	Pipette out 10 mL of the 10^{-9} M solution into the flask and make the volume up to the mark with water

2.11 Conversion of Molarities into ppm

The atomic weight of a simple ionic species is equal to its concentration in ppm at 10^{-3} M.

This can be explained as under:

The atomic weight of fluoride is 19; hence, a 10^{-3} M solution of a fluoride salt contains 19 ppm fluoride. 1 M solution is, therefore, equivalent to 19,000 ppm fluoride.

The atomic weight of Ca = 40; hence, a 10^{-3} M solution of $CaCO_3$ contains 40 ppm calcium and 10^{-2} M solution is equivalent to 400 ppm of calcium.

Or

Atomic weight of Ca = 0 and 1 M = 40 g or 40,000 mg = 40,000 ppm; hence, 10^{-2} M solution is equivalent to $40{,}000 \times 10^{-2}$ ppm = 400 ppm.

2.12 Conversion of ppm to Parts per Hundred (%) (ppm $\times$ 10^{-6} $\times$ 100)

See Table 2.2.

2.13 Preparation of One Molar (1 M) Solutions of Concentrated Acids

See Table 2.3.

Table 2.2 Conversion of ppm to parts per hundred

1 ppm =	1/1,000,000 =	0.000001 =	0.00%
10 ppm =	10/1,000,000 =	0.00001 =	0.00%
100 ppm =	100/1,000,000 =	0.0001 =	0.01%
200 ppm =	200/1,000,000 =	0.0002 =	0.02%
5,000 ppm =	5,000/1,000,000 =	0.005 =	0.50%
10,000 ppm =	10,000/1,000,000 =	0.01 =	1.00%
20,000 ppm =	20,000/1,000,000 =	0.02 =	2.00%

Table 2.3 Preparation of 1M solutions of concentrated acids

Acids	Molarity	Percent	S.G.	Volume of concentrated acid per litre
Acetic acid (CH_3COOH)				
MW = 60.05	17.4 M	99	1.05	58 mL
Hydrochloric acid (HCl)				
MW = 36.46	11.6 M	36	1.18	86 mL
Nitric acid (HNO_3)				
MW = 63.01	16.4 M	69	1.42	61 mL
Sulphuric acid (H_2SO_4)	17.6 M			56 mL/1 (1 M)
	36 N	95	1.84	28 mL/(1 N)
Hydrobromic acid (HBr)				
MW = 20.01	27 M	48	1.49	11.1 mL
Hydriodic acid (HI)				
MW = 127.90	7 M	57	1.70	143 mL
Perchloric acid ($HClO_4$)	11.7 M	70	1.71	85.5 mL
MW = 100.46	9.5 M	61	1.66	106 mL
Phosphoric acid (H_3PO_4)	45 N			22.5 mL (1 N)
MW = 97.99	15 M	88	1.69	68 mL (1 M)
Ammonia solution	14.5 M			
MW = 35.05	15 N	28	0.88	71 mL (1 M)

2.14 Formula to Calculate the Volume of Stock Solution Required to Prepare Solution of Desired Normality

$$\frac{100 \times \text{desired normality} \times \text{molecular weight} \times \text{desired volume}}{1{,}000 \times \text{Basicity} \times \text{percentage of stock solution} \times \text{specific gravity of stock solution}}$$

2.15 Formula to Calculate Volume of Stock Solution Required to Prepare Solution of Desired Percentage

$$\frac{\text{Desired percentage} \times \text{desired volume}}{\text{Percentage of stock solution}}$$

(Where the percentage is weight/volume)

2.16 Formula to Calculate the Dilution Factor Used in Spectroscopic Estimations

$$\frac{\text{Absorbance} \times \text{dilution factor}}{1{,}000}$$

(Where concentration is per litre)

Calculation of dilution factor

$$\frac{1 \times \text{Initial volume} \times \text{volume made for spetrophotometric analysis}}{\text{Weight of sample} \times \text{initial volume taken for dilution}}$$

Example

Weight of sample taken = 5 g
Initial volume made = 50 mL
Initial volume taken for dilution = 1 mL
Volume made for spectrophotometeric analysis = 10 mL

$$\text{Dilution factor} = \frac{1 \times 50\,\text{mL} \times 10\,\text{m}}{5\,\text{g} \times 1\,\text{mL}}$$

2.17 Preparation of Standard Solution of Acids and Bases

The preparation of standard acid solution, viz. H_2SO_4, HCl, HNO_3 and alkalis, viz. KOH, NaOH have difficulty because they cannot be pipetted/weighed very accurately. Moreover alkalis gather moisture while weighing and assay percentage of acids varies. Therefore, oxalic acid, which can be accurately weighed is used as primary standard to standardize NaOH which then is used to standardize acids using indicators as end points. If the strength of one component is known exactly, the strength of the other can be experimentally found by using the under mentioned equation.

$$N_1 \times V_1 = N_2 \times N_2$$

N and V indicate normality and volume of acids and bases respectively.

Reagents: 0.1 N H_2SO_4, 0.1 N HCl, 0.1 N HNO_3, 0.1 N NaOH, 0.1 N oxalic acid, 1% phenolphthalein.

Glassware/equipments: Conical flask (250 mL), burette (25 mL), burette stand, volumetric flask and beakers, funnel, droppers.

Table 2.4 Common solutions used in analytical procedures

Primary standards	*Per cent solutions*
N/7 Na_2CO_3	2% HNO_3 (v/v)
N/10 Oxalic acid	3% KNO_3 (w/v)
2.04 N Na_2CO_3	2% Boric acid (w/v)
	40% NaOH (w/v)
	20% Ammonium molybdate (w/v)
Secondary standards	*Ratio solutions (v/v)*
N/7 H_2SO_4	1:4 H_2SO_4
N/7 HNO_3	1:2 HCl
N/7 NaOH	1:4 Liquor ammonia
N/10 $KMnO_4$	Saturated solution
2.04 N H_2SO_4	Saturated ammonium oxalate
2.50 N NaOH	

Procedure: Standardization of NaOH: Pipette out 10 mL of NaOH into a 250 mL conical flask. Add a drop of indicator phenolphthalein. Fill the burette with 0.1 N oxalic acid and set at zero mark. Titrate till pink colour changes to colourless and note the volume of acid used. Repeat the same process thrice.

Observation: Volume of 0.1 N oxalic used to neutralize 10 mL of NaOH

S.No.	Initial reading	Final reading	Volume used (mL)
1			
2			
3			
Mean volume = a			

Calculation of normality of NaOH
Applying

$$N_1 \times V_1 = N_2 \times V_2$$
$$0.1 \times a = N_2 \times 10$$
$$N_2 = 0.1 \times a/10 \quad \text{say it} = x$$

The primary and secondary standardized solutions of known normally along with different ratio and percentage solutions required for the analysis of various proximate principles are given in Table 2.4.

2.17.1 Normal Solutions

2.17.1.1 Primary Standards

Primary standards are used for the standardization of solutions of unknown strength.

Preparation of N/7 Na_2CO_3

Reagent

- Anhydrous sodium carbonate.
- As sodium carbonate is highly hygroscopic, dry it in hot air oven at 100 ± 2°C for 3–4 h to make it anhydrous.

Calculations

Molecular weight of Na_2CO_3 = 105.989;
i.e. [(2 × 22.9898) + 12.0112 + (3 × 15.9994)]

$$\text{Equivalent weight} = \frac{\text{Molecular weight}}{\text{Acidity}} = \frac{105.989}{2} = 52.9945$$

(Acidity of Na_2CO_3 = 2, based on the reaction,
$Na_2CO_3 + 2HCl \rightarrow 2NaCl + H_2O + CO_2$)
Gram-equivalent weight = 52.9945 g

Therefore,
1,000 mL of 1 N Na_2CO_3 contains 52.9945 g
(or) 1,000 mL of N/7 Na_2CO_3 contains 7.5706 g

Procedure

To prepare 250 mL of N/7 Na_2CO_3, weigh exactly 1.8927 g of oven-dried and desiccator-cooled anhydrous Na_2CO_3 in a clean dry beaker. Dissolve in distilled water and make up the volume by transferring the beaker contents with repeated washings into a 250 mL volumetric flask and mix by gentle shaking. Label it as N/7 Na_2CO_3. It is utilized as primary standard for standardizing N/7 H_2SO_4.

Preparation of N/10 Oxalic Acid

Reagent

Oxalic acid $[(COOH)_2{\cdot}2H_2O]$

Calculation

Molecular weight of $[(COOH)_2{\cdot}2H_2O]$ = 126.0666
i.e. [2(12.0112 + 31.9988 + 1.0080) + 2(2.0159 + 31.9988)]

$$\text{Equivalent weight} = \frac{126.0666}{2} = 63.0333$$

Gram-equivalent weight = 63.0333 g

Therefore,
1,000 mL of 1 N oxalic acid contains 63.0333 g
(or) 1,000 mL of N/10 oxalic acid contains 6.3033 g

Procedure

Weigh exactly 1.5758 g of oxalic acid in a clean dry beaker and make up to 250 mL in a volumetric flask with distilled water, shake gently and label as N/10 oxalic acid. The strength of N/10 oxalic acid can be checked by titrating against N/10 Na_2CO_3 using phenolphthalein as indicator. It can serve as a primary standard for preparing N/10 $KMnO_4$.

Preparation of 2.04 N Na_2CO_3

Reagent

Anhydrous Sodium carbonate

Calculations

Molecular weight = 105.989
Gram-equivalent weight = 52.9945 g

Therefore, 1,000 mL of 1 N Na_2CO_3 contains 52.9945 g
(or) 1,000 mL of 2.04 N Na_2CO_3 contains 108.1088 g

Procedure

To prepare 250 mL, weigh accurately 27.0272 g of oven-dried desiccator-cooled, anhydrous Na_2CO_3 in a clean dry beaker, dissolve in small quantity of distilled water, and make up the volume to 250 mL with repeated distilled water washings in a volumetric flask, shake gently and label. It serves as a primary standard for the preparation of 2.04 N H_2SO_4.

2.17.1.2 Secondary Standards

Preparation of N/7 H_2SO_4

Reagents

1. Sulphuric acid (AR) (Sp. gr., 1.84; Purity, 98%)
2. N/7 Na_2CO_3 solution
3. Methyl orange indicator

Calculations

Molecular weight of H_2SO_4 = 98.0776; i.e. [(2 × 1.0080 + 32.064 + 4(15.9994)]

$$\text{Equivalent weight} = \frac{98.0776}{2} = 49.0388$$

Gram-equivalent weight = 49.0388 g
Basing on Sp. gr. = Mass/volume, 49.0388 g of H_2SO_4 = 49.0388/1.84 = 26.65 mL.
Taking purity as 98%, 26.65 mL of pure H_2SO_4 will be present in 27.19 mL.
(or) 1,000 mL of N/7 H_2SO_4 contains 3.885 mL.

Procedure

Add slowly 4 mL of H_2SO_4 from the side of a beaker containing half the quantity of required distilled water, with the help of measuring cylinder and make up the volume to 1,000 mL by transferring the beaker content to a volumetric flask after cooling and by repeated washings to the beaker. It is standardized against 10 mL of N/7 Na_2CO_3 in a conical flask in the presence of 1–2 drops of methyl orange with the help of a burette. The change of colour from orange–yellow to orange–red will indicate the quantity of sulphuric acid solution of unknown strength consumed to neutralize 10 mL of N/7 Na_2CO_3. Calculate the strength of unknown H_2SO_4 solution using the equation $N_1V_1 = N_2V_2$ to add either distilled water or concentrated sulphuric acid depending upon the increase or decrease in the strength of the solution, respectively. Titration are repeated until alteast three consecutive readings are obtained. The standardized N/7 H_2SO_4 is then stored in a clean reagent bottle after labelling.

On the basis of this procedure, sulphuric acid of different normalities can be prepared as per the requirement.

Preparation of N/7 NaOH

Reagents

1. Sodium hydroxide pellets (AR)
2. N/7 H_2SO_4
3. Phenolphthalein indicator

Calculations

Molecular weight of NaOH = 39.9972; i.e. (22.9898 + 15.9994 + 1.0080)
Equivalent weight = 39.9972/1 = 39.9972
Gram-equivalent weight = 39.9972 g
Therefore, 1,000 mL of 1 N NaOH contains 39.9973 g
(or) 1,000 mL of N/7 NaOH contains 5.7139 g

Procedure

Weigh around 6 g of NaOH quickly (highly hygroscopic) in a clean dry beaker, add small quantity of distilled water and transfer the contents to 1 L volumetric flask after cooling. Make up the volume by giving repeated washings to beaker. Mix the contents by gentle shaking. Find the strength of unknown NaOH by titrating from a burette against 10 mL of N/7 H_2SO_4 in a conical flask in the presence of 1–2 drops of phenolphthalein till the contents turn pink. Add calculated quantity of either distilled water or NaOH as per the observed strength till three consecutive readings are obtained. Store in a clean reagent bottle and label as N/7 NaOH.

Preparation of N/7 HNO_3

Reagents

1. Nitric acid (AR) (Sp. gr., 1.42; Purity, 69%)
2. N/7 NaOH
3. Phenolphthalein indicator

Calculations

Molecular weight of HNO_3 = 63.0129; i.e. [1.0080 + 14.0067 + 3(15.9994)]
Equivalent weight = 63.0129/1 = 63.0129
Gram-equivalent weight = 63.0129 g
Basing on Sp. gr. = Mass/volume,
63.0129 g of HNO_3 = 63.0129/1.42 = 44.3756 mL
Considering purity as 69%, 44.3753 mL of pure HNO_3 will be present in 64.3120 mL
Therefore, 1,000 mL of 1 N HNO_3 contains 64.3120 mL
(or) 1,000 mL of N/7 HNO_3 contains 9.1874 mL

Procedure

Pour slowly about 10 mL of HNO_3 from the sides of the beaker containing half the quantity of required distilled water with the help of a measuring cylinder and make up the volume to 1,000 mL by transferring the contents of the beaker after cooling and with repeated washings. Mix the contents gently and standardize HNO_3 solution of unknown strength by titrating with burette against 10 mL of N/7 NaOH in a conical flask in the presence of 1–2 drops of phenolphthalein till the contents turn colourless. After calculating actual strength of the HNO_3 solution on hand, add required quantity of either distilled water or acid accordingly and recheck the strength by getting three consecutive readings. Store in a clean reagent bottle and label as N/7 HNO_3.

Preparation of N/10 $KMnO_4$

Reagents

1. Potassium permanganate crystals (AR)

2. N/10 oxalic acid
3. 1:4 H_2SO_4

Calculations

Molecular weight of $KMnO_4$ = 158.0376; i.e. [39.102 + 54.9380 + 4(15.9994)]
Gram-equivalent weight in acidic medium = 31.6 g (refer equivalent weight for calculations as already described)

Therefore, 1,000 mL of 1 N $KMnO_4$ contains 31.6 g
(or) 1,000 mL of N/10 $KMnO_4$ contains 3.16 g

Procedure

Dissolve about 3.5 g of $KMnO_4$ in 1,000 mL distilled water in a round bottom flask and boil the contents for 10–15 min to remove the traces of organic matter, if any present in the distilled water to avoid reduction of $KMnO_4$ and allow it to stand in dark coloured (preferably amber coloured) container for a few days. Filter the contents through glasswool. Titrate $KMnO_4$ of unknown strength with burette against 10 mL each of N/10 oxalic acid and 1:4 H_2SO_4 in conical flask, when contents of the flask are at about 70–80°C (roughly when the first bubble appears at the time of heating the contents of flask) till a pink colour persists. Find the strength of unknown $KMnO_4$ and add required quantity of either distilled water (preferably boiled and cooled) or $KMnO_4$ according to the calculated concentration of solution in hand. Check the strength to 1/10 normality till three consecutive readings are obtained and store in a clean dry amber coloured reagent bottle. Label it as N/10 $KMnO_4$.

Preparation of 2.04 N H_2SO_4

Reagents

1. Sulphuric acid (AR) (Sp. gr., 1.84; Purity, 98%)
2. 2.04 N Na_2CO_3
3. Methyl orange indicator

Calculations

Molecular weight of H_2SO_4 = 98.0776
Equivalent weight = 49.0388
Gram-equivalent weight = 49.0388 g
Basing on Sp. gr. = Mass/volume, 49.0388 g of H_2SO_4 = 26.65 mL.

Considering 98% purity,
26.65 mL pure H_2SO_4 will be present in 27.19 mL
Therefore, 1,000 mL of 1 N H_2SO_4 contains 27.19 mL
(or) 1,000 mL of 2.04 N H_2SO_4 contains 55.4676 mL.

Procedure

Gently add 56 mL of H_2SO_4 from the sides of a beaker containing half the required quantity of distilled water. Make up the volume to 1,000 mL in a volumetric flask after cooling by transferring the beaker contents and repeated washings. Standardize against 2.04 N Na_2CO_3 in the presence of 1–2 drops of methyl orange (orange–yellow to orange–red) by addition of required amount of either distilled water or H_2SO_4 depending upon the concentration of unknown solution till three consecutive readings are obtained. Store in a clean dry reagent bottle and label as 2.04 N H_2SO_4.

Preparation of 2.5 N NaOH

Reagents

1. Sodium hydroxide pellets (AR)
2. 2.04 N H_2SO_4
3. Phenolphthalein indicator

Calculations

Molecular weight of NaOH = 39.9972
Equivalent weight = 39.9972
Gram-equivalent weight = 39.9972 g
Therefore, 1,000 mL of 1 N NaOH contains 39.9972 g
(or) 1,000 mL of 2.50 N NaOH contains 99.993 g

Procedure

Dissolve approximately 100 g of NaOH in half of the required quantity of distilled water in a beaker and make up the volume to 1,000 mL in a volumetric flask by transferring the beaker contents after cooling and by repeated washings to the beaker. Standardize by titrating NaOH solution of approximate strength from a burette against 10 mL of 2.04 N H_2SO_4 in conical flask in the presence of 1–2 drops of phenolphthalein till the pink colour appears. Calculate the strength of unknown to add required quantities of either distilled water or NaOH depending upon actual strength. Ten millilitre of 2.04 N H_2SO_4 should neutralize 8.16 mL of 2.50 N NaOH. Repeat the titration until three consecutive readings are obtained. Store in a clean dry reagent bottle and label as 2.50 N NaOH.

2.18 Percentage Solutions

2.18.1 Reagents

1. Nitric acid (AR) (Purity, 67%)
2. Potassium nitrate (AR)

3. Boric acid (AR)
4. Sodium hydroxide
5. Ammonium molybdate (AR)

2.18.2 Preparation

1. 2% HNO_3 (v/v):
   Calculate the quantity of nitric acid required to prepare desired quantity taking purity into consideration and mix the acid slowly from the sides of a beaker and make up the total volume. Cool and store. E.g. To prepare 1,000 mL of 2% HNO_3 take 30 mL of 67% HNO_3 and 970 mL of distilled water.
2. 3% KNO_3 (w/v):
   Mix the required quantity of potassium nitrate in known volume of distilled water and make up the final volume with the help of a measuring cylinder. E.g. To prepare 1,000 mL of 3% KNO_3, dissolve 30 g in approximately 970 mL of distilled water to finally measure 1,000 mL.
3. 2% Boric acid (w/v):
   Dissolve the calculated amount of boric acid in distilled water by constant stirring and slight warming, as boric acid is sparingly soluble at room temperature. E.g. To prepare 1,000 mL, take 20 g boric acid and approximately 980 mL of distilled water. It is required for preparing Tashiro's indicator as described earlier.
4. 40% NaOH (w/v):
   Dissolve weighed quantity of crude or commercial sodium hydroxide flakes in ordinary tap water by constant stirring. Store after cooling. E.g. To prepare 1,000 mL, dissolve 400 g of NaOH flakes in approximately 600 mL of tap water.
5. 20% Ammonium molybdate (w/v):
   Dissolve the required quantity of ammonium molybdate in equal amounts of distilled water and liquor ammonia solution by constant stirring. Store in a clean bottle. E.g. To prepare 100 mL take 20 g of ammonium molybdate and approximately 40 mL each of distilled water and liquor ammonia solution.

2.19 Ratio Solutions

Reagents

1. Sulphuric acid (AR) (Purity, 98%)
2. Hydrochloric acid (AR) (Purity, 35%)
3. Liquor ammonia solution

Preparation

1. 1:4 H_2SO_4 (v/v):
   Add one volume of acid to four volumes of distilled water slowly; cool, label and store in a bottle. E.g. For 1,000 mL of 1:4 H_2SO_4, take 200 mL H_2SO_4 and 800 mL distilled water.

2. 1:2 HCl (v/v):
   Add one volume of acid to two volumes of distilled water slowly; cool, label and store in bottle. E.g. For 1,000 mL of 1:2 HCl, take 333 mL acid and 666 mL distilled water.
3. 1:4 Liquor ammonia (v/v):
   Mix one volume of liquor ammonia solution to four volumes of distilled water, label and store in a bottle. Care should be taken to cool liquor ammonia bottle while opening the cork, especially in summer months, to avoid sudden spurting of accumulated ammonia gas in the bottle. E.g. To prepare 1,000 mL, take 200 mL of liquor ammonia solution and 800 mL distilled water.

2.20 Titration

The process of bringing a measured volume of a solution of known concentration into reaction with the desired constituents (or its equivalent of another substance) is called titration. In other words adding of a solution of known strength to another in order to complete the reaction is known as titration. Since volume of a solution of unknown strength is measured, it is also known as volumetric analysis.

1. Titre
   The titre is the weight of solute contained in a mL of solution or the weight of any substance which will react with or equivalent to 1 mL of solution.
   As described earlier a solution of accurately known concentration is called as standard solution, which may be prepared directly (primary standard) or by standardization (secondary standard) through reaction with a primary standard. The end point of a titration occurs when chemically equivalent amounts of reactants are brought together, as indicated by an abrupt change in colour as shown by an indicator, at the stoichiometric point of the reaction.
2. Stoichiometric point
   The stoichiometric point is an equivalence point at which an equivalent of the reacting substance has been added, irrespective of the type of reaction involved.
3. End point
   The end point in an acid–alkali titration is that point at which the titration is stopped, being shown by the colour change of the particular indicator used.
   The suitability or otherwise of an indicator in any given titration depends upon the pH value at which the indicator shows its specific colour change.
4. Indicator
   Indicator is a substance that indicates the physico-chemical status of a reaction. They may be internal, external or self indicators. They are mostly organic compounds of high molecular weight. When dissolved in water or any suitable solvent, they behave either as a weak acid or weak base. Basic indicator possess a coloured cation and acidic indicator possess coloured anion. The internal structural rearrangement is responsible for colour change. Important indicators used in titrimetric analysis are given below (Table 2.5):

Table 2.5 Common indicators used in titration

Indicator	pH range	Titration	Acidic colour	Basic colour
Methyl orange	3.1–4.4	Strong acid & strong base/weak acid & weak base/strong acid & weak base	Orange red	Orange yellow
Methyl red	4.2–6.2	Strong acid & weak base	Red	Yellow
Phenolphthalein	8.2–10.0	Strong acid & strong base/weak acid & strong base	Colourless	Pink
Bromocresol green	3.6–5.2	Weak acid and weak base	Yellow	Blue

5. Preparation of indicators
   Prepare indicators by dissolving requisite quantities of indicators in the solvents as shown below after thorough mixing followed by filtration through Whatman No. 1 filter paper.
   Methyl orange – 0.1 g in 100 mL distilled water.
   Methyl red – 0.2 g in 100 mL alcohol (95–96%).
   Phenolphthalein – 1.0 g in 110 mL alcohol (95–96%) and 90 mL distilled water.
   Bromo cresol green – 0.04 g in 100 mL of alcohol.
   Tashiro's indicator – add 10 mL each of methyl red and bromocresol green to 1,000 m of 2% boric acid solution (boric acid dissolves in distilled water by slight warming).
6. Types of reactions in titrimetry

   (i) Neutralization reaction:

   The reaction of an acid with base is called neutralization.
   E.g.

$$\begin{array}{llll} \text{Acid} & + \ \text{Base} & & \\ H^+ & + \ OH^- & \longrightarrow & H_2O \\ CH_3COOH & + \ OH^- & \longrightarrow & H_2O + CH_3COO \end{array}$$

   (ii) Acidimetry:

   Acidimetry is a reaction in which the amount of base in a sample is determined by titration with a standard acid.

   (iii) Alkalimetry:

   Alkalimetry is the measurement of acid in a sample by titration with a standard alkali.

   (iv) Precipitation and complexation reactions:

   A reaction which forms precipitate or a complex approaches completeness because of extensive removal of ions from solution.

E.g.

$$Cl + Ag^{+} \rightarrow AgCl \text{ (Insoluble)}$$
$$2CN + Ag^{+} \rightarrow Ag\ (CN)_2 \text{(Complex ions)}$$

(v) Oxidation–reduction or Redox reactions:

Redox reactions involve the transfer of one or more electrons from one to another substance. A reducing agent is a substance that looses electrons and thus becomes oxidized, while on oxidizing agent gains electrons and becomes reduced.

Mathematical relations in titrimetry

1. Volume × normally = gram equivalent weight
   e.g. Half litre of 0.2 N reagent contains, 0.5 × 0.2 = 0.1 GEW of solute
2. Volume of known × normally of known = volume of unknown × normally of unknown
   i.e. $V_1 \times N_1 = V_2 \times N_2$
   If any three of the terms are known, the fourth can be calculated.
3. $\text{Volume} \times \text{Normally} = \dfrac{\text{g substance}}{\text{g equivalent weight of the same substance}}$

Storing and preserving standard solutions:

1. The bottle must be kept tightly stoppered to prevent evaporation loss of solvent to avoid increase in solute concentration.
2. The bottle should be shaken before the withdrawn of a portion of solution to ensure uniform composition of both withdrawn portion and the remainder left in the bottle.
3. Portions of solution once withdrawn should be discarded and never be returned to the bottle to minimize the danger of contamination.
4. Some standard solutions must be protected from atmospheric gases such as sodium hydroxide solution which is effectively diluted by atmospheric CO_2 by carbonic acid formatin.
5. Solution containing potassium permanganate and silver nitrate should be stored in dark glass bottles (amber coloured) or kept in dark when not in use to prevent light catalyzed decomposition. Solutions containing sodium thiosulphate should be protected from bacteria-induced decomposition.
6. Aqueous solutions should always be made using pure water (deionized or double distilled). Solution should remain clear after preparation.
7. Solution should always be properly stored, e.g. some solutions can be stored at room temperature while others require storage at low temperatures. Solutions which are sensitive to light should always be stored in brown bottles.
8. Always note down the MW printed on the bottle and take into account hydration state of the substance, purity, specific gravity, etc.

Chapter 3
Buffers and Their Preparation

Buffers are indispensable in biochemical and molecular studies as they maintain a near constant pH to the media while performing various laboratory operations such as during extraction, isolation and purification of various biomolecules. Selection of an appropriate buffer with optimal pH is important as it may have influence on extractability, stability and even biological functioning. The stability as well as activity of various enzymes in the cell is highly dependent on the pH of the system. Therefore, it is important to provide a stable system in form of buffer which could stabilize the hydrogen ion concentration and negate adverse effects due to change in pH on the enzyme under study. To understand the buffering capacity of a buffer it is important to understand the concept of pH, acid and bases.

3.1 Selection of Buffers

While selecting a buffer the following points need to be taken into account.

It should:

- Possess adequate buffering capacity in the required pH range. Generally buffers are most effective over a range of one pH unit on either side of their p*K*' value, e.g. Tris which has p*K*' value of 8.3 has an effective pH range of 7–9.
- Chemically inert and not react or bind with biomolecules or other components.
- Have high degree of purity and should not contain impurities which may interfere with estimations.
- Be enzymatically and hydrolytically stable.
- Maintain pH that is minimally influenced by temperature, ionic composition and concentration or salt effect of the medium.
- Not be toxic.
- Not absorb light in the visible or ultraviolet regions.

R. Katoch, *Analytical Techniques in Biochemistry and Molecular Biology*,
DOI 10.1007/978-1-4419-9785-2_3, © Springer Science+Business Media, LLC 2011

3.2 Acids and Bases

The modern concept of acids and bases defines acid as proton donor and a base as proton acceptor. On ionization, an acid donates a proton and at the same time a corresponding base (which is capable of accepting a proton) is formed. Such a base is known as conjugate base:

$$\text{Acid} \rightleftarrows H^+ + \text{Base}$$

Acids and bases are classified into two groups: strong and weak acids and bases. Strong acids or bases are those which get completely ionized in solution so that the concentration of free H^+ or OH^- is the same as the concentration of the acid or base. For example,

$$\text{Strong acid: } HCl \rightarrow H^+ + Cl^-$$
$$\text{Strong base: } NaOH \rightarrow Na^+ + OH^-$$

Conversely, weak acids or bases dissociate only to a limited extent and the concentration of free H^+ and OH^- depends on the dissociation constant.

$$\text{Weakacid: } CH_3COOH \rightleftarrows CH_3COO^- + H^+$$

3.3 Concept of pH

By using pH indicators, an approximate value of pH of a solution can be obtained. The indicators dissociate like a weak acid on coming in contact with solution and give range of colours depending on pH. Dissociated and undissociated forms of the indicator have different colours. pH papers coated with indicators of different pH range are commercially available. However, accurate pH can be measured using pH metre which measures e.m.f. of a concentration cell developed from a reference electrode, test solution and a glass electrode sensitive to H^+ ions. The combined electrode constituting glass and reference electrode is dipped into the test solution for accurate measurement of pH.

Henderson–Hasselbalch equation
The Henderson–Hasselbalch equation is simply another way of expressing dissociation constant of an acid

$$HA \rightleftarrows H^+ + A^-$$

$$K\ \text{ion} = K' = \frac{[H^+][A^-]}{[HA]}$$

on rearranging,

$$[H^+] = k' = \frac{[HA]}{[A^-]}$$

Taking negative logarithm on both sides

$$-\log[H^+] = -\log k' - \log\frac{[HA]}{[A^-]}$$

substituting pH for $-\log[H^+]$ and pK' for $-\log k'$, where pk' is $-\log$ for dissociation constant

$$pH = pK' - \log\frac{[HA]}{[A^-]}$$

or

$$pH = pk' + \log\frac{[A^-]}{[HA]}.$$

The above equation is known as Henderson–Hasselbalch equation which can be written in a more general form as

$$pH = pK' + \log\frac{[\text{conjugate base}]}{[\text{conjugate acid}]}.$$

The above relationship is very useful equation from which either pH of the solutions of various concentration ratios of a conjugate acid–base pair of known pK' can be calculated or the ratio of conjugate acid–base of known pK' to obtain a buffer of desired pH can be found out.

Dissociation of water, its ionic product K_w

Water is weak electrolyte, which dissociates only slightly to form H^+ and OH^- ions.

$$H_2O \rightleftarrows H^+ + OH^-$$

The equilibrium constant of this dissociation reaction is 1.8×10^{-16} mole/L at 25°C.

Thus

$$K_{eq} = \frac{[H^+][OH^-]}{[H_2O]} = 1.8 \times 10^{-16}$$

One thousand millilitre of pure water contains 1,000 g of water. Since the molecular weight of water is 18, its molar concentration can be calculated and is equal to 1,000/18 or 55.5 moles/L. Because the concentration of H_2O in dilute aqueous solution is essentially unchanged from that in pure H_2O, this figure may be taken a constant. Thus,

$$\frac{[H^+][OH^-]}{55.5} = 1.8 \times 10^{-16},$$

$$[H^+][OH^-] = 1.8 \times 10^{-16} \times 55.5$$
$$= 1.01 \times 10^{-14},$$

or

$$K_W = 1.01 \times 10^{-14}$$

where K_W is ion product of water which expresses the relationship between the concentration of H^+ and OH^- ions in aqueous solutions. The equation may be used to calculate the concentration of H^+ in pure water. Let x be the concentration of H^+ on dissociation of water. Since production of one H^+ is associated with generation of one OH^-, the concentration of OH^- will also be x.

Therefore,

$$[H^+][OH^-] = 1.01 \times 10^{-14}$$

$$x^2 = 1.01 \times 10^{-14}$$

or

$$x = 1.01 \times 10^{-7}\ \text{moles/L}.$$

The conventional method of expressing concentration of H^+ ions is by means of pH, which is defined as negative logarithm of H^+ ion activity.

Thus,

$$\text{pH} = \log\frac{1}{aH^+} = -\log aH^+,$$

where aH^+ is defined as the activity of H^+. In this context no distinction is made between activities and concentration. Hence,

$$\text{pH} = \log\frac{1}{H^+} = -\log[H^+].$$

It is important to note that the pH is logarithmic function, thus, when the pH of a solution decreases by 1 unit from 5 to 4, the H^+ ion concentration increases by tenfold, i.e. from 10^{-5} to 10^{-4} M.

Applying the term pH to the ion product of water

$$[H^+][OH^-] = 1.0 \times 10^{-14},$$
$$\log[H^+] + \log[OH^-] = \log(1.0 \times 10^{-14}),$$
$$-\log[H^+] - \log[OH^-] = 14,$$
$$\text{pH} + \text{pOH} = 14,$$
$$\text{or}\quad \text{pH} = 7.$$

Hence pH of pure water is 7.0.

3.4 Buffer System

A buffer system is one that resists a change in pH on the addition of acid or alkali and constitutes of conjugate acid–base pair. Most commonly, the buffer solution consists of a mixture of a weak acid and its conjugate base, e.g. a mixture of acetic acid and sodium acetate is a buffer solution. A well-known physiological buffer system is the carbonate–bicarbonate system of blood.

The basis for functioning of a buffer with regard to its ability to resist change in pH can be illustrated by the following example. Suppose we have a buffer containing 5 mL of 0.1 M sodium acetate and 4 mL of 0.1 M acetic acid. The effect of pH on addition of 1 mL of 0.1 N HCl to this buffer is expressed as

Before addition of acid

$$\begin{aligned}
\text{Total volume of buffer is } (5\,\text{mL} + 4\,\text{mL}) &= 9\,\text{mL},\\
\text{Concentration of acetate} &= 5/9 \times 0.1\,\text{M} = 0.5/9\,\text{M},\\
\text{p}K'\text{of acetic acid} &= 4.76.\\
\text{Thus, pH} &= 4.76 + \frac{\log 0.5/9}{0.4/9}\\
&= 4.76 + (0.097) = 4.86.
\end{aligned}$$

On addition of acid

HCl provides H^+ which combines with the acetate ion to give acetic acid. Therefore, it reduces the amount of acetate ions present and increases the amount of undissociated acetic acid leading to an alteration in the salt/acid ratio and hence to a change in pH.

$$\begin{aligned}
\text{The final volume of buffer after acid addition} &= 10\,\text{mL}\\
\text{Conc. of acetate after acid addition} &= 5/10 \times 0.1\,\text{M} - 1/10 \times 1.0\,\text{M}\\
&= 0.04\,\text{M}.\\
\text{Conc. of acetic acid after acid addition} &= 4/10 \times 0.1\,\text{M} + 1/10 \times 0.1\,\text{M}\\
&= 0.05\,\text{M}.\\
\text{Thus, pH} &= 4.76 + \log 0.04/0.05\\
&= 4.76 + (-0.097)\\
&= 4.66.
\end{aligned}$$

Thus, the pH changes marginally from 4.86 to 4.66 on addition of 1 mL of 0.1 N HCl. However, if the some amount of the acid is added to pure water its pH will change from 7.0 to 2.0.

3.5 Preparation of Buffers

Protocols for the preparation of most of the commonly used buffers are available in books and manuals. The example showing calculations for preparing a buffer of required pH are shown below:

For preparing 1 L of 0.1 M acetate buffer of pH 5.22 (pK_a of acetic acid is 4.74) we need to determine the required ratio of conjugate base (acetate ion) to weak acid (acetic acid) in this buffer solution by implication of previously discussed Henderson–Hasselbalch equation.

$$\mathrm{pH} = \mathrm{p}K' + \log \frac{[\mathrm{CH_3COO^-}]}{[\mathrm{CH_3COOH}]}$$

Putting the values

$$5.22 = 4.76 + \log \frac{[\mathrm{CH_3COO^-}]}{[\mathrm{CH_3COOH}]}$$

or

$$\log \frac{[\mathrm{CH_3COO^-}]}{[\mathrm{CH_3COOH}]} = 5.22 - 4.76 = 0.48$$

or

$$\frac{[\mathrm{CH_3COO^-}]}{[\mathrm{CH_3COOH}]} = \mathrm{antilog}\, 0.48 = 3.$$

Therefore, in this particular buffer solution the molar ratio of CH_3COO^- to CH_3COOH should be 3. In other words, 75% of the buffer components should be present as the conjugate base CH_3COO^- and 25% of the components as acetic acid. Since 1 L of 0.1 M acetate buffer should contain 0.1 mole of acetate plus acetic acid, the solution should have 0.075 moles of acetate ions and 0.025 moles of acetic acid. Further we need to calculate sodium acetate (g) required to obtain 0.075 moles of acetate ions and to obtain 0.025 moles of acetic acid 1.5 g of acetic acid would be used. Mix these solutions after their preparation and make the final volume to 1 L with water to obtain 1 L of buffer of the desired pH and concentration.

3.6 Preparation of Some Commonly Used Buffers

Buffers can be made in stock solutions and these are diluted before use. Preparation of some of the buffers frequently used in biochemical studies is given below. Unless otherwise stated, by following these procedures 0.1 M buffer can be obtained.

1. Acetate buffer
   Stock solutions:

   A: 0.2 M acetic acid (11.15 mL/L)
   B: 0.2 M sodium acetate (16.4 g to sodium acetate or 27.2 g of sodium acetate $4H_2O$ per litre)

   x mL of A + *y* mL of B, diluted to a total volume of 100 mL.

x	*y*	pH	*x*	*y*	pH
46.3	3.7	3.6	25.5	24.5	4.6
44.0	6.0	3.8	14.8	35.2	5.0
41.0	9.0	4.0	10.5	39.5	5.2
36.8	13.2	4.2	8.8	41.2	5.4
30.5	19.5	4.4	4.8	45.2	5.6

2. Phosphate buffer
   Stock solutions:

   A: 0.2 M monobasic sodium phosphate (27.8 g on 1 L)
   B: 0.2 M dibasic sodium phosphate (53.65 g of $Na_2HPO_4 \cdot 7H_2O$ or 71.7 g $Na_2HPO_4 \cdot 12H_2O$ in 1 L)

   x mL of A + *y* mL of B, diluted to a total volume of 200 mL.

x	*y*	pH	*x*	*y*	pH
93.5	6.5	5.7	65.5	43.5	6.7
90.0	10.0	5.9	39.0	61.0	7.0
85.0	15.0	6.1	16.0	84.0	7.5
77.5	22.5	6.3	5.3	94.7	8.0
68.5	31.5	6.5			

3. Tris (hydroxymethyl) aminomethane buffer or Tris buffer
   Stock solutions:

   A: 0.2 M Tris (hydroxymethyl) aminomethane (24.2 g/L)
   B: 0.2 M HCl
   50 mL of A + *y* mL of B diluted to a total volume of 200 mL. 0.05 M Tris–HCl buffer will be obtained.

y	pH	*y*	pH
5.0	9.0	26.8	8.0
8.1	8.8	32.8	7.8
12.2	8.6	38.4	7.6
16.5	8.4	41.4	7.4
21.9	8.2	44.2	7.2

4. Boric acid–borax buffer
   Stock solutions:

   A: 0.2 M solution of boric acid (12.4 g in 1,000 mL)
   B: 0.05 M solution of borax (19.05 g in 1,000 mL; 0.2 M in terms of sodium borate)

   50 mL of A, *x* mL of B, diluted to a total of 200 mL.

x	pH
2.0	7.6
3.1	7.8
4.9	8.0
7.3	8.2
11.5	8.4
17.5	8.6
22.5	8.7
30.0	8.8
42.5	8.9
59.0	9.0
83.0	9.1
115.0	9.2

5. Carbonate–bicarbonate buffer
   Stock solutions:

   A: 0.2 M solution of anhydrous sodium carbonate (21.2 g in 1,000 mL)
   B: 0.2 M solution of sodium bicarbonate (16.8 g in 1,000 mL)

   x mL of A, *y* mL of B, diluted to a total of 200 mL

x	*y*	pH
4.0	46.0	9.2
7.5	42.5	9.3
9.5	40.5	9.4
13.0	37.0	9.5
16.0	34.0	9.6
19.5	30.5	9.7
22.0	28.0	9.8
25.0	25.0	9.9
27.5	22.5	10.0
30.0	20.0	10.1
33.0	17.0	10.2
35.5	14.5	10.3
38.5	11.5	10.4
40.5	9.5	10.5
42.5	7.5	10.6
45.0	5.0	10.7

6. Citrate buffer
   Stock solutions:

   A: 0.1 M solution of citric acid (21.01 g in 1,000 mL)
   B: 0.1 M solution of sodium citrate (29.41 g $C_4H_5O_7Na_3{\cdot}2H_2O$ in 1,000 mL)

   x mL of A, *y* mL of B, diluted to a total of 100 mL.

x	*y*	pH	*x*	*y*	pH	*x*	*y*	pH
46.5	3.5	3.0	33.0	17.0	4.0	18.0	32.0	5.2
43.7	6.3	3.2	31.5	18.5	4.2	16.0	34.0	5.4
40.0	10.0	3.4	28.0	22.0	4.4	13.7	36.3	5.6
37.0	13.0	3.6	25.5	24.5	4.6	11.8	38.2	5.8
35.0	15.0	3.8	23.0	27.0	4.8	9.5	41.5	6.3
			20.5	29.5	5.0	7.2	42.8	6.2

7. Glycine-HCl buffer
   Stock solutions:

   A: 0.2 M glycine (15.01 g in 1,000 mL)
   B: 0.2 N HCl

   25 mL of A, *x* mL of B diluted to a total of 100 mL.

x	pH
22.0	2.2
16.2	2.4
12.1	2.6
8.4	2.8
5.7	3.0
4.1	3.2
3.2	3.4
2.5	3.6

8. Glycine–NaOH buffer
   Stock solutions:

   A: 0.2 M solution of glycine (15.01 g in 1 L)
   B: 0.2 M NaOH

   50 mL of A + *x* mL of B, diluted to a total of 200 mL.

x	*y*
4.0	8.6
6.0	8.8
8.8	9.0
12.0	9.2
16.8	9.4
22.4	9.6
27.2	9.8
32.0	10.0
38.6	10.4
45.5	10.6

Chapter 4
Techniques in Biochemical Evaluation

The different techniques employed in biochemistry plays a crucial role in different analysis. This chapter provides us the knowledge about some of the crucial techniques involved in different analysis.

4.1 Spectrophotometry

This is one of the important technique in biochemical analysis. There are three fundamental applications of this technique:

(i) If the absorbancy index (a_s) at a specific wavelength is known, the concentration of a compound can be readily determined by measuring the optical density at that wavelength. With a large a_s, as in nucleotides, very small quantities of the absorbing material (2–4 μg) can be accurately measured.
(ii) The course of a reaction can be determined by measuring the rate of formation or disappearance of a light-absorbing compound. Thus, NADH absorbs strongly at 340 nm, whereas the oxidized form (NAD+) has no absorption at this wavelength. Therefore, reactions involving the production or utilization of NADH (or NADPH) can be assayed by this technique.
(iii) Compounds can frequently be identified by determining their characteristic absorption spectra in the ultraviolet and visible regions of the spectrum.

There are two fundamental laws associated with the technique of spectrophotometry:

Lambert's and Beer's laws – Lambert's law states that the light absorbed is directly proportional to the thickness of the solution being analysed:

$$A = \log_{10}\frac{I_o}{I} = a_s b,$$

R. Katoch, *Analytical Techniques in Biochemistry and Molecular Biology*,
DOI 10.1007/978-1-4419-9785-2_4, © Springer Science+Business Media, LLC 2011

where I_o is the incident light intensity, I is the transmitted light intensity; a_s is the absorbancy index characteristic of the solution, b is the length or thickness of the medium, and A is the absorbancy.

Beer's law states that the amount of light absorbed is directly proportional to the concentration of solute in solution:

$$\log_{10}\frac{I_o}{I} = a_s c$$

and the combined Beer–Lambert law is

$$\log_{10} I_o/I = a_s b\, c.$$

If b is held constant by employed standard cell or cuvette, the Beet–Lambert law reduces to the form

$$A = \log_{10} I_o/I = a_s c.$$

The absorbancy index a_s is defined as A/Cb, where C is the concentration of the substance in grams per litre and b is the distance in centimeters travelled by light in the solution. The molar absorbancy index a_m is equal to a_s multiplied by the molecular weight of the substance.

All spectrophotometers have the following essential parts (Fig. 4.1):

(a) A source of radiant energy (*L*).
(b) A monochromator, which is a device for isolating monochromatic light or narrow bands of radiant energy.

The typical spectrophotometer consists of either a grating or a prism (B), which is used to disperse the radiant energy into a spectrum, together with an exit slit C, which selects a narrow portion of the spectrum. The cuvette, and the emergent light – a light-tight unit; the incident light strikes the cuvette, and the emergent light passes into photocell, which converts the emerging light energy into a signal of measurable electrical energy.

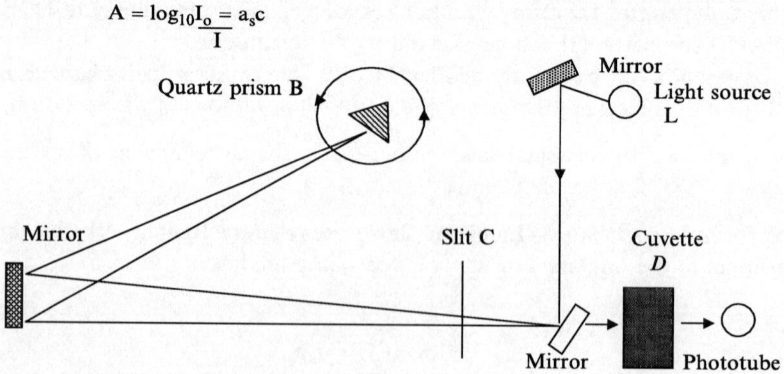

Fig. 4.1 Construction of a typical spectrophotometer

4.2 Measurement of pH with Glass Electrode

The most effective way of accurately measuring the pH value of a biochemical system is to employ a pH metre with a glass electrode. The potential of the glass electrode (E_g) relative to the external reference electrode (E_{ref}) is related to the pH as follows:

$$\mathrm{pH} = \frac{E_g - E_{ref}}{0.0591} \quad \text{at } 25°\mathrm{C}.$$

The typical glass electrode assembly consists of

$$\underset{\text{Silver} - \text{Silver chloride electrode}}{\mathrm{Ag, AgCl(S), HCl(0.1\,M)}}||\text{Glass membrane}||\underset{\text{Calomelhalf} - \text{cell}}{\text{Solution X}|\mathrm{KCl(Sat), Hg_2Cl_2(S), Hg}}$$

When two solutions of different H^+ ion concentrations are separated by a thin glass membrane, potential difference related to differences in pH of the two solutions is obtained. A typical glass electrode is illustrated in Fig. 4.2.

The potential difference ($E_g - E_{ref}$) is carefully measured either with a potentiometer type of pH metre or with a direct-reading pH metre (line-operated) consisting normally of a simple triode amplifier using the negative feedback principle. It should be noted that the results obtained are in terms of activity (a_H) rather than concentration of hydrogen ion [H^+].

The glass electrode pH metre is the preferred tool, as it is an extremely sensitive and stable instrument The glass electrode should be carefully washed after each pH determination, particularly after dealing with protein solutions, since proteins may

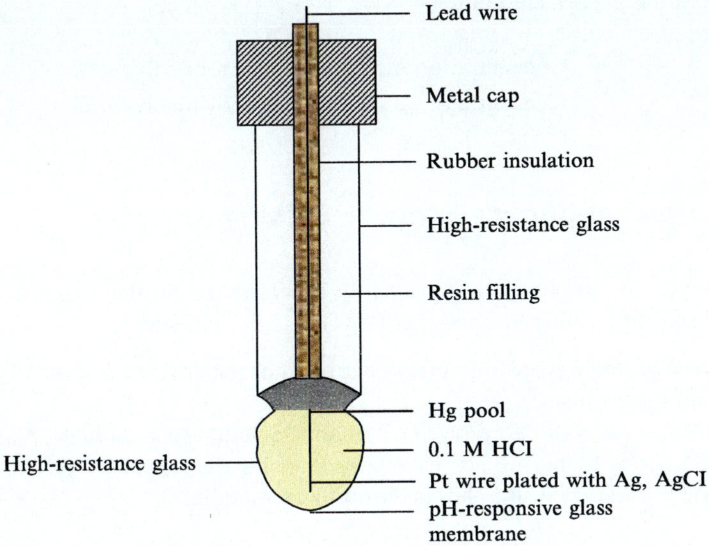

Fig. 4.2 Construction of a typical glass electrode

absorbed on the glass membrane surface, resulting in errors. In non-aqueous solutions a partial dehydration of the glass membrane may occur with changes in the potential difference, which may also lead to errors. Poorly buffered solutions should be thoroughly stirred during measurements, since a thin layer of solution at the glass solution interface may not reflect the true activity of the rest of the solution.

4.3 Chromatography

Generally chromatography involves all those techniques in which various components in a mixture get separated during their passage through a porous media due to the differences in their migration rates. During the process, the sample is applied at one end of a porous support which holds the stationary phase (which may be a solid or a liquid) and the mobile phase (a liquid or a gas) is then made to flow over it. The various constituents in a sample get separated essentially due to differences in their partition or distribution behaviour between the stationary phase and the mobile phase. Different compounds vary in their distribution coefficients in a given biphasic system and this constitutes the underlying principle for their separation in the different types of chromatographic techniques. In simple terms, movement of those substances in the sample which have greater affinity for stationary phase is retarded since they tend to spend more time in this phase. On the other hand, the compounds with relatively higher affinity for mobile phase travel at a much faster rate as these are carried along with the flowing mobile phase. This eventually results in the separation of various components in the mixture.

The partition or distribution coefficient (K_d), is defined as the ratio of concentration of a compound in two phases (say A and B) at equilibrium. Hence, a situation where both the phases are liquids

$$K_d = \frac{\text{Concentration of the compound in solvent A}}{\text{Concentration of the compound in solvent B}}$$

4.3.1 Types of Chromatographic Techniques

Chromatographic techniques may broadly be classified on the basis of nature of support used for holding the stationary phase. These include:

(i) *Paper chromatography*: In this system a filter paper sheet is used as a support for stationary phase.
(ii) *Thin layer chromatography* (TLC): In this technique a glass plate, plastic sheet or a piece of metal foil serves as a support for the stationary phase, which is applied in the form of a thin layer on these materials.

(iii) *Column chromatography*: Here stationary phase is packed into a tubular glass, polypropylene or metal columns.

The various chromatographic techniques have also been classified into the following types depending upon the forces or the interacting phenomenon between the solute molecules and the stationary phase:

(a) Partition chromatography
(b) Adsorption chromatography
(c) Ion-exchange chromatography
(d) Molecular sieve chromatography
(e) Affinity chromatography

The different chromatographic techniques are described in detail.

4.3.1.1 Paper Chromatography

In paper chromatography, the sample to be analysed is applied in the form of a spot near one of the edges of a Whatman filter paper. This filter paper is then kept in an atmosphere already saturated with water vapours to form a thin film of water around cellulose fibres of the paper, which then acts as a stationary phase. An appropriate solvent system, which functions as a mobile phase, is then allowed to flow over the sample spot. On coming in contact with mobile phase, the various components of the sample get partitioned between the stationary and the mobile phases. Those constituents having a higher affinity for the stationary phase move less rapidly as compared with those having higher affinity for the mobile phase. Evidently the components will get well separated from each other if their K_d values are sufficiently different. Satisfactory separation of components of interest can hence be achieved by judicious selection of the mobile phase.

Depending upon the direction of flow of mobile phase, the two commonly used systems are

1. Ascending paper chromatography
2. Descending paper chromatography

In the ascending method, the solvent is kept at the base of the chamber and the edge of the paper where the sample has been applied is immersed in the solvent taking care that the sample spots do not get dipped in the solvent but remain just above the surface of the solvent. The solvent moves up or ascends the paper by capillary action and the separation of different components occurs on the basis of differences in their partition coefficients. In the descending method, the paper is hung in such a way that the side where the sample has been spotted dips in a trough, which is fitted at the top of the chamber, and contains the mobile phase and this solvent travels down the paper under the force of gravity.

Identification of Components

After development, the filter paper is dried in air. The location of the compounds under investigation is carried out by making use of their specific chemical, physical or biological property. For instance, if these components form coloured complex or product with a particular reagent, they can then be conveniently located by spraying the chromatogram with that reagent. In case the compounds absorb UV light or show UV fluorescence, the paper can be examined under strong UV light in darkness. The UV absorbing compounds would appear as dark spots while UV fluorescent compounds would show a characteristic fluorescence under UV light. If in metabolic studies, a radioactive precursor has been used, then the products derived from it can be detected from the radioactive zones or spots on the chromatogram (Fig. 4.3).

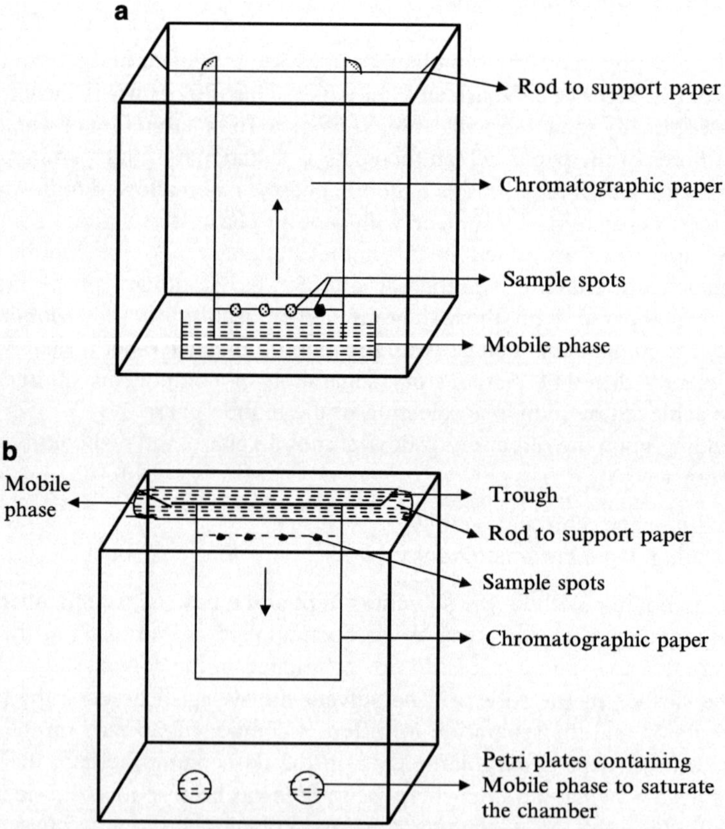

Fig. 4.3 Paper chromatography, arrow indicates the direction of flow of mobile phase. (**a**) Ascending (**b**) Descending paper chromatography

Identification

The separated compounds are identified on the basis of their R_f value, which denotes relative to front. The R_f value is calculated as follows:

$$R_f = \frac{\text{Distance travelled by the component from base line}}{\text{Distance travelled by the solvent from base line}}.$$

Since a number of factors such as composition of solvent system, temperature, pH, mode of development, grade of filter paper and flow rate influence the R_f value, it is always advisable to run authentic standards along with the sample.

Modifications for Development of Chromatograms

Sometimes a single run with one solvent system in a single dimension is not sufficient to obtain satisfactory separation of the components. A number of modifications has been employed to achieve better resolution of such overlapping or very closely located compounds.

(i) *Two-dimensional chromatography*: In this method, the chromatogram is first developed in one solvent system in one direction. After air drying, the paper is again developed in a second solvent system in a direction perpendicular to the previous run. The compounds having similar R_f values in first solvent system might have different mobilities in the second solvent system and hence they get separated. However, the limitation of bidimensional chromatography is that only one sample can be spotted or applied for analysis on each filter paper.

(ii) *Reverse phase chromatography*: In all the above cases, the aqueous phase of the solvent system is used to saturate the internal atmosphere of the chamber, and water absorbed by the cellulose fibres of the paper serves as the stationary phase. Thus, the stationary phase is hydrophilic, whereas the mobile phase is hydrophobic in nature. For separation of hydrophobic substances like fatty acids, the filter paper is treated with lipophilic compounds like silicon grease, kerosene oil, paraffin, Vaseline, etc. while the mobile phase is hydrophilic in nature. Since the characteristics of stationary and mobile phases are converse of that in the conventional paper chromatography, this modified form is referred to as "reverse phase chromatography".

(iii) *Multiple developments*: In this procedure, the paper is developed with the same solvent system several times in the same direction. The chromatogram is air-dried between the successive developments. This mode of development of chromatogram can be used advantageously when the sample contains mixture of components some of which migrate quite fast and get well separated from each other and a group of other components which remain clustered near origin due to their low and closely similar R_f values. These slow-moving components get further apart with each successive run resulting in their better separation.

(iv) *Continuous development*: It is also known as over-run chromatography. In this case the chromatogram is developed in the same solvent system in the same direction continuously for a long time even after the solvent has started dripping down from the other end of paper. Care, however, must be taken that the fastest moving compound does not get eluted out of the paper. This approach is useful for achieving better separation of compounds having very low and similar mobilities. Evidently R_f value cannot be calculated as the solvent front cannot be calculated and determined due to overflow of the solvent. The identification is hence done on the basis of R_g value which is calculated as follows:

$$R_g = \frac{\text{Distance moved by an unknown compound}}{\text{Distance moved by a standard compound}}.$$

4.3.1.2 Thin Layer Chromatography

During TLC separations, instead of paper, the supporting material is a glass plate, a plastic sheet or a piece of metal foil. A thin layer of the stationary phase is laid over this inert support. The layer may be as thin as 250 μm for analytical separations and as thick as 2–5 mm for preparative separations. A binding agent such as calcium sulfate or gypsum may be incorporated into the chromatographic media to facilitate firm adhesion of adsorbent to the plate. Most widely used adsorbent in TLC is silica gel "G" which contains 13% $CaSO_4$. Another commercially available media is Silica Gel "H" which does not contain $CaSO_4$. Since Silica Gel "G" has binder, its slurry should not be kept for a long time and it should be spread immediately after preparation of its suspension.

For development, the plate on which the sample spots have been applied is placed in an air tight glass jar containing the solvent. The location and identification of separated components is carried out in the same manner as in the case of paper chromatography. Also like paper chromatography, TLC is amenable to two-dimensional mode of development.

In TLC, depending upon the nature of the chromatographic media used, the separation can be achieved by partitioning, adsorption, ion exchange or molecular sieving phenomenon and the separation can be achieved within an hour. This technique is relatively more sensitive since lower concentrations of compounds in the mixture can be successfully separated and detected. Corrosive agents like H_2SO_4 and high temperatures can be used to locate the separated compound which is not possible in paper chromatography (Fig. 4.4).

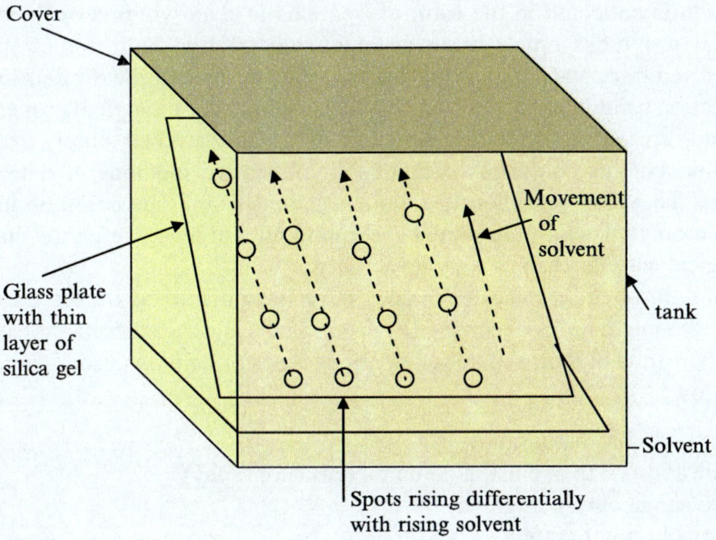

Fig. 4.4 Thin layer chromatography using silica gel

4.3.1.3 Column Chromatography

In column chromatography, separation is achieved by passage of the sample through vertically fixed tubular glass or polypropylene column which is packed with an appropriate chromatography media. Usually, commercially available columns have a porous sintered plate fused at their base which prevents the stationary phase from flowing out of the column. This sintered base is positioned as near the base as possible in order to minimize the dead space to reduce the chances of post-column mixing of the separated compounds. Alternatively, a simple glass burette with a plug of glass wool at the base can be used as a column. At the base there is a small capillary tubing through which the effluent from the column flows into the test tubes in which fractions are collected. At the top of the column a solvent reservoir with a delivery system is fitted. The stationary phases used in column chromatography are water insoluble, porous, solid particles, and the resolution of sample components occurs depending upon the principle underlying the separation phenomenon. The flow rate and resolution characteristics are influenced by the size and shape of the stationary phase. Large and coarse particles have higher flow rate but give comparatively poor resolution while finer particles with large surface to volume ratio have slower flow rate but greater resolution efficiency. Generally, particles of 100–200 mesh are used for routine analysis but for high resolution the smaller particles of 200–400 mesh are used.

For fractionation and separation of components, the sample is loaded on top of the column and eluted with an appropriate buffer. The effluent emerging from base of the column is collected in the form of fractions of fixed volume or fixed time in individual test tubes either using an automatic fraction collector or manually. The collected fractions are then analysed for the presence of the desired substance. The detection technique depends on physical, chemical or biological property of the compound. Presence of coloured compounds can be identified simply from visual observation but for colourless compounds, alternative methods of detection are employed. They can be either the colour reactions or may be based on its unique physical property such as ultraviolet absorption, fluorescence, refractive index, etc. or biological activity such as enzyme activity.

Column chromatographic techniques have been classified on the basis of the nature of the interactions occurring between solutes and the stationary phase which ultimately results in their separation. Various types of column chromatography are given below.

- Adsorption chromatography
- Gel filtration (size exclusion column chromatography)
- Ion exchange chromatography
- Affinity chromatography
- Gas chromatography
- High performance (pressure) liquid chromatography

Adsorption Chromatography

Adsorption is a phenomenon in which compounds are held onto the surface of a solid adsorbent, having specific adsorption sites, through weak non-ionic interactions such as vander Waal's forces and hydrogen bonding. Different compounds bind with varying strengths and hence can selectively be desorbed. For good resolution, selection of right type of the adsorbent and the eluent or mobile phase is essential. Some of the commonly used adsorbents include charcoal, silica, alumina, hydroxyaptite, etc. Eluent influences quality of separation since polarity of the mobile phase influences the adsorption considerably. Non-polar solvents favour maximum adsorption which decreases with increase in polarity of the solvent. In general, the polar solvents are preferred for the substances having polar or hydrophilic groups and non-polar solvents for substances having hydrophobic or non-polar groups. For example, alcoholic solvents containing –OH group substances ; acetone or ether for substances with carbonyl groups and hydrocarbons such as toluene or hexane for non-polar substances. For gradient elution, mixture of the polar and non-polar solvents of different ratios can be used to obtain eluent of varying polarities.

Gel Filtration (Size Exclusion Column Chromatography)

The chromatographic media used in this technique are porous, polymeric organic compounds with molecular sieving properties. These are cross-linked polymers which swell considerably in water forming a gel of a three-dimensional network of pores. The size of the pore is determined by degree of cross-linking of polymeric chains. Different solutes in a mixture get separated on the basis of their molecular size and shape during their passage through a column packed with the swollen gel particles. The terms "exclusion chromatography", "gel filtration" and "molecular sieve" chromatography are used for this separation process. The large molecules in sample are unable to penetrate through the pores into the gel and thus remain excluded while the small molecules enter into the gel beads. Obviously the volume of solvent accessible to large molecules is very much less (V_o), whereas small molecules, which can freely penetrate into the gel have access to solvent inside (V_o) the spherical beads.

Different Applications of Gel Filtration

The major applications of gel filtration phenomenon have been described below:

- *Desalting or group separation*: In a number of experiments as during preparation of nucleic acids and protein purification, inorganic salts or other low-molecular weight reagents are extensively used, which often have to be removed from the final preparation. This can effectively be achieved by passing preparation through a column of Sephadex G-10 or G-25. The high-molecular weight macromolecules like proteins and nucleic acids remain excluded from the gel particles and are recovered from the column immediately after void volume, while the movement of low-molecular weight compounds, such as salts, is considerably impeded due to their entry into the beads. This method of desalting is relatively faster than the process of dialysis and hence is particularly suitable for desalting of labile compounds.
- *Concentration of dilute solution*: Dry Sephadex (G-25, coarse) when added to a dilute solution of high-molecular weight substance absorbs water during swelling. The high-molecular weight substances, however, remain in solution due to their complete exclusion. Because of reduced volume, the concentration of the higher-molecular weight substances in solution increases without any effect on pH and ionic strength. The added beads are then removed from the solution by centrifugation.
- *Purification of materials*: The availability of a large variety of gel media (dextran, polyacrylamide, agarose) with varying fractionation range has extended the use of this technique for separation as well as for purification of biological materials ranging from low-molecular weight compounds like sugars, polymers like proteins and nucleic acids to superstructures such as viruses. Generally, for purification or fractionation purposes, the gel selected is such that the compound to be purified falls within the fraction range of the gel.
- *Molecular weight estimation*: As mentioned earlier, separation of various components in the sample is based on the differences in their molecular size. Larger the molecular weight of a compound, lesser will be its elution volume. In fact, this method has been exploited for determination of molecular weight of

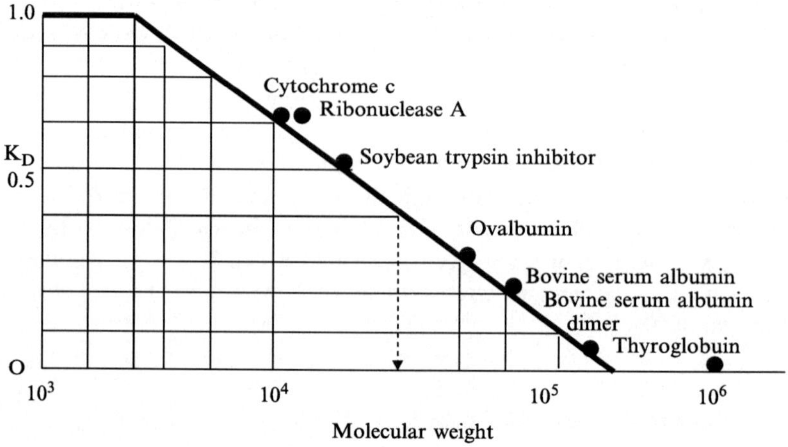

Fig. 4.5 Relationship between K_d and the logarithm of the molecular weight of proteins as determined by column gel filtration on Sephadex G-150

Fig. 4.6 A graph showing relation between molecular weight and elution volume during gel filtration. A–E are standard proteins of known molecular weight and X* is the protein whose molecular weight is being determined

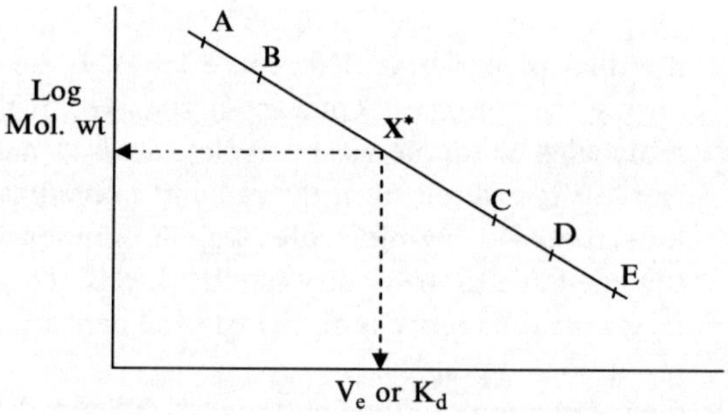

a solute, particularly proteins, from their elution characteristics (V_e or K_d value). A plot of log molecular weight vs. elution volume or K_d gives a linear curve. Hence, by using proteins of known molecular weights, a calibration curve can first be prepared and form the elution volume or K_d value, the molecular weight of protein of interest can readily be obtained. This method of molecular weight determination is simple, inexpensive and does not require homogeneously purified preparation of the protein or compound of interest and is non-destructive (Figs. 4.5 and 4.6).

Ion Exchange Chromatography

Ion exchange chromatography is a type of adsorption chromatography in which retention of a solute occurs due to its reversible electrostatic interaction with the oppositely charged groups on an ion exchanger. Hence, this technique is useful for separation of compounds which bear a net electric charge such as proteins, amino acids, nucleic acids, etc. Ion exchangers are prepared from either certain synthetic resins which are insoluble porous organic molecules or naturally occurring biopolymers such as cellulose to which various groups known as fixed ions are covalently attached. These fixed ions are balanced by equal and oppositely charged ions from the solution referred to as counter ions. Depending upon the nature of the counter ions, these ion exchangers are of two types: *cation exchangers* in which the counter ions are cationic or positively charged and *anion exchangers* which have negatively charged counter ions. Counter ions are mobile and can be easily exchanged by other similarly charged molecules in the sample. Nature of the resin matrix remains unchanged during this exchange process. Generally, resin-based ion exchangers are used for separation of low-molecular weight biomolecules and cellulosic ion exchangers are more suitable for isolation of marcromolecules such as proteins and nucleic acids. Ion exchangers have to be precycled to get an appropriate charge and their complete swelling. In case of anion exchanger, it is treated first with alkali and then with acid and finally washed with water till it is neutral. Conversely, cation exchangers is first given the acid treatment and then alkali treatment and finally washed with water till neutral. After precycling, the ion exchanger is packed into a column and is equilibrated with the counter ion by passing 2–3 bed volumes of the starting buffer of a required pH. In case of anion exchanger, Tris–HCI buffer and for cation exchanger K^+ or Na^+ buffer is used.

Affinity Chromatography

Purification by affinity chromatography is different from all other forms of chromatography in the sense that this technique does not make use of the differences in the physical properties (like solubility, adsorption, molecular weight and ionic properties) of the molecules to be separated, rather it exploits one of the unique and fundamental properties of biopolymers, i.e. the specificity of their interaction with other biomolecules. Therefore, affinity chromatography is a type of adsorption chromatography in which the substance to be isolated is specifically and reversibly bound to a complementary binding substance (ligand) immobilised on an insoluble chromatographic bed material (matrix). The other substances in the mixture remain unbound and are washed away while the substance of interest (the bound substance) is subsequently recovered by displacement from the ligand either by specific (affinity) elution or by non-specific (change in pH or ionic concentration) elution.

Purification by affinity chromatography is often of the order of several thousandfold recoveries of the material which are generally very high. The operation of this technique needs following information:

- Structure and biospecificity of the compound to be isolated
- Various biospecific ligands which can be covalently attached to the chromatographic bed material (matrix)
- Whether the ligand retains its biospecific binding affinity for the substance of interest after its attachment to the matrix

- Different methods for selectively desorbing the bound substance in an active form after washing away the unbound substances

Before selecting a ligand, two attributes must be kept in mind:

1. The ligand must exhibit specific and reversible binding with the substance to be purified.
2. It must have chemically reactive functional groups which allow it to be attached to the matrix without destroying its binding activity with the substance of interest.

It is possible to select a ligand which displays absolute specificity and binds exclusively with one particular compound only. It is also possible to select a ligand which displays group specificity, e.g. for the purification of enzymes a substrate analogue, or an inhibitor, a cofactor or an activator can be used as ligand while a specific receptor or a carrier protein can be used for the purification of vitamins and hormones. Antigen–antibody interactions can be exploited for the purification of either of these. Similarly, for nucleic acid purification a complementary base sequence or histones and for lectins, cell surface receptors or polysaccharides can be successfully employed as ligands. If several functional groups are available, the ligand should be coupled with the matrix via the group like $-NH_2$, –COOH, –SH and –OH.

An ideal chromatographic bed material (matrix) to which the ligand is covalently bound must possess the following attributes:

1. It must possess suitable groups to which ligand can be covalently coupled. Many groups may be introduced into matrix to couple ligands. They may be nucleophilic as NH_2, SH, OH or electrophilic such as activated acid chlorides, carbonyls activated by carbodiimide, iothiocyanate or diazonium salts.
2. It must remain unchanged under the conditions of attachment of ligand.
3. During the binding of the macromolecule and its subsequent displacement from ligand, it must retain its physical and chemical stability.
4. It must not exhibit non-specific adsorption.
5. It should have an open pore structure.

The most commonly used matrices are cross-linked dextrans (e.g. Sephacryl), agarose (e.g. Sepharose), polyacrylamide gel (Bio gel P), polystyrene, cellulose, porous glass and silica. Sepharose is a bead-form of agarose gel which displays virtually all features required of a successful matrix for immobilizing biologically active molecules. The hydroxyl groups on the sugar residues can be easily derivatized for covalent attachment of a ligand.

Using Spacer Arm During Affinity Separations

Generally, it is observed that the active site of the biological substance, e.g. enzyme is located deep within the molecule and the attachment of the ligand directly to the matrix, sometimes, interferes with its ability to bind the macromolecule due to

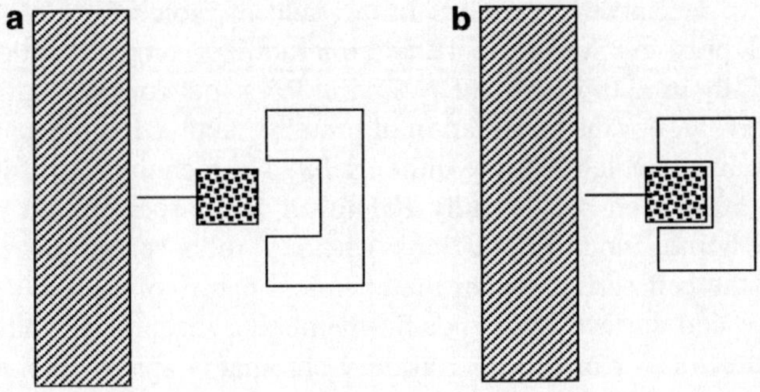

Fig. 4.7 Separation by affinity chromatography

steric hinderances between the matrix and the substance to be bound to the ligand. This interference can be prevented by interposing a "spacer arm" between the matrix and the ligand to facilitate effective binding as given in the figure.

The length of the spacer arm is one of the important factors. If it is too short, the arm is ineffective and the ligand fails to bind the substance in the sample. If it is too long, non-specific effects become pronounced and reduce the selectivity of separation. The optimum length of the spacer arm is generally 6–10 C-atoms or their equivalent. Chemical nature of the spacer arm (e.g. whether hydrophobic or hydrophilic) is also critical for the success of separation. There are two approaches which are generally followed for coupling of the ligand. In one case, spacer arm is first linked to the matrix followed by the coupling of the ligand, whereas the second approach involves the binding of the spacer arm to the ligand which is then linked to the matrix. It is the first approach which is more convenient and is preferred over the second one (Fig. 4.7).

For attachment of the ligand with the matrix, the matrix is given preliminary treatment with cyanogens bromide at pH 11. This causes activation of the matrix and the molecules containing primary amino groups could then easily be coupled to CNBr activated matrices. Different spacer arms including 1,6-diamino hexane, 6-amino hexanoic acid, and 1,4-bis(epoxy-propoxy)butane have been used to which the ligand can be attached by conventional organo synthetic procedures involving the use of succinic anhydride and a water soluble carbodiimide. A number of supports of agarose, dextran and polyacrylamide type are commercially available with a variety of spacer arms and ligands.

Applications of Affinity Chromatography

Affinity chromatography occupies a unique place in separation technology since it is the only technique which enables purification of almost any biomolecule on the basis of its biological function. The principle of affinity chromatography has been extended to purify a large number of enzymes, other proteins including immunoglobulins and receptor proteins and nucleic acids and so has contributed considerably to recent developments in the field of molecular biology. Poly (U) Sepharose 4B, poly (A) Sepharose 4B and immobilized single-stranded DNA have been successfully used to isolate mRNA, viral RNA and complementary RNA and DNA, respectively. For the purification of proteins involved in nucleic acid metabolism, immobilized nucleotides are quite useful. The technique of affinity chromatography has also been successfully employed for the separation of a mixture of cells into homogenous populations where it relies either on the antigenic properties of the cell surface or on the chemical nature of exposed carbohydrate residue on the cell surface on a specific membrane receptor–ligand interactions. Useful modifications or methods of affinity chromatography technique have also been developed.

Gas Chromatography

Principle

Gas chromatography (GC) as the name suggests, is particularly suited for the separation of gases and volatile liquids or solids in their gaseous state. The compounds of low polarity are best separated by GC. The technique is highly sensitive, reproducible and has high speed of resolution. When the stationary phase is an active solid such as silica, the method is referred to as gas solid chromatography (GSC). However, if the stationary phase is a liquid such as polymers of silicone coated onto the surface of an inert granular solid then the technique is known as gas-liquid chromatography (GLC). The stationary phase, whether a solid or a liquid coated as thin film on surface of a solid support, is packed in a glass or stainless steel column which is narrow, coiled, 1–3-m long and with 2–4 mm internal diameter. An inert carrier gas (mobile phase) such as nitrogen, helium or argon is made to flow through the column. The temperature of the column is maintained high in an oven to keep the compound to be separated in their volatile state. These volatilized compounds get partitioned between the liquid or solid stationary phase and the gaseous mobile phase and hence get separated because of differences in their partition coefficients. After leaving the column, the separated compounds pass through the detector, sensed and recorded by the recorder.

Capillary columns of internal diameter 0.03–1 mm and length up to 100 mm made of glass or steel are used for performing GC. Two types of capillary columns viz. wall coated open tubular columns (WCOT) and support coated open tubular columns (SCOT) are available for this purpose. In WCOT, as the name suggests, walls of the capillary column are coated with the stationary phase. Since the stationary phase is a liquid and is directly coated on the walls of the capillary column, only a small amount of the stationary phase is present in this system. Accordingly, only a small amount of the sample can be applied on to the WCOT column. In SCOT, the stationary phase is in the form of a thin layer on surface of a solid support which in turn is packed into the capillary column. Hence, the capacity

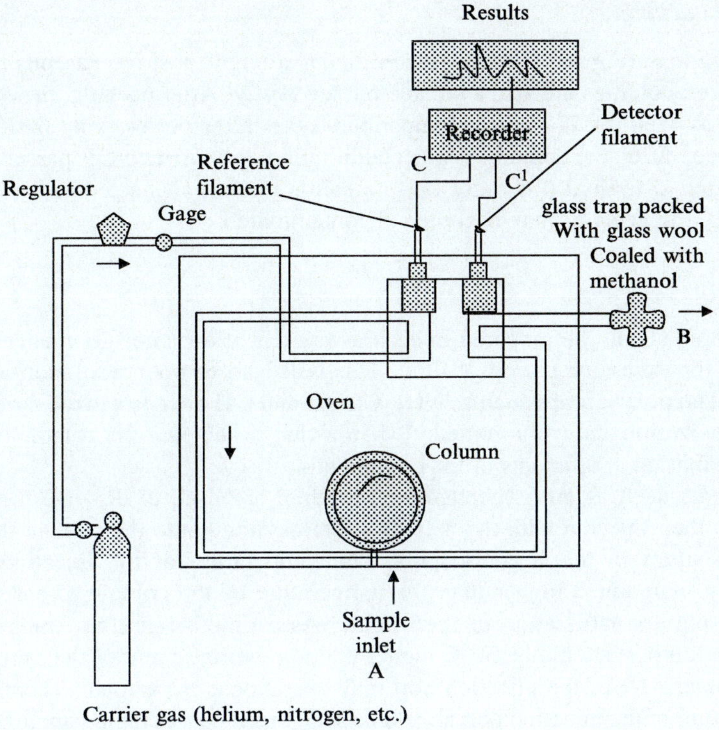

Fig. 4.8 Construction of gas chromatography system

of SCOT system is much higher than that of WCOT and a larger amount of sample can be applied (Fig. 4.8).

Solid Support and Stationary Phase

The purpose of the solid support is to provide a large uniform, inert surface area for holding a thin layer of the liquid stationary phase. The support should be inert, and should have high mechanical strength, large surface area, regular shape and uniform size. The most commonly used support is celite, the OH groups of which are modified by silanization with hexamethyl disilazane to minimize interaction with the sample. The correct choice of the stationary phase is perhaps the most important parameter in GC. Ideally, the stationary phase must be non-volatile and thermostable at temperature used for analysis. It should be chemically inert towards the solutes of interest at the column temperature. The high boiling point organic compounds such as polyethylene glycols, methyl-phenyl and methyl-vinyl silicon gums, esters of adipic, succinic and phthalic acid, polyesters, polyethylene glycols are used as liquid stationary phases. The operating temperature must be compatible with the phase chosen. At very high temperature, the phase may get volatilized and cause excessive column bleeding which may contaminate the detector.

Column Packing

The columns are generally dry packed under a slight positive gaseous pressure. Pre-packed columns are also available but are costly. After packing, the column is kept in an oven for 24–48 h at temperature near the upper working limit. This is done to condition the column. While conditioning, the carrier gas is passed through the column at normal flow rates but the column is not connected to the detector, otherwise the detector may become gel contaminated.

Sample Preparation

The sample should be prepared in such a way that it does not get retained on the column for excessive period of time. This will lead to poor resolution and peak tailing. The polar groups such as NH_2, COOH and OH are derivatized by methylation, silanization and trifluoromethyl silanization to increase the volatile character and distribution coefficients of the compounds.

Solvents such as ether, heptane or methanol are used to dissolve the sample which is then injected with the help of a microsyringe onto the column through a rubber septum in the injection port. The temperature of the injection port is generally maintained higher than the temperature of the column to ensure rapid and complete volatilization of the sample. As a thumb rule, the temperature of the injection port should be 50°C higher than the boiling point of the sample. Too high temperature of the injection port may decompose the sample. Therefore, the temperature of the injection port should be such that it causes rapid vapourization of the sample without decomposing it.

Carrier Gas

The primary function of the carrier gas is to carry the volatile components through the column. The gas used should be inert and should not react either with sample or with stationary phase. Its secondary purpose is to carry the separated components to the detector so that it is suitable for detector use. It should be readily available in extra pure form and be inexpensive.

Normally, nitrogen, helium and argon are the three most commonly used carrier gases. The column temperature must be high enough so that analysis can be accomplished in a reasonable length of time. The retention time doubles with every 30°C decrease in column temperature. Lower the temperature better is the resolution and longer is the analysis time. Therefore, a balance has to be struck between the peak retention time and resolution. Chromatographic separation can be achieved isothermically where a constant temperature is employed or by temperature programming where the temperature is increased gradually.

Detectors

Characteristics such as selectivity, sensitivity, response, noise, minimum detectable quantity and linear range, should be given consideration while making choice of the detector. The detector should be simple to operate, inexpensive and as far as possible insensitive to changes in flow rate and temperature. Commonly used detectors during GC operations are:

- *Flame ionization detector* (FID): Since it responds to almost all organic compounds, it is the most widely used detector. It has a wide linear response range (10^6) and can detect as low a concentration as 1 ng. The detector consists of two electrodes. One of the electrodes is the jet of the flame which is produced by introducing a mixture of hydrogen and air into the detector, while the other electrode is made of brass or platinum wire which is mounted near the tip of the flame. When the carrier gas carrying the sample components emerges from the column, the sample signal is recorded by the recorder. An FID has an upper temperature limit of 400°C and the minimum quantity which can be detected by this detector is of the order of 5×10^{-12} g.
- *Nitrogen phosphorus detector* (NPD): NPD responds efficiently to detect the compounds containing N and/or P. It shows poor response for the compounds which possess neither of these elements. This detector has an upper temperature limit of 300°C, narrow linear response range of 10^4 and detection limits of 10^{-11} g s^{-1}. The principle of operation is same as that of FID but the NPD has the sodium salt fused onto the electrode system. NPD is widely used in the analysis of organophosphorus pesticides.
- *Electron capture detector* (ECD): The compounds which have the capacity to capture the electrons are best detected by this detector. Here a radioactive source (^{63}Ni) ionizes the column gas and produces electrons which give a current across the electrodes to which suitable voltage is applied. When the carrier gas carrying the electron capturing substance emerges from the column, it captures the ionized electrons. This results in the drop of the current which is traced on a chart paper by the recorder. The detector has upper temperature limit of 300°C, high detection sensitivity (10^{-12} g s^{-1}) but much lower linear range (10^2–10^4). The detector is best suited for the halogen-containing compounds such as pesticides DDT, dieldrin and aldrin.

Amplifiers and Recorders

When components leave the column and pass through the detector discussed above, small and weak electrical signals are produced which are amplified by an amplifier before they are fed to the recorder. Recorders generally consist of two basic parts viz. a chart paper and a pen which moves on the chart paper and traces the signals being activated from the amplifier in the form of peaks.

High Performance (Pressure) Liquid Chromatography

High performance liquid chramotography or high pressure liquid chromatography (HPLC) also works on the basis of partitioning, adsorption, ion exchange or molecular sieving phenomena. The conventional column chromatography suffers from two major flaws as it is generally a time consuming process and quality of resolution is poor. This is mainly because of the fact that in conventional column chromatography the mobile phase percolates through the column under the force of gravity or by small pressure applied by peristaltic pump. This accounts for the slow flow rate, which in addition to extending the time required for elution of the sample creates the problem of peak broadening through diffusion phenomenon resulting in poor resolution. In general, resolution of individual components can be improved by decreasing the particle size of stationary phase. However, in conventional column chromatography, this is not feasible because the use of finer gel material will further lower the permeability of the column contributing to the resistance to flow of mobile phase, which can be overcome by use of high pressure. Therefore, the stationary phases of smaller particle size, which can withstand high pressures, have been developed and have facilitated the development of a new chromatographic technique called HPLC, which gives faster and superior resolution with sharp peaks.

Components of HPLC

The basic HPLC equipment consists of the following components: (i) solvent reservoir, (ii) pump, (iii) damping device, (iv) pressure gauge, (v) sampling device, (vi) column, (vii) detector, (viii) fraction collector and (ix) recorder. The essential features of HPLC equipment are shown in Fig. 4.9.

The pump delivers the solvent from the reservoir at a constant flow which is smoothed out by means of damping device. From the damping unit, the mobile

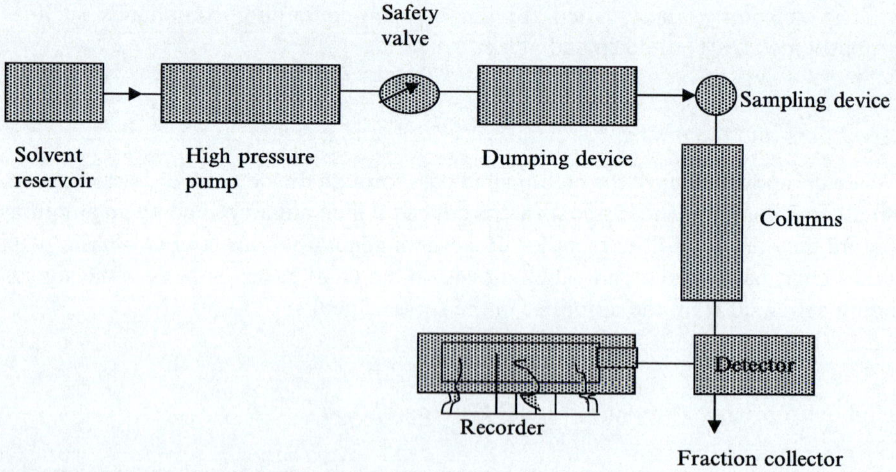

Fig. 4.9 Essential components of HPLC

flows via sample injector into the column. The inlet pressure of the column is monitored with a manometer. After leaving the column, the sample components are monitored by the detector and their tracings drawn by the recorder.

Solvent reservoir: The solvent used in HPLC must be of high purity as any traces of impurities or suspended material can seriously affect the column efficiency and can interfere with the detection system. Solvents, as supplied commercially, contain substantial amount of dissolved air. Formation of air bubbles can seriously interfere with satisfactory separation by HPLC because the air bubbles affect the column efficiency and also the solute detection. Thus, some of the conventional solvent reservoirs are equipped with degassifier or the solvent can also be degassed (by heating, stirring, subjecting it to vacuum, ultrasonic vibrations or bubbling helium gas) before pouring it into the reservoir. Highly pure solvents are available commercially, but even with such solvents it is advisible to introduce a 1–5-μm microfilter prior to the pump to prevent any particulate impurities from entering into the column.

Pumps: In HPLC, the column is quite narrow and is packed with superfine particles. There is high resistance to flow of solvent and high pressures are, therefore, required to achieve satisfactory and constant flow rates. Therefore, a good pumping system which delivers pulse free solvent flow up to 20 mL/min at pressures to the extent of 300–400 atmospheres is one of the most important features in HPLC. All materials in the pump should be chemically resistant to all the solvents used in HPLC. Various pumping systems operate on the principle of either constant pressure or constant displacement.

Constant pressure pumps facilitate delivery of the solvent at a constant pressure. A gas at high pressure is introduced into the pump which forces, in turn, the solvent from pump into the column. Constant pressure is maintained throughout, which causes a decrease in permeability of the column with time which in turn results in decreased flow rates. Such pumps do not compensate for this decrease in flow rate and so provide uniform and pulseless solvent flow.

The second type of pumps are the constant displacement pumps, which displaces a constant amount of the solvent from the pump into the column and so maintains a constant flow rate irrespective of the changing conditions within the column. These pumps produce small pulse of flow between two displacements and so pulse dampners are usually introduced between the pump and the column to smoothen the flow and to minimize the pulsing effect. Two commonly used constant displacement pumps are:

(i) Motor driven syringe type pump.
(ii) Reciprocating pump which delivers a fixed and constant volume of the solvent onto the column at each stroke.

Sample injection: Sample can be introduced into the column either by a syringe injection through a septum of an injection port or by a sample loop from which it is swept into the column by the eluent. The sample is loaded directly on top of the column to avoid appreciable mixing of the sample with the eluent. In the syringe injection mode, the sample is injected with help of a microsyringe (which can withstand high pressures) directly onto the column bed. While loading the sample, the system should not be under pressure. Hence, before applying the sample, the pump is turned off and when the pressure is dropped near atmospheric pressure, the sample is introduced. After the sample has been injected, the pump is switched on again. This procedure is known as stop flow injection.

The second type of system is loop injection. Here the sample is introduced with the help of a metal loop of fixed small volume. The loop is filled with the sample and by appropriately adjusting the sample valve, the solvent from the pump is channelled through the loop. The sample is thus flushed by the solvent from the loop whose outlet opens directly at top of the column bed.

HPLC columns: Since glass tubing cannot withstand pressures in excess of 70 atm, stainless steel precision bored columns with an internal mirror finish for efficient packing, are normally used. These straight columns of 15–50 cm length and 1–4 mm diameter can withstand very high pressures of up to 5.5×10^7 Pa and are relatively corrosion resistant. At the end of the column, homogenously porous plugs of stainless steel or Teflon are used to retain the packing material and to ensure the uniform flow of the solvent through the column. At times, repeated application of impure samples may result in clogging and the loss of resolving power of the column. To prevent this, a short column of length 1–2 cm and internal diameter equal to that of analytical column is generally introduced between the injector and the analytical column. This short column is called guard column and is packed with material with which analytical column is packed. The guard column retains the solid particles in the sample before it enters the main column. The guard columns can be replaced at regular intervals.

Stationary phases: One of the basic requirements for HPLC is that the packing material which serves as stationary phase or support for stationary phase should be pressure stable, and withstand the operating pressure applied during separation (Table 4.1). Three forms of column packing materials are available based on the nature of the rigid solid structure:

(i) Totally porous materials or microporous supports: In these supports the micropores ramify through particles which are generally 5–10 μm in diameter.
(ii) Porous layer beads or pellicular supports: These are superficially porous supports where a thin, porous, active layer is coated onto a solid core such as impervious glass beads. The thickness of the porous layer is generally 1–3 μm. The size of glass beads used is between 25 and 50 μm.
(iii) Bonded phases: The stationary phase is chemically bonded to an inert support such as silica. The type of particular stationary phase, with their commercial names, which can be used for different types of chromatographic separations are listed in Table 4.1.

Detectors: Detectors are the devices which continuously monitor changes in the composition of the eluent coming out of the column. Most commonly used

Table 4.1 HPLC stationary phases

Chromatographic separation principle	Commercial name	Type of stationary phase	Name of support
Partition	ULTRA Pak TSK ODS	Octadecylsilane	Porous
	ULTRA Pak TSK NH_2	Alkylamine	Porous
	Bondapak-C18/Corasil	Octadecylsilane	Pellicular
	μ Bondapk-NH_2	Alkylamine	Porous
Adsorption	Corasil	Silica	Pellicular
	Partisil C8	Octylsilance	Porous
	Pellumina	Alumina	Pellicular
	Micropak	Alumina	Microporous
Exclusion	Superose	Agarose	Soft gel
	Fractogel TSK	Polyvinylchloride	Semi-rigid gel
	Bio-Glas	Glass	Rigid solid
	Styragel	Polystyrene divinyl benzene	Semi-rigid gel
Ion exchange	Perisorb-KAT	Strong acid	Pellicular
	Partisil-SAX	Strong base	Porous
	Micropak-NH_2	Weak base	Porous
	Partisil-SCX	Strong acid	Porous

detectors are refractive index detector, UV detector, electrochemical and fluorometric detectors.

1. *Refractive index detector* (RID): Refractive index (RI) of dilute solutions changes proportionally with solute concentration. This relationship is exploited for quantitative detection of solutes in the column eluate. The relationship between the change in RI and solute concentration is only moderately dependent on the type of solute, making this a quite universal, yet not very sensitive detection principle, RID can, therefore, be applied to general purpose. This detector suffers from many defects including low sensitivity, tendency to be affected by temperature or flow speeds and incompatibility for being used in gradient elution unless chosen solvents are identical RI. It measures the bulk RI of sample eluent system. Hence, any substance whose RI differs sufficiently from that of eluent can be detected. To attain adequate sensitivity, the temperature of the eluent and measuring cell is held constant to ±0.001°C. Variation in flow rates also interfere with response of differential refractometer. Hence, very good damping is essential for the pumps producing pulsating flow.
2. *UV-VIS-absroption detectors*: The basis of quantitative absorbance photometry is Lambert–Beer's law, i.e. the absorbance of a solution is proportional to the concentration of the absorbing solute, the light path length and the extinction coefficient. Fixed wavelength detectors utilize lamps which emit light of a few discrete wavelengths. The most common of these lamps is the low pressure mercury lamp emitting over 90% of its light at 254 nm. Lower wavelength lamps such as zinc lamps (214 nm) and cadmium lamps (229 nm) are available. The combination of the lamp and a filter determines the fixed operating wavelength of the detector. Variable wavelength (VW) detectors use a light source

with a continuous emission spectrum and a continuously adjustable (narrow) band filter, called monochromator. The most common light source for these detectors is the deuterium lamp whose usable emission spectrum ranges from about 190 nm to about 350 nm, with an intensity maximum between 220 and 240 nm. Above 300 nm, the output intensity is low, therefore, some VW at wavelength above 350 nm. Specially designed VW detectors have an additional or optional tungsten lamp, which can be used as detectors for HPLC have been introduced which allow automatic rapid change of wavelength setting within 1–2 s or less across their entire wavelength span, which typically ranges from about 190 nm to about 600 nm.

3. *Electrochemical detectors*: Methods used in HPLC based on electroanalysis can be classified as bulk property and solute property electrochemical detectors. Bulk property electrochemical detectors respond to a change in an electrochemical property of the bulk liquid flowing through the measuring cell , whereas the solute property electrochemical detectors respond to a change in voltage (potentiomety) or current (voltammetry or coulometry) when an analyte passes through the cell.
4. *Fluorescence detectors*: The quantity of fluorescent light emitted from excited molecules in dilute solutions is proportional to the intensity of excitation source, illuminated volume of the sample solution, quantum efficiency of fluorescence of sample and the concentration of the solute to be detected. Ideally, fluorescence radiation, as a result of suitable excitation of the sample molecules, is measured against a dark background. Therefore, the main source of detector noise is dark current noise of the photodetector, which is mainly determined by temperature. As rule for every 10°C increase in temperature the dark current noise of the photodiode gets doubled.

4.4 Centrifugation

Centrifugation is one of the common step during different biochemical procedures. The term "*g*" is commonly used to specify the gravitational force exerted on the homogenate being centrifuged. It is defined as the gravitational force acting on a 1 g mass at distance *r* (cm) from the axis of rotation. It can be readily calculated from the formula

$$F = \frac{S^2 r}{89,500},$$

where *F* is the relative centrifugal force (*g*), *r* is the radial distance (cm) from the center or axis of rotation and *S* is the speed of rotation of the rotor (rpm). Thus, in the given figure, *F* has been considered 6,200 × *g* at the bottom of the centrifuge tube (Fig. 4.10).

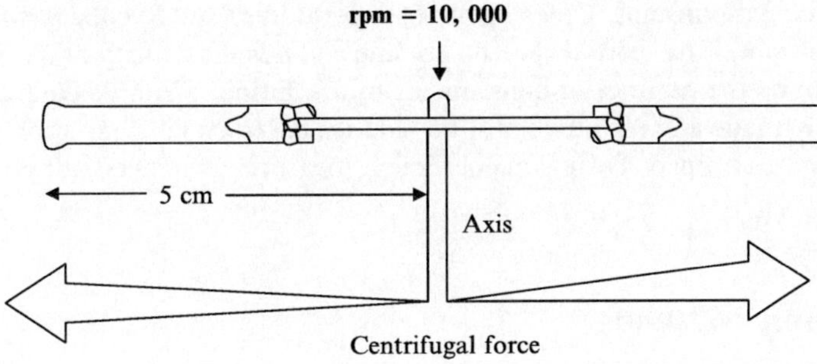

$$F = \frac{(10{,}000)^2 5}{89{,}500} = 6200 \times g$$

Fig. 4.10 Centrifugal force from the axis of rotation

4.5 Ultracentrifugation

This instrument can measure certain properties of a molecule, such as molecular weight, shape, size and density, as well as the number of components in a protein solution. The ultracentrifuge subjects a small volume of solution (less than 1 mL contained in a quartz cell) to a carefully controlled centrifugal force, and records, by means of optical and photographic systems, the movement of the macromolecules in the centrifugal field. The solute molecules, which are initially uniformly distributed throughout the solution in the cell, are forced toward the bottom of the cell by the centrifugal field. This migration leaves a region at the top of the cell that is devoid of solute and contains only solvent molecules. The migration also leaves a region in the cell where the solute concentration is uniform. A boundary is set up in the cell between solvent and solution in which concentration varies with distance from the axis of rotation. The measurement of the boundary's movement, which represents the movement of the protein molecules, is the basis of the analytic method. By the data obtained, namely, the sedimentation rate, the Svedberg unit (S) can be calculated. A Svedberg unit, named in honour of T. Svedberg, the Swedish pioneer in the field, is defined as the velocity of the sedimenting molecule per unit of gravitational field or 1×10^{-13} cm/s/dyne/g. Typical S values are 4.4 for bovine serum albumin, 1.83 for cytochrome c, and 185 for tobacco mosaic virus. With a knowledge of the diffusion coefficient, molecular weights can be readily calculated. The basic equation relates S and molecular weight:

$$\text{mol wt} = \frac{RTS}{D(1 - Vp)},$$

where R is the gas constant, T the absolute temperature, S the Svedberg unit, D the diffusion constant, V the partial specific volume and p is the density of the solution.

To determine the number of components in a solution, a simple centrifugation can be readily made and the number of boundaries based on concentration gradient peaks can be determined. Diffusion coefficient measurements need not be made.

4.6 Isotopic Methods

Radioisotopes. The most useful radioisotopes, ^{14}C, ^{35}S, ^{32}P and ^{3}H, are β-ray emitters; that is, when the nuclei of these atoms disintegrate, one of the product is an electron which moves with energies characteristic of the disintegrating nucleus. The β rays interact with the molecules through which they traverse, causing dissociation, excitation or ionization of the molecules. It is the resultant ionization property which is used to measure quantitatively the amount of radioisotope present.

Units. A curies is the amount of emitter which exhibits 3.7×10^{10} disintegrations/s (dps). More common units are a millicurie, mc (10^{-3} curie), and a microcurie, μc (10^{-6} curie).

Specific activity. This is defined as disintegrations/minute per unit of substance (mg, μmole, etc.).

Dilutions factor. The factor is defined as

$$\frac{\text{Specific activity of precursor fed}}{\text{Specific activity of compound isolated}}.$$

This factor is used frequently to express the precursor relation of a compound in the biosynthesis of a second compound. Thus, in the sequence $A \rightarrow B \rightarrow C \rightarrow D$, the dilution factor for $C \rightarrow D$ would be small, whereas for A it would be large. Therefore, a small dilution factor would indicate that compound C fed to a tissue has better precursor relationship to the final product that compound A with a large dilution factor.

Percentage of incorporation. This is also useful to compare the proximity of a precursor in the biosynthesis of a second compound. If labelled, compound A is administered to an experimental system and some of the radioactivity is incorporated into compound D, the percentage of incorporation expressed as curies (or microcuries) in D is divided by curies (or microcuries) in A × 100 [(Ci of D/Ci of A) × 100].

Stable isotopes. Stable isotopes of several of the biologically important elements are available in enriched concentrations and therefore may be used to "tag" or label compounds. As an example, deuterium, the hydrogen atom with mass of 2, is present in most H_2O to the extent of 0.02% only. The remainder of the hydrogen

Table 4.2 Some common radioisotopes

Element	Radiation	Half-life	Energy of radiation (meV) [a]
^{3}H	β−	12.1 years	0.0185
^{14}C	β−	5,100 years	0.156
^{32}P	β−	14.3 years	1.71
^{35}S	β−	87.1 days	0.169

[a] Million electron volts

atoms has a mass of 1. This concentration of 0.02% is known as the normal abundance of deuterium. It is possible to obtain heavy water in which 99.9% of the hydrogen atoms are deuterium. The concentration of a heavy isotope is usually measured as atom% excess; this is the amount, in percent, by which the isotope exceeds its normal abundance. Thus, the two stable isotopes of nitrogen are $_{7}N^{14}$ and $_{7}N^{15}$, which have a normal abundance of 99.62 and 0.38%, respectively. If a sample of nitrogen gas contain 4.00%$_{7}N^{15}$ (and 96.00%$_{7}N^{14}$), the concentration of $_{7}N^{15}$ in this sample is said to be 3.62 atom% excess. Other stable isotopes that are available in enriched concentrations and therefore may be used as tracers in biochemistry are $_{8}O^{17}$, $_{8}O^{18}$, $_{6}C^{13}$, $_{16}S^{33}$ and $_{16}S^{34}$; the normal abundance of these isotopes can be found in any chemical handbook. The principles underlying the use of stable isotopes are similar to those employed with radioisotopes (Table 4.2).

Measurements. Liquid scintillation counting is the most popular technique for measuring radioisotopes. The technique is based on the use of a scintillation solution containing fluors and a multiplier phototube. The scintillation solution converts the energy of the radioactive particle into light; the multiplier phototube responds to the light by producing a charge which can be amplified and counted by a scaling circuit.

In liquid scintillation counting, the radioactive substance is usually dissolved in a suitable organic solvent containing the fluor. Alternately, the radioactive sample, which can consist of filter paper containing the sample, which can consist of filter paper containing the sample, is suspended or immersed in the scintillation fluid. Under these conditions the energy of the radioactive particle is first transferred to the solvent molecules, which may then ionize or become excited. It is the electronic excitation energy of the solvent which is transferred to the fluor (solute). When the excited molecules of the solute return to their ground state, they emit quanta of light that are detected by the phototube. One problem associated with this technique is the quenching of the light output by coloured substances in the sample. In addition, the fluor molecules may be quenched if foreign substances absorb their excitation energy before it is released as light. Methods are available for determining the amount of quenching exhibited by the radioactive sample. Scintillation counting is particularly useful for determining the weak β particles of tritium (^{3}H) and carbon-14(^{14}C). The efficiency of counting these particles can be as high as 50% and 85%, respectively.

Chapter 5
Carbohydrate Estimations

Carbohydrates defined as aldehydes or ketones of polyhydric alcohols, which also include those biopolymers, yield these compounds on hydrolysis. They occur in animals, plants as well as microorganisms and serve diverse structural and metabolic roles. Sugars such as glucose are among the major sources of energy whereas starch and glycogen function as storage polysaccharides in plants and animals, respectively. In addition, carbohydrates are structural components of cell walls, connective tissues in animals and exoskeletons of invertebrates. These are also constituents of vital biomolecules like nucleic acids, coenzymes such as NAD(P) FAD, etc.

5.1 Classification of Carbohydrates

Carbohydrates referred to as saccharides (Greek: Sakcharon, meaning sugar) are classified as given below:

1. *Monosaccharides*
   These carbohydrates are a single unit of polyhydroxy aldehydes or ketones and cannot be hydrolysed into simpler compounds. Simple sugars like glucose, arabinose, ribose, fructose, galactose, etc. are examples of monosaccharides.
2. *Oligosaccharides*
   They contain 2–10 U of monosaccharides (monomers) linked to each other via. glycosidic linkages, e.g. sucrose, lactose, maltose, raffinose, stachyose, etc.
3. *Polysaccharides*
   Polysaccharides are composed of ten or more monosaccharide units linked to each other with glycosidic linkages, e.g. starch, glycogen, cellulose, etc.

R. Katoch, *Analytical Techniques in Biochemistry and Molecular Biology*,
DOI 10.1007/978-1-4419-9785-2_5, © Springer Science+Business Media, LLC 2011

5.2 Qualitative Tests for Carbohydrates

During biochemical investigations it may become necessary to establish whether a given sample, particularly of a purified preparation, contains carbohydrates or not. Several rapid tests based on specific colour reactions typical for their group are available:

Name of test	Application
Molisch's test	General test for carbohydrates
Anthrone test	General test for carbohydrates
Iodine test	For glycans (starch, glycogen)
Barfoed's test	To distinguish between monosaccharides from reducing disaccharides
Seliwanoff's test	For ketoses
Fehling's test	For reducing sugars
Benedict's test	For reducing sugars
Picric acid test	For reducing sugars
Mucic acid test	For galactose
Bial's test	For pentoses

5.2.1 *Molisch's Test*

Principle

This is a general test for all carbohydrates. Conc. H_2SO_4 hydrolyses glycosidic bonds to yield monosaccharides which in the presence of an acid get dehydrated to form furfural and its derivatives. These products react with sulphonated α-naphthol to give a purple complex. Polysaccharides and glycoproteins also give a positive reaction (Fig. 5.1).

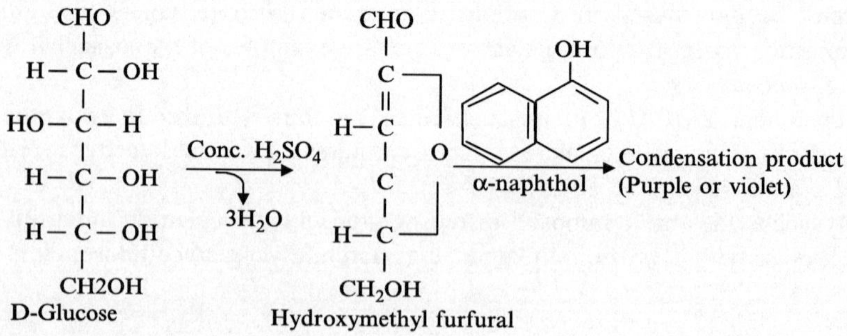

Fig. 5.1 Reaction for molisch's test

Reagents

1. Conc. H_2SO_4.
2. α-naphthol: 5% (*w*/*v*) in ethanol (prepare fresh).

Procedure

Add 2–3 drops of α-naphthol solution to 2 mL of the test solution. Taking precautions, gently pipette 1 mL conc. H_2SO_4 along the side of the test tube so that the two distinct layers are formed. Observe any colour change at the junction of two layers. Appearance of purple colour indicates the presence of carbohydrates.

Precautions

1. α-naphthol solution is unstable and should be prepared fresh.
2. Conc. H_2SO_4 should be added along the sides of the test tubes causing minimal disturbance to the contents in the tube.

5.2.2 *Anthrone Test*

Principle

Anthrone reaction is another test for carbohydrates. In this the furfural produced reacts with anthrone to give bluish green coloured complex (Fig. 5.2).

Materials and Reagents

1. Boiling water bath.
2. Conc. H_2SO_4.
3. 0.2% (*w*/*v*) anthrone solution in conc. H_2SO_4.

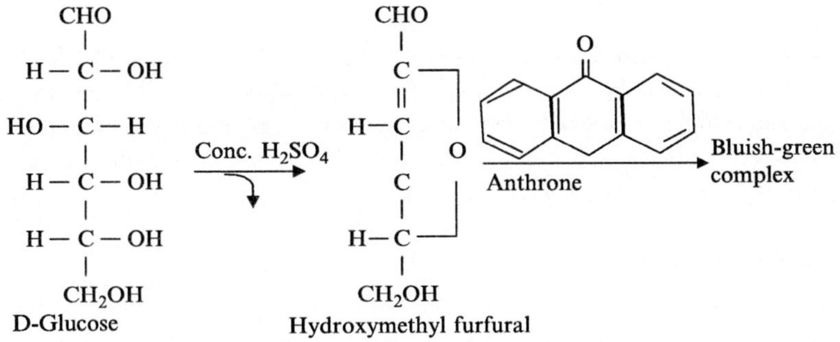

Fig. 5.2 Reaction for Anthrone test

Procedure

Add 0.5–1 mL of the test solution to about 2 mL of anthrone reagent and mix thoroughly. Observe whether the colour changes to bluish green. If not, examine the tubes again after keeping them in boiling water bath for 10 min.

5.2.3 *Iodine Test*

Principle

Iodine forms coloured adsorption complexes with polysaccharides. Starch gives blue colour with iodine, while glycogen reacts to form reddish brown complex. Hence, it is a useful, convenient and rapid test for detection of amylase, amylopectin and glycogen.

Reagents

1. Iodine solution: Prepare 0.005 N iodine solution in 3% (*w*/*v*) potassium iodide solution.
2. 1% Test solutions of glucose, sucrose, starch, glycogen, cellulose, etc.

Procedure

Take 1 mL of the sample extract or test solution in a test tube. Add 4–5 drops of iodine solution to it and mix the contents gently. Observe if any coloured product is formed. Note the colour of the product.

5.2.4 *Barfoed's Test*

Principle

This test is used for distinguishing monosaccharides from reducing disacchardes. Monosaccharides usually react in about 1–2 min while the reducing disaccharides take much longer time, between 7 and 12 min, to get hydrolysed and then react with the reagent. Brick red colour is obtained in this test, which is due to the formation of cuprous oxide (Fig. 5.3).

Material and Reagents

1. Boiling water bath.
2. Barfoed's reagent: Dissolve 13.3 g of copper acetate in 200 mL water and add 1.8 mL of glacial acetic acid to it.

$$(CH_3COO)_2Cu + 2H_2O \rightarrow 2CH_3COOH + Cu(OH)_2$$

Cupric acetate — Cupric hydroxide

$$Cu(OH)_2 \rightarrow CuO + H_2O$$

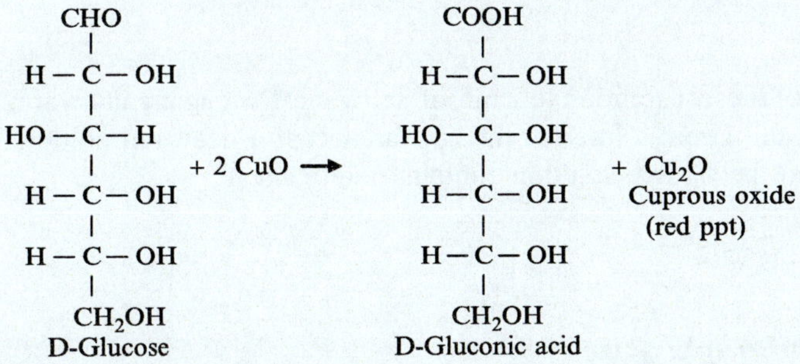

Fig. 5.3 Reaction for Barfoed's test

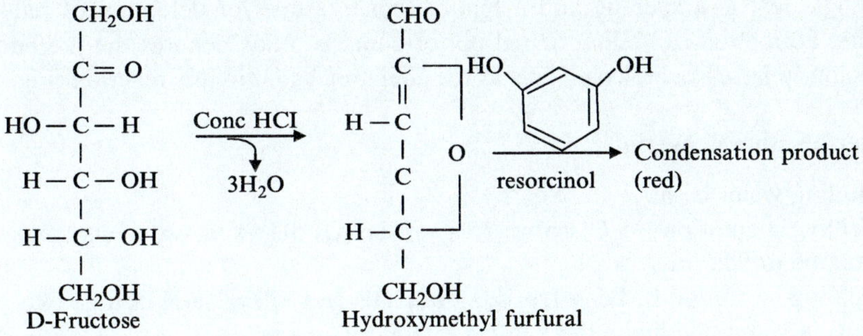

Fig. 5.4 Reaction for Seliwanoff's test

Procedure

Take 2 mL of Barfoed's solution in a test tube and add 1 mL of sample solution to it. Keep the test tubes in a boiling water bath. A briskly boiling water bath should be used for obtaining reliable results. Look for the formation of brick red colour and also note the time taken for its appearance.

5.2.5 Seliwanoff's Test

Principle

This test is used to distinguish aldoses from ketoses. Ketoses undergo dehydration to give furfural derivatives which then condense with resorcinol to form a red complex. Prolonged heating will hydrolyse disaccharides and other monosaccharides which also eventually give colour (Fig. 5.4).

Materials and Reagents

1. Boiling water bath.
2. Seliwanoff's reagent: 0.05% (*w*/*v*) resorcinol in 3 N HCl.

Procedure

Add 1 mL of the test solution to 2 mL of Seliwanoff's reagent and warm in a boiling water bath for 1 min. Note for the appearance of a deep red colour. This would indicate that the sample solution contains a keto sugar.

5.2.6 Fehling's Test

Principle

Fehling's test is a specific and a highly sensitive test for detection of reducing sugars. Formation of yellow or red ppt of cuprous oxide denotes the presence of reducing sugars. Rochelle salt acts as the chelating agent in this reaction (Fig. 5.5).

Materials and Reagents

1. Boiling water bath.
2. Fehling's solution A: Dissolve 35 g of $CuSO_4{\cdot}5H_2O$ in water and make the volume to 500 mL.
3. Fehling's solution B: Dissolve 120 g of KOH and 173 g Na-K tartrate (Rochelle salt) in water and make the volume to 500 mL.
4. Fehling's reagent: Mix equal volumes of Fehling's solution A and B. These solutions must be mixed immediately prior to use.

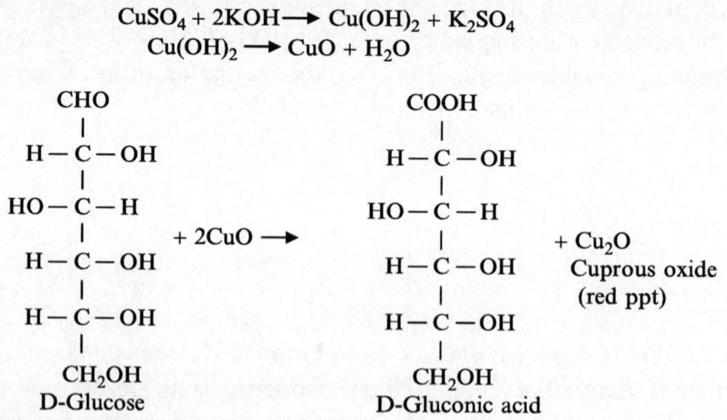

Fig. 5.5 Reaction for Fehling's test

Procedure

Add 1 mL of Fehling's reagent (Reagent No. 4) to 1 mL of aliquot of the test solution. Mix thoroughly and place the test tubes in vigorously boiling water bath. Look out for the formation of red ppt of cuprous oxide which would indicate the presence of reducing sugars in the test solution.

5.2.7 *Benedict's Test*

Principle

Benedict's test is more convenient and this reagent is more stable. In this method sodium citrate functions as a chelating agent. Presence of reducing sugars results in the formation of red ppt of cuprous oxide.

$$Na_2CO_3 + 2H_2O \rightarrow 2NaOH + H_2CO_3$$

$$2NaOH + CuSO_4 \rightarrow Cu(OH)_2 + Na_2SO_4$$

$$Cu(OH)_2 \rightarrow CuO + H_2O$$

$$D\text{-}Glucose + 2CuO \rightarrow D\text{-}gluconic\ acid + Cu_2O\ \text{(Red ppt)}$$

Materials and Reagents

1. Boiling water bath.
2. Benedict's reagent: Dissolve 173 g of sodium citrate and 100 g of anhydrous Na_2CO_3 in 600 mL of host H_2O. Dilute to 800 mL with water.
3. Dissolve 17.3 g of $CuSO_4{\cdot}5H_2O$ in 100 mL hot water. Cool and dilute to 100 mL.
4. Add Reagent No. 2 to Reagent No.3 slowly with constant stirring. Make the final volume to 1 L.

Procedure

Add 0.5–1 mL of the test solution or sample extract to 2 mL of Benedict's reagent (Reagent No. 4). Keep the test tubes in a vigorously boiling water bath. Observe for the formation of red precipitates whose appearance would suggest the presence of reducing sugars in the given solution.

5.2.8 *Picric Acid Test*

Principle

This test is another test for detection of reducing sugars. The reducing sugars react with picric acid to form a red coloured picramic acid (Fig. 5.6).

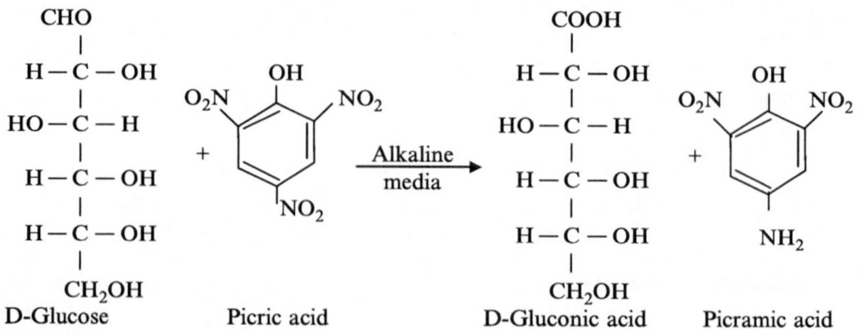

Fig. 5.6 Reaction for Picric acid test

Materials and Reagents

1. Boiling water both.
2. Saturated picric acid: Dissolve 13 g of picric acid in distilled water, boil and cool.
3. 10% Na_2CO_3.

Procedure

Add 1 mL of saturated picric acid to 1 mL of sample solution followed by 0.5 mL 10% Na_2CO_3. Heat the test tubes in a boiling water bath. Appearance of red colour indicates the presence of reducing sugars in the sample solution.

5.2.9 *Mucic Acid Test*

Principle

This test is specific for galactose. This sugar can be distinguished from other monosaccharides by its reaction with conc. HNO_3. Oxidation of other monosaccharides yields soluble dicarboxylic acids whereas galactose produces insoluble mucic acid.

Materials and Reagents

1. Boiling water bath
2. Soild galactose
3. Solid glucose
4. Conc. HNO_3

Procedure

Take about 50 mg galactose and 50 mg glucose separately in different test tubes. Add 1 mL of water and 1 mL conc. HNO_3 to each tube. Heat the tubes in boiling

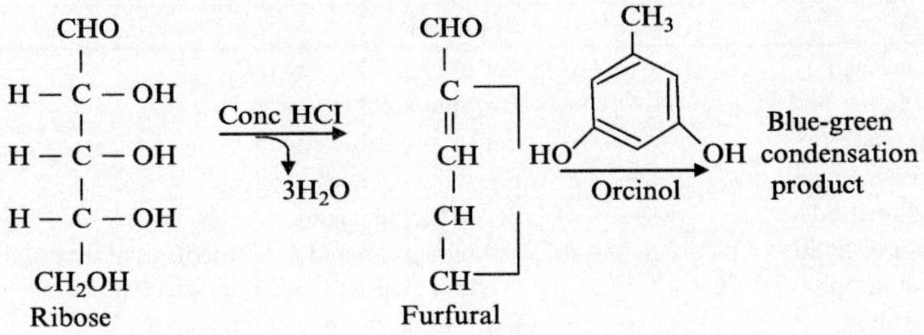

Fig. 5.7 Reaction for Bial's test

water bath for 1.5 h. Add 5 mL water and keep them over night. Insoluble mucic acid will be formed in the case of galactose but not glucose or other sugars.

5.2.10 Bial's Test

Principle

This test is used for the determination of pentose sugars. Reaction is due to the formation of furfural in the acid medium which condenses with orcinol in the presence of ferric ions to give a blue–green coloured complex, which is soluble in butyl alcohol (Fig. 5.7).

Materials and Reagents

1. Boiling water bath.
2. Dissolve 1.5 g of orcinol in 100 mL of conc. HCl and add 20–30 drops of 10% ferric chloric solution to it.

Procedure

To 2 mL of Bial's reagent add 4–5 drops of test solution. Heat in a boiling water bath. Observe for the formation of blue–green coloured complex.

5.2.11 Quantitative Determination of Carbohydrates

There are several methods for the quantitative determination of various types of carbohydrates. For accurate determination of a particular sugar, it may be necessary to separate out the various carbohydrates in a mixture or a tissue extract by chromatographic techniques and then estimate them individually. Some of the methods which are employed for determination of different groups or types of sugars are:

Method		Applicable
Anthrone method	⟶	For total soluble sugars
Ferricyanide method	⟶	For total soluble sugars
Dubois method	⟶	For total soluble sugars
Somogyi-Nelson method	⟶	For reducing sugars
Picric acid method	⟶	For reducing sugars
High pressure liquid chromatography (HPLC)	⟶	For separation and determination of monosaccharides, disaccharides, and oligosaccharides
Enzyme methods	⟶	For estimation of specific sugars

In HPLC method, sugars in an appropriately processed sample are first separated from each other by HPLC and their presence as well as the amount in the column effluent is continuously monitored and recorded by a suitable detector such as RI (refractive index) detector. The identity of sugars in individual peaks is established by comparison with elution profiles of authentic samples of different sugars.

Amount of some of the sugars in a sample can also be determined enzymatically. One main advantage of this technique is that prior separation of individual sugars is not essential since some of the enzymes exhibit absolute substrate specificity amount of glucose can conveniently and accurately be determined, even if the sample contains other closely related sugars like galactose, mannose etc., by using glucose oxidase.

Chapter 6
Estimation of Lipids

Lipids are non-polar organic biomolecules which are totally or nearly insoluble in water but are quite soluble in non-polar organic solvents like ether, chloroform or benzene. They serve as major structural components of the membranes and also form a protective coating on many organisms. Some of the vitamins and hormones are lipids. There are several different classes of lipids but all of them derive their distinct properties due to the hydrocarbon nature of their structure. Lipids have been classified in several ways but the most acceptable classification is the one based on the structure of their backbone. Based on this, they are divided into two groups:

1. Complex lipids.
2. Simple lipids.

Complex lipids are esters of fatty acids. Among various forms of complex lipids such as acylglycerides, phospholipids, sphingolipids and waxes, fatty acids are covalently joined by an ester linkage to a trihydroxy alcohol, glycerol, or its derivative. Since the complex lipids yield soap on alkaline hydrolysis, they are also called saponifiable lipids. Simple lipids, on the other hand, do not contain fatty acids and are therefore called non-saponifiable lipids. They include compounds such as terpenes, sterols, etc.

6.1 Extraction, Separation and Estimation of Lipids from Oil Seed

Principle

Lipids are soluble in some organic solvents. This property of specific solubility in non-polar solvents is used for extracting lipids from tissues. In biological materials, the lipids are generally bound to proteins and they are, therefore, extracted with either

R. Katoch, *Analytical Techniques in Biochemistry and Molecular Biology*,
DOI 10.1007/978-1-4419-9785-2_6, © Springer Science+Business Media, LLC 2011

a mixture of ethanol and diethyl ether or a mixture of chloroform and methanol. Inclusion of methanol or ethanol in the extraction medium helps in breaking the bonds between the lipids and proteins.

Materials and Reagents

- Iodometric flasks
- Separatory funnel
- Oil seeds (sunflower, peanuts or soybean)
- Anhydrous sodium sulphate
- Chloroform: methanol mixture (2:1)
- 1% sodium chloride

Procedure

1. Take 1 g of the oil seed and grind it in the presence of 5 g of anhydrous sodium sulphate in a pestle and mortar. A small amount of acid washed sand may be used as an abrasive if the seed material is tough.
2. Add 20 mL of chloroform–methanol mixture to it and transfer it to an air tight glass stoppered iodometric flask. Shake the content of the flask on a mechanical shaker for 1 h and then filter it through a glass-sintered funnel. Repeat the extraction of the residue twice and pool the filtrates.
3. Remove the solvent from the residue by distilling under vacuum. Since the residue left after drying contains crude lipids, extract it again with 10 mL of chloroform–methanol mixture containing 1 mL of 1% sodium chloride.
4. Take the pooled fractions in a separatory funnel, shake it thoroughly and allow it to stand for 5 min. The lipids will be recovered in the lower chloroform layer while soap, glycerol and other water insoluble impurities move into the upper layer.
5. Drain out the lower layer and treat the upper layer again 3–4 times with 5–10 mL of chloroform–methanol mixture to extract any residual lipid from it.
6. Collect the lipid containing fractions in a pre-weighed beaker.
7. Evaporate the solvent by keeping the beaker in warm water bath (50°C) with a constant blowing of a slow stream of nitrogen gas over the surface.
8. Record the weight of the beaker and determine the amount of crude lipids in the sample by subtracting the weight of empty beaker.
9. Express the results in terms of % crude lipid in the given sample of the oil seed.

   - The sample should not be exposed to high temperature or light as some lipids get polymerized or decomposed on exposure to light, heat and oxygen.

6.2 Determination of Saponification Value of Fats and Oils

Principle

Hydrolysis of fat with an alkali results in the formation of salts of fatty acids (also called soap) and glycerol. This process is called *saponification*. From the amount of potassium hydroxide utilized during hydrolysis, the saponification value of a given fat sample can be calculated. The saponifiaction value is defined as mg of KOH required to saponify 1 g of the given fat (Fig. 6.1).

It may be recalled that three molecules of KOH are consumed for saponification of each molecule of tracylglycerol irrespective of chain length of fatty acid. Each gram of a triacylglycerol with shorter chain fatty acids will contain larger number of molecules of the triacylglycerol and will thus require much more KOH. *The saponification value is therefore an indication of average molecular weight of the fatty acids in an acylglyceride.*

The procedure involves refluxing of known amount of fat or oil with a fixed but an excess of alcoholic KOH. The amount of KOH remaining after hydrolysis is determined by back titrating with standardized 0.5 N HCl and the amount of KOH utilized for saponification can therefore be calculated.

Materials and Reagents

1. Reflux condenser
2. Boiling water bath
3. Burette
4. Test compounds (tristearin, coconut oil, butter)
5. Fat solvent: A mixture of 95% ethanol and ether (1:1 *u*/*v*)
6. 0.5 N alcoholic KOH: Prepare 0.5 N solution of KOH by dissolving 28.05 g of KOH pellets in 20 mL water and make the volume to 1 L with 95% ethanol
7. 1% Phenolphthalein solution in 95% alcohol
8. 0.5 N HCl

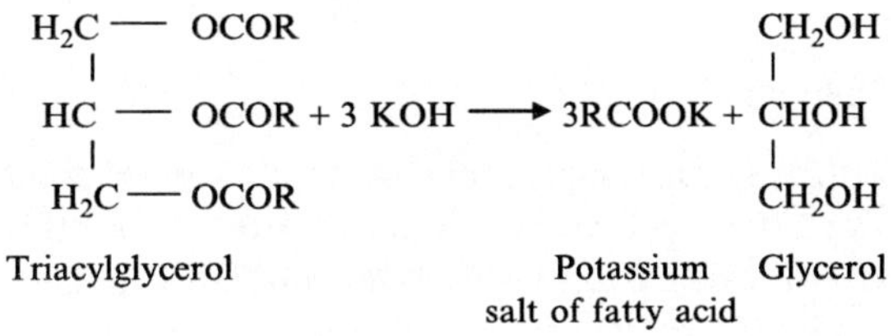

Fig. 6.1 Saponification reaction of triacylglycerol

Procedure

1. Weigh accurately 1 g of the fat sample in a conical flask and dissolve it in about 3 mL of the fat solvent (Reagent No. 5).
2. Add 25 mL of 0.5 N alcoholic KOH, attach a reflux condenser to it and reflux the contents on boiling water bath for 30 min.
3. Cool to room temperature and add a few drops of phenolphthanlein into the flask.
4. Titrate the contents of the flask with 0.5 N HCl till the pink colour disappears.
5. Similarly, run a blank by refluxing 25 mL of 0.5 N alcoholic KOH without any fat sample.

Calculation

0.5 N KOH in blank $= x$ mL
0.5 N KOH in test sample $= y$ mL
Titre value for sample $= (x-y)$ mL

$$\text{Saponification value} = \frac{28.05 \times \text{titre value}}{\text{Weight of sample (g)}}.$$

The multiplication factor of 28.05 in the above equation is included since 1 mL of 0.5 N KOH contains 28.05 mg of KOH.

Precautions

1. As alcohol is highly inflammable therefore precaution is required during heating.
2. During refluxing, effective cooling of condenser is required so that alcohol does not get evaporated during saponification.

6.3 Determination of Acid Value of Fats and Oils

Principle

Different fat samples may contain varying amount of fatty acids. In addition, the fats often become rancid during storage and this rancidity is caused by chemical or enzymatic hydrolysis of fats into free acids and glycerol. The amount of free fatty acids can be determined volumetrically by titrating the sample with potassium hydroxide. The acidity of fats and oils is expressed as its acid value or number which is defined as mg KOH required for neutralizing the free fatty acids present in 1 g of fat and oil. The amount of free fatty acids present or acid value of fat is a useful parameter which gives an indication of the age and extent of its deterioration.

Materials and Reagents

1. Burette.
2. Conical flasks.
3. Test compounds (olive oil, butter, margarine, etc.).
4. 1% phenolphthalein solution 95% alcohol.
5. 0.1 N Potassium hydroxide: Weigh 5.6 g of KOH and dissolve it in distilled water and make the final volume to 1 L. Standardize this solution by titrating known volume of 0.1 N oxalic acid (prepared by taking 630 mg oxalic acid in 100 mL water) using phenolphthalein as an indicator till a permanent pink colour appears. Calculate the actual normality (N_2) of KOH solution from equation $N_1V_1 = N_2V_2$ where N_1 and V_1 are normality and volume of oxalic acid taken for titration and V_2 is the volume of KOH solution used.
6. Fat solvent (95% ethanol: ether 1:1, *u*/*v*).

Procedure

1. Take 5 g of fat sample in a conical flask and add 25 mL of fat solvent (Reagent No. 6) to it. Shake well and add a few drops of phenolphthalein solution and again mix the contents thoroughly.
2. Titrate the above solution with 0.1 N KOH until a faint pink colour persists for 20–30 s.
3. Note the volume of KOH used.
4. Repeat the steps 1–3 with a blank which does not contain any fat sample.

Calculations

0.1 N KOH solution used for blank = x mL
0.1 N KOH solution used for sample = y mL
Titre value for sample = $(y-x)$ mL

$$\text{Acid value (mg KOH/g fat)} = \frac{\text{Titre value} \times \text{Normality of KOH} \times 56.1}{\text{Weight of sample (g)}}.$$

1 mL of 1 N KOH contains 56.1 mg of KOH. Hence, factor of 56.1 is incorporated in the numerator in the above equation to obtain weight of KOH from the volume of 0.1 N KOH solution used during this titration.

6.4 Determination of Iodine Number of Fat

Principle

The most important analytical determination of an oil/fat is the measurement of its unsaturation. The generally accepted parameter for expressing the degree of carbon to carbon unsaturation of fat, oil or their derivatives is iodine value. Iodine value or

Fig. 6.2 Reaction for the estimation of Iodine number of fatty acid

$$CH_3(CH_2)_7CH = CH(CH_2)_7COOH + ICI$$
Oleic acid
$$\rightarrow CH_3(CH_2)_7CH\text{—}\underset{\text{I}}{CH}\text{—}\underset{\text{CI}}{CH}(CH_2)_7COOH$$
$$ICI + KI \rightarrow KCI + I_2$$
$$I_2 + 2Na_2SO_3 \rightarrow 2NaI + Na_2S_4SO_6$$

iodine number is defined as grams of iodine absorbed by 100 g of fat. It is a useful parameter in studying oxidative rancidity of triacylglycerols since, higher the unsaturated, greater is the possibility of rancidity.

Estimation of iodine number is based on the treatment of a known weight of fat or oil with a known volume of standard solution of iodine monochloride, and then determining the amount of unused iodine monochloride from iodine liberated, on addition of excess of KI. The released iodine is titrated against 0.1 N sodium thiosulphate solution using starch as an indicator (Fig. 6.2).

Materials and Reagents

1. Stoppered bottles.
2. Burette (25 mL).
3. Test compounds: 2% solution of corn oil, olive oil and butter in chloroform.
4. Wij's solution: Dissolve 8.5 g of iodine and 7.8 g of iodine trichloride separately in 450 mL of acetic acid each. Mix both the solution and make the volume upto 1 L.
5. 0.1 N sodium thiosulphate: Dissolve 24.82 g of $Na_2S_2O_3$. $5H_2O$ in 1 L of water. To check its normality, take 20 mL of 0.1 N potassium dichromater, add 10 mL of 15% KI and then 5 mL of HCl. Dilute to 100 mL with water and titrate with thiosulphate solution till the yellow colour appears. Now add a few drops of starch solution (Reagent No. 7) and continue the titration till the blue colour disappears. Note the volume of thiosulphate solution and calculate its exact normality ($N_1V_1 = N_2V_2$, where N_1 is the normality, V_1 is the volume of dichromate solution taken for titration and V_2 is the volume of thiosulphate solution used).
6. 10% Potassium iodide solution: Dissolve 10 g of KI crystals in water and make up the volume to 100 mL.
7. 1% Starch indicator: Take 1 g starch and dissolve it in 100 mL water, boil for a min, cool and centrifuge to get a clear solution.

Procedure

1. Take 10 mL of fat solution into stoppered bottles and add 25 mL of Wij's solution. Shake thoroughly and allow it to stand in dark for 1 h.
2. Similarly, prepare a blank in which fat solution is replaced by chloroform.
3. After the reaction time of 1 h in dark, rinse the stopper and neck of the bottle with 50 mL of water and add 10 mL of potassium iodide solution.
4. Titrate the liberated iodine with standard sodium thiosulphate solution till the content of the flask becomes pale yellow in colour.

5. Add a few drops of starch solution and continue to titrate it further with sodium thiosulphate solution till the blue colour disappears.

Calculations

The difference between the blank and test readings gives the amount of 0.1 N sodium thiosulphate required to react with an equivalent volume of iodine. One litre of 0.1 N iodine solution contains 12.7 g of iodine. The iodine number can thus be calculated as follow:

Volume of 0.1 N sodium thiosulfate used for blank = x mL
Volume of 0.1 N sodium thiosulfate used for sample = y mL

$$\text{Iodine number} = \frac{(x - y) \times 12.7}{1{,}000} \times \frac{100}{\text{wt. of sample (g)}}.$$

Precautions

1. The bottles must be shaken thoroughly throughout the titration to ensure that all the iodine is expelled from the chloroform layer.

6.5 Solubility Test for Lipids

Triaclyglycerols with small chain fatty acids are somewhat soluble in water but those containing non-polar long chain fatty acids are insoluble and they form emulsions in water. All triacylglycerols are soluble in diethyl either, chloroform and benzene. They are slightly soluble in cold methanol, ethanol and acetone but their solubility increases on warming. Understanding of their solubility characteristics is helpful in developing efficient procedures for extraction of various lipids from the biological materials.

Materials and Reagents

1. Fatty acids (butyric, palmitic and oleic acids)
2. Fats and oils (butter, olive oil, cod liver oil, phospholipids, etc.)
3. Solvents (water, acetone, ethyl alcohol, chloroform, diethyl either, etc.)

Procedure

1. Take small amount of different lipids in various test tubes and add water. Shake well and check their solubility.
2. Note the change, if any, in solubility on warning the above tubes in water bath at 50°C for 5 min.
3. Repeat the solubility test using different solvents.
4. Record the observation regarding solubility of these lipids and conclude about their solubility characteristics.

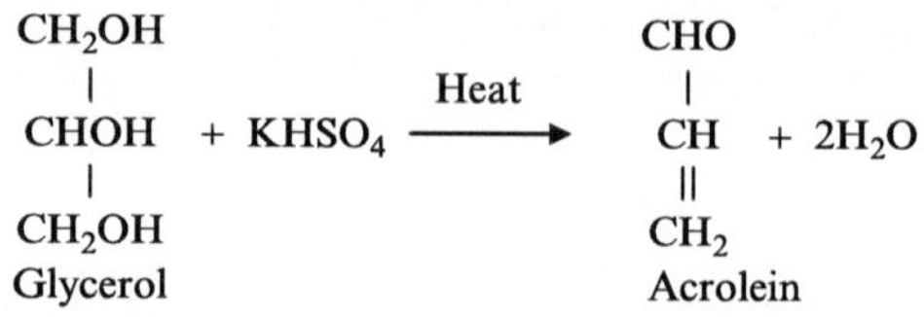

Fig. 6.3 Reaction for Acrolein test

6.6 Acrolein Test for Glycerol

Principle

When glycerol, either in free form or as an ester of fatty acids, is heated with potassium hydrogen sulphate till it gets dehydrated to an unsaturated aldehyde called acrolein. Acrolein can be identified by its characteristic pungent smell (Fig. 6.3).

Materials and Reagents

1. Lipid samples (butter, olive oil, stearic acid, glycerol, etc.)
2. Anhydrous potassium hydrogen sulphate

Procedure

1. Take approximately 1.5 g of potassium hydrogen sulphate in a test tube, and add five drops of the liquid test sample or an approximately equivalent weight of the test sample if it is solid. Cover the test sample completely by adding more of solid potassium hydrogen sulphate on top of it.
2. Heat the test tube slowly on burner and note the odour of the fumes evolved from the tube.

6.7 Qualitative Test for the Presence of Fatty Acids by Titrimetric Method

Principle

The presence of non-esterified fatty acids in a given sample can be determined by titrating it with an alkali using phenolphthalein as an indicator.

Materials and Reagents

1. Lipid samples (butter, olive oil, stearic acid dissolved in 50% alcohol)
2. Phenolphathalein: Prepare 1% solution in alcohol
3. 0.1 N NaOH
4. Erlenmeyer flasks

Procedure

1. Take 10 mL of 0.1 N NaOH in an Erlenmeyer flask and add a drop of phenolphthalein solution, which will give permanent pink colour.
2. To this, with pipette add the test solution drop by drop with constant shaking of the flask.
3. Disappearance of pink colour indicates the presence of free fatty acids in the given test compound.
4. Repeat this with other lipid samples.

6.8 Test for Unsaturation of Fatty Acids in Lipid Sample

Principle

The fatty acids present in animal fats are usually fully saturated, whereas those occurring in vegetable oils contain one or more double bonds in their hydrocarbon chain. A semiquantitative estimate about the degree of unsaturation of lipid samples can be deduced since halogens are readily added across the double bonds and this reaction results in decolorization of bromine water or iodine solution (Fig. 6.4).

Materials and Reagents

1. Test solutions (olive oil, corn oil, coconut oil, oleic acid, etc.).
2. Bromine water: Add 5 mL of bromine to 100 mL of water. Shake the mixture and keep it in dark bottle.

Procedure

1. Take approximately 5 mL of test solution in a test tube and slowly add bromine water dropwise and shake the tube after each addition.
2. Keep on adding bromine water till it fails to get decolorized and retains its colour.
3. Note the amount of bromine water added.
4. Repeat the experiment with other test solutions.

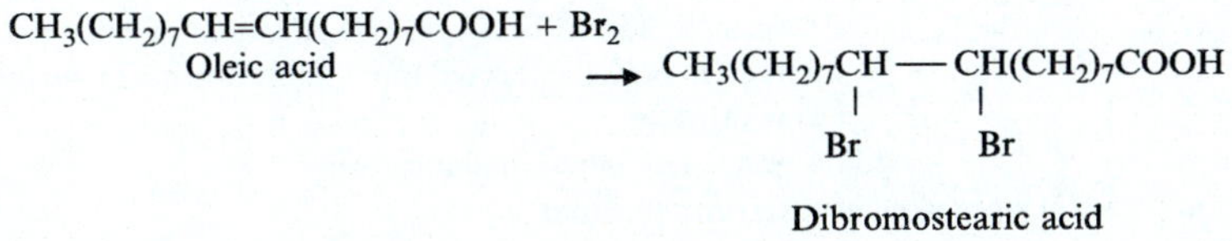

Fig. 6.4 Reaction for estimation of unsaturation in oleic acid

6.9 Separation of Different Lipid Fractions by Thin Layer Chromatography (TLC)

The qualitative and quantitative analysis of non-polar and polar lipids can be effectively done after separating the components in these fractions by thin layer chromatography (TLC). It is a chromatographic technique in which different components in a sample get separated during their passage along a very thin layer, usually of 0.20–0.25 mm thickness, of a suitable chromatographic media which is spread as a uniform layer on a glass plate (see Sect. 9.4 for details of TLC). The chromatographic material used for this experiment is Silica Gel-G. Different lipids are adsorbed onto activated Silica Gel-G with varying degrees of strength. Those components, which are not adsorbed or adsorbed with lesser strength, tend to move faster along with the mobile phase (solvent system), whereas those which are held more firmly by the adsorbent (stationary phase) travel slowly. This differential distribution of various components between the mobile and stationary phases determines the rate of migration of different compounds in a mixture and results in their separation from each other.

Materials and Reagents

Activated TLC plates	Glass plates (20 × 20 cm) onto which a 0.2-mm thick layer of silica gel-G has been layered with the help of a spreader. The prepared thin layer plates are dried at room temperature and then activated at 110°C for 45 min before use
Thin layer chromatographic tanks	
Developing mixtures	Chloroform:methanol:water (65:25:4) for separation of phospholipids and galactolipids from polar lipids
	Toluene:ethyl acetate:ethanol (2:1:1) for separation of galactolipids from phospholipids
	Hexane:diethylether:acetic acid (80:20:1) for separation of various neutral lipids
Spraying reagents for location	Sulphuric acid: 50% H_2SO_4 in water of spots on TLC plates (*u/v*)
	Perchloric acid: 20% perchloric acid in water (*u/v*)
	Ferric chloride spray: dissolve 50 mg $FeCl_3$. $6H_2O$ in 90 mL water and add 5 mL each of glacial acetic acid and conc. sulphuric acid
	Nihydrin spray: 0.2 g ninhydrin dissolved in 100 mL ethanol
	Anthrone spray: dissolve 0.2 g anthrone in 100 mL conc. H_2SO_4
	Molybdate spray: prepare a solution containing 16 g ammonium molybdate in 120 mL water. To 80 mL of this solution add 40 mL conc. HCl and 10 mL mercury. Shake it for 30 min. To this solution add 200 mL conc. H_2SO_4 and remaining 40 mL of ammonium molybdate, cool the mixture and make the volume to 1 L with water
	Iodine spray: 1% iodine in chloroform
Various lipid fractions prepared as discussed previously	
Standard lipids: make 0.2% solution of cholesterol, tristearin, phosphatidyl ethanolamine, palmitic acid, stearic acid, lecithin, galactosyldiacylglycerol	

Procedure

1. Apply 10–20 μL fraction of the lipid sample in the form of a spot at a distance of 2 cm starting from the left bottom edge of the activated TLC plate.
2. Develop the plates in the appropriate developing mixture (depending upon the types of lipids to be separated) in an air tight chromatographic glass tank till the solvent front moves up to 4 cm below the top edge of the glass plate.
3. Take out the plates, dry them for 5 min in air and spray the plates with the required detection reagent. The specificity and use of different spraying reagents is given below.
4. Locate the position of the lipid spots on the glass plate and measure the distance travelled by the individual lipid component. Calculate their R_f values and compare them with those of standards.

Spraying Reagents

1. Sulphuric acid spray: This reagent forms charred black spots with all the lipids. Spray the plates with H_2SO_4 and keep the plates in an oven at 120°C for 30 min. Mark the spots on the plate with pencil.
2. Iodine vapours: This is also a general detection reagent for all lipids. On keeping the plates in a jar or development tank containing a trough filed with iodine crystals, yellowish or brown spots appear and then fade away after some time on exposure of plates to air.
3. Perchloric acid spray: Heat the developed chromatogram at 100°C for 15 min and spray with perchloric acid. Note the brown spots which are formed by the lipids.
4. Ferric chloride spray: Spray the developed plates with ferric chloride and heat them at 100°C for 2–3 min in an oven. Lipid samples of cholesterol and cholesteryl esters will produce red or violet spots.
5. Ninhydrin spray: For detection of lipids containing amino groups, spray the developed plates with ninhydrin reagent and heat them in oven at 100°C for 5 min. Violet spots will indicate their presence and location on the plates.
6. Anthrone spray: Spray the chromatogram with anthrone reagent and heat the plate in oven at 100°C for 10 min. Formation of green spots will be due to its reaction with glycolipids and violet spots due to sulpholipids.
7. Molybdate spray: Phospholipids are identified as violet spots on spraying the developed plates with molybdates reagent or iodine solution. Heat the plates in an oven at 100°C for 5 min after spraying them with molybdate reagent.

6.10 Separation and Identification of Lipids by Column Chromatography

Principle

The extracted lipids are fractionated into neutral, non-polar and polar lipids by adsorbing them on solid adsorbent in non-polar solvents. They are then eluted stepwise with solvents of increasing polarity. The individual components in neutral and polar fractions obtained by this method may further be separated by TLC.

Materials and Reagents

1. *For column chromatography*

   (a) Lipid preparation (as discussed)
   (b) Silicic acid
   (c) Chloroform
   (d) Acetone
   (e) Methanol
   (f) Glass column (2.5 cm diameter × 15 cm length)
   (g) Glass wool

2. *For estimation of galactose*

   (a) Anthrone reagent: Dissolve 0.2 g anthrone in 100 mL of conc. H_2SO_4.
   (b) Standard galactose solution: Dissolve 25 mg galactose in water and make the volume to 250 mL. This solution contains 100 μg galactose/mL.

3. *For estimation of phosphorus*

   (a) Digestion mixture: Sulphuric acid and 72% Perchloric acid (9:1, *u/v*)
   (b) Aminonaphthol suphonic acid (ANSA) reagent: Dissolves 3.43 g of sodium metabisulphite, 0.063 g of 1-amino-2-naphthol-4-sulfonic acid and 0.125 g of sodium, filter through Whatman No. 1 filter paper and make the volume to 25 mL with water.
   (c) 0.26% ammonium heptomolybdate.
   (d) ANSA-molybdate reagent: Mix 22 mL of ANSA reagent (Reagent No 11) and 500 mL ammonium heptamolybdate just before use. This reagent is stable for 1 h only.
   (e) Standard phosphorus: Dissolve 34 mg of KH_2PO_4 in water and make the volume to 250 mL. This solution contains 1 μmol of phosphorus/mL.

Procedure

1. Wash silicic acid thrice with double distilled water to remove fine particles and activate it at 125°C for 14 h.
2. Suspend the activated silicic acid in chloroform and carefully pour the slurry into the glass column having glass wool plug at its bottom. Allow the adsorbent to settle and add more slurry till a column with a bed height of 9 cm is obtained.
3. Wash the packed column with three column volumes of chloroform.
4. Apply lipid preparation (dissolved in chloroform) to the silicic acid column and elute it with 200 mL of chloroform at constant flow rate of 2 mL/min. The neutral lipids, i.e. triacylglycerols, diacylglycerols, sterols and sterol esters get eluted during this step. The different lipid components in these fractions may

further be separated by TLC The polar lipids, which remain adsorbed in the column during previous step are then eluted with 120 mL of acetone. These fractions contain glycolipids and sterol glycolipids.

5. The residual glycolipids and all the phospholipids are finally eluted from the column with 280 mL of methanol. Collect fractions of 5 mL each and examine fractions for the presence of galactose (Step 6) and phosphorus (Step 7).
6. Take 1 mL of the fraction in a test tube and place it in a water bath at 70°C to evaporate acetone. Add 3 mL of diluted sulphuric acid (1 mL of H_2SO_4 diluted to 13 mL with water) and keep the sample for hydrolysis in boiling water bath for 1.5–2.0 h to release galactose. Allow to cool and extract the liberated fatty acids in 5 mL chloroform. The aqueous layer is used for estimation galactose. Take 1 mL of the aqueous reagent (Reagent No.8) and mix the contents. For reagent blank use 1 mL of water in place of the aqueous phase of the hydrolyzed lipid optical density at 625 nm against reagent blank after cooling them under running tap water. Prepare a standard curve as described above over a range of 0–100 μg of galactose. Determine the amount of galactose in the sample tubes from the standard curve.
7. Pipette 1 mL of the phospholipid fractions (containing 0.5–3.0 μg phosphorus) in test tubes and evaporate methanol by heating them in water bath at 70°C. Now add 0.4 mL of digestion diacid mixture (Reagent No. 10) and heat at 180°C for about 45 min to release phosphorus from phospholipids. When the solution in test tubes becomes clear, remove the tubes. After allowing them to cool, add 9.6 mL of ANSA-molybdate reagent (Reagent No. 13) and keep them in boiling water bath for 10 min. Simultaneously prepare a reagent blank by replacing acid hydrolysate with water and also a standard curve using standard solution of KH_2PO_4 (Reagent No. 14) containing 0–1.0 μmol of phosphorus. Cool the tubes and note the optical density at 640 nm against the reagent blank. Determine the amount of phosphorous in each fraction from the standard curve.

6.11 Estimation of Fatty Acids by Gas–Liquid Chromatography (Morrison and Smith, 1964)

Analysis of complex fatty acid mixtures can be carried out by gas–liquid chromatography (GLC). In this technique the fatty acids are first converted into a volatile form usually methyl esters. The esters of fatty acids are identified by comparing with a set of standard fatty acid esters and then quantified (Fig. 6.5).

Reagents

1. Sodium methoxide (0.5 N). Dissolve sodium methoxide in dried methanol A.R.
2. BF_3 – methanol reagent (Sigma)
3. Hexane/Petroleum ether, 40–60°C (Spectroscopic grade)
4. Sodium sulphate (anhydrous)

All solvents should be moisture-free.

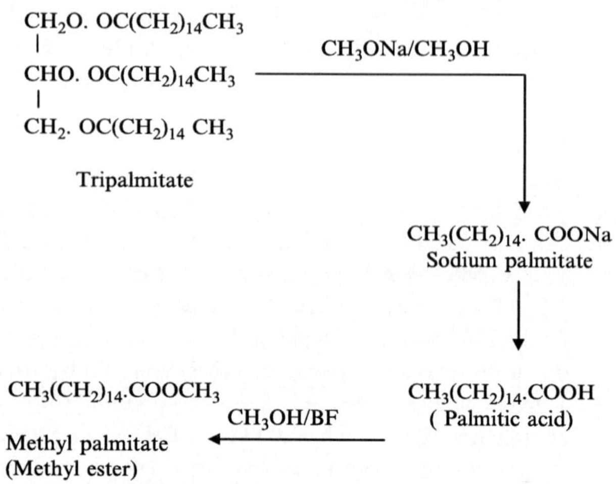

Fig. 6.5 Reaction showing formation of methyl ester of fatty acid

GLC conditions

Column	10% DEGS (diethylene glycol succinate) on chromosorb P or W (60–80 mesh)
Detector	Flame ionization detector (FID)
Carrier gas	Nitrogen/argon with a flow rate of 40–50 mL/min. Convenient flow rates of hydrogen (0.5 kg/cm^2) and air (0.8 kg/cm^2) are also used with the detector as a fuel
Column/oven temperature	170–200°C
Detector temperature	230°C
Injector port temperature	230°C
Recorder speed	1 cm/min
Generally the injector and detector temperatures are kept 50°C above column temperature	

Procedure

1. Preparation of methyl esters

   (a) Grind 0.1 g of oilseed with 4 mL of 0.5 N sodium methoxide in instalment of 2 mL each in a pestle-mortar. Transfer the contents to an air tight screw capped 15 mL vial. Wash the pestle-mortar twice with 1 mL of 0.5 N Sodium methoxide and transfer into the vial.

   or

   Take 20 mg oil (one or two drops of oil) along with 5 mL of 0.5 N sodium methoxide in a screw capped vial.

   (b) Keep the vial in a boiling waterbath for 10–15 min.

   (c) Cool to room temperature, add 1–2 drops BF_3-methanol reagent and again heat for 5 min.

   (d) Cool and add 1–2 mL of hexane, shake and wait till hexane layer separates out.

2. Fatty acid analysis

   (a) Take appropriate aliquot (2–3 μL) of the hexane layer containing methyl esters and inject into pre-conditioned gas chromatograph.
   (b) Measure the retention time and identify individual fatty acids by comparing with the reaction time of standard methyl esters. Esters appear in the order of increasing number of carbon atoms and of increasing unsaturation for the same number of carbon atoms. The area of each peak should be calculated by measuring peak height and width at half height. After computing total peak area for the sample, calculate per cent area under peak that would give per cent of respective fatty acid.

$$\text{Peak area} = \text{Peak height} \times \text{width at half height.}$$

Precautions

1. Do not switch on the gas chromatograph without the flow of carrier gas.
2. Solvents used should be of high purity.
3. Add some anhydrous sodium sulphate to the solvents to remove moisture traces.
4. BF_3 is a carcinogenic compound. Do not allow it to come in contact with skin or hand. Do not pipette it out by mouth.
5. In step (a) 4, sometimes the hexane layer does not separate out. Add 2+ or 3 drops of water and tap the vial to separate the layer. Pipette out the upper layer and remove moisture by adding a pinch of anhydrous sodium sulphate.
6. Do not load more than 2–3 μL of methyl ester in the column at a time.
7. Any unesterified liquid component will spoil the column in most cases and hence esterification must be checked before injecting into GLC.
8. If proper resolution of the peaks is not there, wait for sometime and find out the reasons for it. May be flame has extinguished or carrier gas, hydrogen or air has exhausted. Then load the sample again.
9. Sodium methoxide is generally in powder form. If it has lumps then keep it in a Petri dish in an oven for 1–2 h at 110°C to remove moisture.
10. In case of power failure, continue the flow of carrier gas and after resumption of power, allow the instrument to run for about 1 h to remove the sample already loaded in the column.
11. While switching off the instrument, first switch off injector, detector and oven but continue the flow of carrier gas till the oven temperature is brought down to 50°C.
12. To activate the column, allow it to condition at 200–210°C for 4–5 h (In case of DEGS bleeding temperature is 225°C) to get good resolution.
13. Strictly avoid introduction of moisture in the column through sample, or solvent or wet glassware.

Chapter 7
Qualitative and Quantitative Estimations of Amino Acids and Proteins

Proteins are present in the living organisms. They form the structural and functional basis of the cell, which is the smallest unit of life. The estimation of proteins and sometimes their constitution of amino acids become necessary for various biochemical and molecular experimentations. Different methods are available for the estimation of proteins and amino acids. Some of the simple colour reactions of proteins which could be used as identification tests are given below.

7.1 Qualitative Tests for Proteins

See Table 7.1.

7.2 Nitrogen Estimation by Micro-Kjeldahl Method (FAO 1970)

Nitrogen is the major element next to carbon, hydrogen and oxygen found in living things. Nitrogen occurs in amino acids, purine and pyrimidine bases, vitamins, aminosugars, alkaloids, compound lipids, etc.; however, the major nitrogen source is proteins. In most proteins, nitrogen constitutes 16% of the total make-up and hence, the total nitrogen content of a sample is multiplied by 6.25 to arrive at the value of the crude protein. By and large micro-kjeldahl technique is adopted to estimate the total nitrogen content in a variety of samples ranging from microbila cells to meat. The procedure described here is highly suitable for food samples such as cereals and pulses flour. In addition, procedures for non-protein nitrogen, protein nitrogen and amino nitrogen are also presented.

R. Katoch, *Analytical Techniques in Biochemistry and Molecular Biology*,
DOI 10.1007/978-1-4419-9785-2_7, © Springer Science+Business Media, LLC 2011

Table 7.1 Qualitative test for proteins

Procedure	Colour of the reaction product	Mechanism
Ninhydrin test		
To 4 mL of the solution which should be at neutral pH add 1 mL of 0.1% freshly prepared ninhydrin solution. Mix the contents and boil for a couple of minutes. Allow to cool	Violet or purple colour	The ninhydrin test is answered by amino acids and proteins. The formation of a complex called Rheumann's purple due to the condensation of two molecules of ninhydrin with one molecular of ammonia acid is responsible for the violet colour. The α-amino group is the reactive group
Biuret test		
To 2 mL of the test solution and 2 mL of 10% NaOH add two drops of 0.1% $CuSO_4$ solution	Violet or pink colour	Compounds with two or more peptide bonds give a violet colour with alkaline copper sulphate solution
Xanthoproteic test		
To 5 mL of the solution add 1 mL of conc. HNO_3. Boil the contents. After cooling add excess 40% NaOH	On adding acid, yellow colour with be noticed, the when NaOH is added deep orange colour will develop	The yellow colour is due to the nitro derivatives of the aromatic amino acids present in the protein. The sodium salts of nitro derivatives are orange in colour
Sulphur test		
To 2 mL of the solution add 2 mL of 40% NaOH and ten drops of 2% lead acetate solution. Boil for a minute and cool	Black precipitate	The sulphur in sulphur containing amino acids of the proteins in presence of NaOH, is changed into Na_2S which forms black lead sulphide when reacted with lead acetate
Glyoxylic test/Hopkins-Cole test for tryptophan		
Add 2 mL of glacial acetic acid to 2 mL of the test solution. Then add about 2 mL of conc. H_2SO_4 carefully down the sides of the test tube. Observe the colour change at the junction of the two liquids	Violet ring is formed at the junction.	The indole group of tryptophan reacts with the glyoxylic acid released by the action of conc. H_2SO_4 on acetic acid to give a purple colour
Sakaguchi test		
To 5 mL of the solution on ice add 1 mL of 10% NaOH solution and 1 mL of 0.02% α-naphthol solution. After few minutes add ten drops of alkaline hypobromide solution	Intense red colour	The guanidine group of arginine reacts with α-naphthol to form a bright red colour complex

(continued)

Table 7.1 (continued)

Procedure	Colour of the reaction product	Mechanism
Modified Millon's test		
To 1 mL of solution add 1 mL of 10% mercuric sulphate in 10% sulphuric acid. Boil gently for half a minute	Yellow precipitate	The yellow precipitate is due to the precipitation of protein. Mercury combines with tyrosine of the protein
Cool under a tap and add a drop of 1% $NaNO_2$ solution and warm gently	Red colour develops	The red colour is due to reaction of the precipitate with the nitrous acid

Principle

The nitrogen in protein or any other organic material is converted to ammonium sulphate by H_2SO_4 during digestion. This salt, on steam-distillation, liberates ammonia which is collected in boric acid solution and titrated against standard acid. Since 1 mL of 0.1 N acid is equivalent to 1.40 1 mg N, calculation is made to arrive at the nitrogen content of the sample.

Materials

– Kjeldahl flasks: 30 mL hard glass flasks
– Digestion rack: Commercial heating apparatus
– Distillation apparatus: Glass distillation apparatus assembly
– Sulphuric acid specific gravity (Sp gr) 1.84
– Mercuric oxide
– Potassium sulphate
– Sodium hydroxide–Sodium thiosulphate solution: Dissolve 600 g of NaOH and 50 g of $Na_2S_2O_3{\cdot}5H_2O$ in distilled water and make to 1 L
– Indicator solution: Methyl red 0.2 g/100 mL ethanol, methylene blue 0.2 g/100 mL ethanol. For mixed indicator, two parts of methyl red solution are added to one part of methylene blue solution
– Boric acid 4% solution
– Standard HCl or H_2SO_4, 0.02 N
– Boiling chips

Procedure

1. Weigh 100 mg of the sample (containing 1–3 mg nitrogen) and transfer to a 30-mL digestion flask.
2. Add 1.9 ± 0.1 g potassium sulphate and 80 ± 10 mg mercuric oxide and 2 mL conc. H_2SO_4 to the digestion flask. If sample size is larger than 20 mg dry weight, 0.1 mL H_2SO_4 should be added for each 10 mg dry material.

3. Add boiling chips and digest the sample till the solution becomes colourless. The time of digestion will vary with regard to the size of the sample, temperature and the mode of digestion.
4. After cooling the digest, dilute it with a small quantity of distilled ammonia-free water and transfer the distillation apparatus. When the nitrogen content of the sample is high, the digest can be made up to a known volume and an aliquot may be transferred to the distillation flask. The Kjeldahl flask should be rinsed with successive small quantities of water.
5. Place a 100-mL conical flask containing 5 mL of boric acid solution with a few drops of mixed indicator with the tip of the condenser dipping below the surface of the solution.
6. Add 10 mL of sodium hydroxide–sodium thiosulphate solution to the test solution in the apparatus.
7. Distill and collect the ammonia on boric acid (at least 15–20 mL of distillate should be collected).
8. Rinse the tip to the condenser, and titrate the solution against the standard acid until the first appearance of violet colour, the end point.
9. Run a reagent blank with an equal volume of distilled water and subtract the titration volume from that of sample titre volume.

Calculation

The nitrogen content of the sample can be calculated based on any one of the following formulae as the case may be

$$\text{Ng/kg} = \frac{(\text{mL HCl} - \text{mL blank}) \times \text{normality} \times 14.01}{\text{Weight (g)}},$$

$$\text{Ng/kg} = \frac{(\text{mL HCl} - \text{mL blank}) \times \text{normality} \times 14.01 \times \text{final volume}}{\text{Weight (g)} \times \text{aliquot volume}}.$$

Non-protein nitrogen

1. Extract a known quantity of powdered material (100 mg) with ice cold 10% TCA (10 mL). (Proteins are precipitated while non-protein nitrogen gets extracted).
2. Centrifuge, wash the precipitate with TCA, pool all the supernatants and make up to a known volume (25 or 50 mL).
3. Take an aliquot and distill as described earlier.
4. Titrate against the standard acid, and calculate the nitrogen content. This gives the percentage of non-protein nitrogen.

Protein nitrogen

Multiplying total nitrogen value with 6.25 will give the crude protein content, which also includes non-protein nitrogen. To get true protein content, deduct the non-protein nitrogen from the total nitrogen and then multiply with the factor.

Amino nitrogen

1. Estimate the total free amino acid content as per the procedure given elsewhere in this book.
2. Multiply the percentage equivalent of leucine with 14/131 to get the percentage of amino nitrogen. If any amino acid other than leucine is used as standard, introduce the molecular weight of that amino acid in the denominator.

Precautions

- Care must be taken so as to get a representative and homogeneous sample.
- When greater quantity of sample (500 mg) is used and the digest is diluted before an aliquot is transferred to the distillation set, care should be taken so that the actual quantity of sulphuric acid so transferred does not exceed the capacity of the 10 mL of $NaOH–Na_2S_2O_3$ solution. The solution being distilled out should always be strongly alkaline.
- If the test solution is still yellow-coloured even after prolonged digestion, addition of a few drops of perchloric acid or hydrogen peroxide will ensure complete oxidation and to get a colourless solution.
- Appropriate factors should be calculated for the acid normalities prepared in the laboratory based on the following

$$1\ \text{mL}\ 0.1\ \text{N acid} = 1.401\ \text{mg N}$$

- A known concentration of ammonium sulphate solution can be distilled as a standard check.
- In general, the nitrogen content is multiplied by the factor 6.25 to arrive at the percentage of crude protein which is based on the assumption that nitrogen constitutes 16% of a protein. However, the nitrogen percent varies with the amino acid composition of the proteins. For more refined expression of protein percentage in samples, different factors are used. These factors were arrived at by the amino acid composition. Some such factors are given below.
- The method is meant for the conventionally operated distillation apparatus and does not attempt to give the methodology for automatic distillation sets that are commercially available. In such cases, the research worker has to follow the supplier's manual for specific procedures (Table 7.2).

Table 7.2 Conversion factors for estimation of crude protein

Item	Conversion factor
Wheat (whole)	5.83
Wheat flour	5.70
Wheat bran	6.31
Rice	5.95
Rye, barley and oats	5.83
Groundnut	5.46
Soya	5.71
Sesame, safflower and sunflower	5.30
Milk and cheese	6.38
Other foods	6.25

7.3 Sample Preparation for Amino Acid Estimation

Amino acids may be determined colorimetrically, or by chromatography techniques or in an amino acid analyzer. For estimating the amino acid composition of foodstuff, feed or any protein, it has to be first hydrolyzed for further estimations. The free amino acids may be extracted from the tissues or foods in ethanol. If the extract sample contains higher amounts of salts and/or sugars they have to be removed before estimation of amino acids.

1. Free amino acids

Materials

1. Ethanol
2. 0.01 N HCl

Procedure

1. Weigh accurately sufficient quantity of the sample, which should have 2–6 μmole of each amino acid.
2. Extract with warm (60°C) 70% ethanol or 0.01 M phosphate buffer, pH 7.0, 3–6 times. Use extractant 5 times the weight of the sample for each extraction.
3. Pool the extracts after filtration or centrifugation and evaporate in a rotary vacuum evaporator to dryness.
4. Take the residue in 1–10 mL of 0.01 N HCl or in a suitable sample diluting buffer.

2. Hydrolysis of proteins

Materials

1. 6 N HCl
2. 0.01 N HCl

Procedure

1. Weigh accurately sufficient quantity of the sample containing about 10 mg protein in a thick-walled heavy glass tube having a constriction at 7.0 cm from top.
2. Add 5 mL of 6 N HCl and place the tube in liquid nitrogen.
3. Evacuate the frozen sample to 0.01 mmHg.
4. Thaw the sample by shaking and slight warming to allow any dissolved air to bubble out.
5. Freeze the contents and evacuate to 0.005 mmHg again.
6. Seal the tube at the constriction using a flame till the tube is under evacuation.
7. Place the tube in a hot air oven at 110 ± 1°C for 22 h to hydrolyze proteins.
8. Break open the seal and evaporate the contents in a rotary evaporator to remove hydrochloric acid.

9. Add water and repeat evaporation 2 times.
10. Take the residue in 1–10 mL 0.01 N HCl or 50 mm citrate buffer (pH 2.2).

However, some amino acids may be degraded partly or completely during acid hydrolysis. Glutamine and asparagines are converted to their respective acids. Cysteine may be converted to cysteic acid. Serine and threonine are very rapidly hydrolysed to form the protein and are lost to some extent. Tryptophan is also partially destroyed during acid hydrolysis.

3. Deproteinization

Materials

1. 1% Picric acid
2. Dowex 1 × 8 Cl^- Resin
3. 0.02 N HCl
4. 0.01 N HCl

Procedure

1. Add 50 mL of 1% picric acid to 10 mL of the sample.
2. Centrifuge at 3,000 rpm for 10 min and reserve the supernatant.
3. Place the supernatant in a glass column (2 × 10 cm) containing Dowex 1 × 8 Cl^- resin to a height of 3 cm.
4. Immediately after the sample sinks into the resin wash the sides of the column with 3 mL of 0.02 N HCl.
5. Repeat washing 5 times.
6. Collect all the eluate and evaporate it to dryness.
7. Take the residue in 0.01 N HCl.

Centrifugation of the sample for 30 min at 2,000 g is a better method for deproteinization. The ultrafiltrate may be used for analysis. Picric acid does remove some amino acids.

4. Desalting

Materials

1. Dower 5 × 8 (H^+) or Amberlite CG-120 (Na^+)
2. 2 N NH_4OH
3. 0.01 N HCl

Procedure

1. Load 10–20 mL of sample over the column of resin (2 × 10 cm) and allow it to be absorbed by the resin.
2. Elute the amino acids with ammonium hydroxide.
3. Collect the eluate (about 100 mL) and evaporate to remove ammonia.
4. Wash the residue twice with water and repeat evaporation.
5. Take the amino acid residue in 0.01 N HCl.

7.4 Estimation of Total Free Amino Acids (Moore and Stein 1948; Misra et al. 1975)

Amino acids are the basic building blocks of proteins. Apart from being bound as proteins, amino acids also exist in the free form in many tissues and are known as free amino acids. They are mostly water soluble in nature. Very often in plants during disease conditions, the free amino acid composition exhibits a change and hence, the measurement of the total free amino acids gives the physiological and health status of the plants.

Principle

Ninhydrin, a powerful oxidizing agent, decarboxylates the alpha-amino acids and yields an intensely coloured bluish purple product which is colourmetrically measured at 570 nm.

Ninhydrin + alpha-amino acid → Hydrindantin + Decarboxylated amino acid
+ Carbon dioxide + Ammonia

Hydrindantin + Ninhydrin + Ammonia → Purple coloured product + water

Materials

1. Ninhydrin: Dissolve 0.8 g stannous chloride ($SnCl_2 \cdot 2H_2O$) in 500 mL of 0.2 M citrate buffer (pH 5.0). Add this solution to 20 g of ninhydrin in 500 mL of 2-methoxyethnol.
2. 0.2 M Citrate buffer pH 5.0.
3. Diluent solvent: mix equal volumes of water and *n*-propanol.

Procedure

Extraction of amino acid
Weigh 500 mg of the plant sample and grind it in a pestle and mortar with a small quantity of acid-washed sand. To this homogenate, add 5–10 mL of 80% ethanol. Filter or centrifuge. Save the filtrate or the supernatant. Repeat the extraction twice with the residue and pool all the supernatants. Reduce the volume if needed by evaporation and use the extract of the quantitative estimation of total free amino acids. If the tissue is tough, use boiling 80% ethanol for extraction.

Estimation

1. To 0.1 mL of extract, add 1 mL of ninhydrin solution.
2. Make up the volume to 2 mL with distilled water.
3. Heat the tube in a boiling water bath for 20 min.
4. Add 5 mL of the diluent and mix the contents.

5. After 15 min read the intensity of the purple colour against a reagent blank in a colorimeter at 570 nm. The colour is stable for 1 h.
6. Prepare the reagent blank as above by taking 0.1 mL of 80% ethanol instead of the extract.

Standard

Dissolve 50 mg leucine in 50 mL of distilled water in a volumetric flask. Take 10 mL of this stock standard and dilute to 100 mL in another volumetric flask for working standard solution. A series of volume from 0.1 to 1 mL of this standard solution gives a concentration range 10–100 μg. Proceed as that of the sample and read the colour.

Draw a standard curve using absorbance vs. concentration. Find out the concentration of the total free amino acids in the sample and express as percentage equivalent of leucine.

7.5 Estimation of Proline (Chinard 1952)

Proline is a basic amino acid found in high percentage in basic proteins. Many workers have reported a several-fold increase in the proline content under physiological and pathological stress conditions. *Free proline is known to play a role in plants under stress conditions*. Though the molecular mechanism has not yet been established for the increased level of proline, one of the hypotheses refers to breakdown of proteins into amino acids and conversion to proline for storage.

Principle

During selective extraction with aqueous sulphosalicylic acid, proteins are precipitated as a complex. Other interfering materials are also presumably removed by absorption to protein-sulphosalicylic acid complex. The extracted proline is made to react with ninhydrin in acidic conditions (pH 1.0) to form the chromophore (red colour) and read at 520 nm.

Materials

1. Acid ninhydrin: Warm 1.25 g ninhydin in 30 mL glacial acetic acid and 20 mL 6 M phosphoric acid, with agitation until dissolved. Store at 4°C and use within 24 h.
2. 3% Aqueous sulphosalicyclic acid.
3. Glacial acetic acid.
4. Toluene.
5. Proline.

Procedure

1. Extract 0.5 g of plant material by homogenizing in 10 mL of 3% aqueous sulphosalicyclic acid.
2. Filter the homogenate through filter paper (Whatman No. 2).
3. Take 2 mL of filtrate in a test tube and add 2 mL of glacial acetic acid and 2 mL acid-ninhydrin.
4. Heat it in the boiling water bath for 1 h.
5. Terminate the reaction by placing the tube in ice bath.
6. Add 4 mL toluene to the reaction mixture and stir well for 20–30 s.
7. Separate the toluene layer and warm to room temperature.
8. Measure the red colour intensity at 520 nm.
9. Run a series of standard with pure praline in a similar way and prepare a standard curve.
10. Find out the amount of praline in the test sample from the standard curve.

Calculation

Express the praline content on fresh-weight-basis as follows:

$$\mu\text{moles per g tissue} = \frac{\mu\text{g proline/mL} \times \text{mL toluene}}{115.5} \times \frac{5}{\text{g sample}},$$

where 115.5 is the molecular weight of proline.

7.6 Lysine Estimation in Grains (Mertz et al. 1975)

The dye-binding method for lysine estimation has been successfully adopted by researchers to identify high lysine lines in barley, wheat, etc. in mass screening programme.

Principle

The acidic dye acrilane orange G binds to the basic amino acid-lysine and arginine in the proteins. In high lysine lines, arginine and lysine are elevated and hence, the dye-binding capacity is high. However, the general increase in protein content also increases the dye-binding capacity. To avoid the effect of protein change, the dye-binding value is divided by the protein value. Hence, higher dye-binding capacity per unit of protein than the normal lines identifies high lysine lines.

Materials

Buffered dye solution: Dissolve 2 g of acrilane orange G, 15.84 g citric acid, 2.98 g $Na_2HPO_4.2H_2O$ and 0.3 g thymol in 1 L of distilled water at 80°C. When measured in a spectrophotometer, this buffered solution should show an absorbance of 0.65/mL at 580 nm.

Procedure

1. Weigh 200 mg of ground and defatted whole kernel powder and transfer into a 100 mL conical flask.
2. Add 15 mL of buffered dye solution and agitate to equilibrate the dye with the reactive groups of the sample. This can be done for 5 min with a blender or homogenizer. Alternatively, the solution can be kept in standard shaker for 30 min.
3. Filter or centrifuge the suspension.
4. Collect and dilute the filtrate or supernatant 200-fold with water.
5. Read at 480 nm against the original dye solution, diluted similarly 200-fold.

Calculation

Dye-binding capacity of the sample is the difference between the absorption units of the sample and the original dye. Divide this value by the total protein content of the sample determined by micro-Kjeldahl method to obtain the dye-binding capacity per unit protein. This procedure is recommended only to compare the lysine content in progenies in a breeding programme and not to get an absolute value of lysine in individual samples. Hence, the weight (200 μg), volume of dye, and dilution factor should be constant for all the samples.

7.7 Estimation of Lysine in Cereal Grains (Mertz et al. 1975)

Lysine is a limiting amino acid in the cereal grains, therefore, estimation of lysine is very crucial for screening varieties.

Principle

The protein in the grain sample is hydrolysed with a proteolytic enzyme, papain. The alpha-amino groups of the derived amino acids are made to form a complex with copper. The ε-amino group of lysine, which does not couple with copper, is made to form ε-dinitropyridyl derivative of lysine with 2-chloro-3, 5-dintropyridine. The excess pyridine is removed with ethyl acetate and the colour of ε-dintropyridyl derivative is read at 390 nm.

Reagents

1. Solution A: 2.89 g $CuCl_2.2H_2O$ in 100 mL water.
2. Solution B: 13.6 g $Na_3PO_4.2H_2O$ in 200 mL water.
3. Sodium borate buffer 0.05 M, pH 9.0.

4. *Copper phosphate reagent*: POUR solution A (100 mL) to B (200 mL) with swirling, centrifuge and discard the supernatant. Resuspend the pellet 3 times in 15 mL borate buffer and centrifuge after each suspension. After third washing, resuspend the pellet in 80 mL of borate buffer. The reagent must be prepared fresh for every 7 days.
5. 3% Solution of 2-chloro-3,5-dinitropyridine in methanol. Prepare fresh just prior to use.
6. 0.05 M Sodium carbonate buffer, pH 9.0.
7. *Amino acid mixture*: grind in a mortar 30 mg alanine, 50 mg glutamic acid, 60 mg aspartic acid, 20 mg cystein, 300 mg glutamic acid, 40 mg glycine, 30 mg histidine, 30 mg isoleucine, 80 mg leucine, 30 mg mehionine, 40 mg phenylalanine, 80 mg proline, 50 mg serine, 30 mg threonine, 30 mg tyrosine and 40 mg value. Dissolve 100 mg of this mixture in 10 mL of sodium carbonate buffer (0.05 M, pH 9.0).
8. Dissolve 400 mg technical grade papain in 100 mL 0.1 M sodium acetate buffer (pH 7.0).
9. 1.2 N HCl.
10. Ethyl acetate.

Procedure

1. To 100 mg of defatted grain sample add 5 mL of papain solution and incubate overnight at 65°C. Cool to room temperature, centrifuge and decant the clear digest.
2. To 1 mL digest taken in a centrifuge tube add 0.5 mL carbonate buffer and 0.5 mL copper phosphate suspension.
3. Shake the mixture for 5 min in a vortex mix and centrifuge.
4. To 1 mL supernatant add 0.1 mL pyridine reagent, mix well and shake for 2 h.
5. Add 5 mL of 1.0 N HCl and mix.
6. Extract 3 times with 5 mL ethyl acetate and discard the ethyl acetate (top) layer.
7. Read the absorbance of aqueous layer at 390 nm.
8. Prepare a blank with 5 mL papain alone repeating steps 1–7.
9. Dissolve 62.5 mg lysine monohydrochloride in 50 mL carbonate buffer (1 mg lysine/mL). Pipette out 0.2, 0.4, 0.6, 0.8 and 1 mL and make up to 1 m with carbonate buffer. Add 4 mL papain to each tube and mix. Pipette out 1 mL from each and add 0.5 mL of amino acid mixture and 0.5 mL of copper phosphate suspension. Carry out steps 3–7. The standard curve represents absorbance value for 40, 80, 120, 160 and 200 μg lysine.

Calculation

Prepare a standard curve from the readings of the standard lysine. Subtract the absorbance of the blank from that of the sample and calculate the lysine content in the aliquot from the graph.

$$\text{Lysine content of the sample} = \frac{\text{Lysine value from graph in } \mu\text{g} \times 0.16}{\text{Percent N in the sample}}$$

$$= \text{g per 16 g N}$$

7.8 Estimation of Methionine in Food Grains (Horn et al. 1946)

Methionine is one of the essential, sulphur-containing amino acids. Although it is present in many food proteins, *methionine is the limiting amino acid in most of the grain legumes.*

Principle

The protein in the grain is first hydrolyzed under mild acidic condition. The liberated methionine gives a yellow colour with nitroprusside solution under alkaline condition and turns red on acidification. Glycine is added to the reaction mixture to inhibit colour formation with amino acids.

Materials

1. 2 N hydrochloric acid.
2. 10 N NaOH (40%).
3. 10% Sodium nitroprusside.
4. 3% Glycine.
5. Orthophosphoric acid (Sp.gr. 1.75).
6. Standard methionine: Dissolve 100 mg of DL-Methionine in 4 mL of 20% HCl and dilute with water to 100 mL.

Procedure

1. Weigh 0.5 g of defatted sample into 50 mL conical flask. Add 6 mL of 2 N HCl and autoclave at 15 lb pressure for 1 h.
2. Add a pinch of activated charcoal to the hydrolysate (autoclaved sample) and heat to boil. Filter when hot and wash the charcoal with hot water.
3. Neutralize the filtrate with 10 N NaOH to pH 6.5. Make up the volume to 50 mL with water after cooling to ambient temperature.
4. Transfer 25 mL of the made up solution into a 100-mL conical flask.
5. Add 3 mL of 10% NaOH followed by 0.15 mL sodium nitroprusside.
6. After 10 min add 1 mL of glycine solution.
7. After another 10 min add 2 mL *ortho*-phosphoric acid and shake vigorously.
8. Read the intensity of red colour after 10 min at 520 nm against a blank prepared in the same way but without nitroprusside.

9. Standard curve: Pipette out 0, 1, 2, 3, 4 and 5 mL of standard methionine solution and make up to 25 mL with water. Follow steps 5–8 to develop the colour in the standards. The 0 level serves as the blank.

Calculation

Draw a standard curve and calculate the methionine content from the graph.

Methionine content in the sample = (methionine content from the graph × 4) mg per g

Methionine is usually expressed as percentage of protein or g per 16 g N.

$$\text{Methionine content in the sample} = \frac{\text{Methionine content from the graph} \times 6.4}{\text{Percentage of N in the sample}}$$

$$= \text{g per 16 g N}$$

7.9 Estimation of Tryptophan (Mertz et al. 1975)

Cereals like Maize and sorghum are deficient in tryptophan in addition to lysine. Methods to determine these essential amino acids are useful to identify nutritionally superior types.

Principle

The indole ring of tryptophan gives an orange-red colour with ferric chloride under strongly acidic condition. The colour intensity is measured at 545 nm.

Reagents

1. *Papain solution*: dissolve 400 mg technical grade papain in 100 mL 0.1 N sodium acetate buffer, pH 7.0. Prepare the solution fresh every day.
2. *Reagent A*: dissolve 135 mg $FeCl_3.6H_2O$ in 0.25 mL water and dilute to 500 mL with glacial acetic acid containing 2% acetic anhydride.
3. *Reagent B*: 30 N H_2SO_4.
4. *Reagent C*: mix equal volumes of reagent A and B about 1 h before use.
5. *Standard tryptophan*: dissolve 5 mg tryptophan in 100 mL water (50 μg/mL).

Procedure

1. Weigh 100 mg of air-dried, powdered and defatted grain sample into a sample test tube.
2. Add 5 mL papain solution, shake well and close the tube.
3. Incubate at 65°C overnight.
4. Cool the digest to room temperature, centrifuge and collect the clear supernatant.
5. To 1 mL supernatant add 4 mL reagent C.
6. Mix in a vortex mixer and incubate at 65°C for 15 min.

7. Cool to room temperature and read the orange-red colour at 545 nm.
8. Set a blank with 5 mL papain alone and repeat steps 3–7.
9. Pipette out 0, 0.2, 0.4, 0.6, 0.8 and 1 mL standard tryptophan and make up to 1 mL with water. Develop colour following the steps 5–7.

Calculation

Draw a standard curve. Subtract the absorbance value of the blank from that of the sample and calculate tryptophan content from the graph.

$$\text{Tryptophan in grain sample} = \frac{\text{Tryptophan value from graph in } \mu\text{g} \times 0.096}{\text{Percent N in the sample}}$$

$$= \text{g}/16\,\text{g N}$$

7.10 In Vitro Protein Digestibility (Satterlee et al. 1979)

The nutritive value of a protein depends primarily on its capacity to supply needs of nitrogen and essential amino acids. Although the chemically determined amino acid composition is used to measure the quality of a protein, the biological availability of these amino acids is the real measure of the quality of the proteins. The availability of amino acids depends upon the extent of digestibility of proteins by the proteolytic enzymes of the alimentary tract. Digestibility of a protein can be assessed using rats which is termed in vivo digestibility. It can also be measured using proteolytic enzymes which is termed in vitro protein digestibility. The results obtained with the later procedure agreed well with in vivo protein digestibility as measured in the rats.

Principle

Proteolytic enzymes are used to digest the protein, and the pH change due to the release of amino acids at a fixed time interval is measured by using the formula:

$$\%\,\text{Digestibility} = 234.84 - 22.56X,$$

where X is the pH after 20 min incubation, the in vitro digestibility could be calculated.

Materials

1. Powdered sample which passes through 80-mesh screen.
2. Glass distilled water.
3. *Three enzyme solution*: 1.6 mg trypsin, 3.1 mg chymotrypsin and 1.3 mg peptidase per millilitre in glass distilled water.
4. *Bacterial protease solution*: 7.95 mg protease (type IV from *Streptomyces griseus*) per millilitre in glass distilled water.

Procedure

1. Add 10 mL of glass distilled water to the powdered sample (amount of sample is adjusted so as to contain 6.25 mg protein/mL).
2. Allow the sample to hydrate for at least 1 h (not longer than 25 h) at 5°C.
3. Equilibrate the sample and three enzyme solution, pH 8.0 at 37°C.
4. Add 1 mL of three enzyme solution to the sample suspension and stir at 37°C.
5. After 10 min from the time of addition of three enzyme solution (still stirring) add 1 mL of the bacterial protease solution.
6. Immediately transfer the solution to 55°C water bath.
7. Nine minutes after adding the bacterial enzyme, transfer the solution back to 37°C water bath (in total 19 min after the addition of the three enzyme solution).
8. Measure the pH of the hydrolysate exactly 10 min after the addition of bacterial enzyme. This is called the 20 min pH.

Calculation

In vitro protein digestibility is calculated using the following equation:

$$\%\,\text{Digestibility} = 234.84 - 22.56X,$$

where X is the pH after 20 min incubation.

Note
First run a control (HNRC sodium caseinate) before each set of samples and this must have a 20 min pH of 6.42 ± 0.05. This control is needed to ensure the presence of proper enzyme activity prior to running any samples.

7.11 Protein Efficiency and Net Protein Ratios (Pellet and Young 1980)

The nutritive value of a protein is best determined by in vivo experiments with rats which give more practical values. Animal experiments have been widely used to evaluate protein quality, along with the chemical techniques, since there are no other better alternatives, rat is used as the experimental animal because the results obtained with rat system have been found to agree many a times with the results obtained with human volunteers.

Measuring the growth rate of young animals fed with a test food over a period of time offers the simplest method to evaluate the nutritive value of proteins. The determination of protein efficiency and net protein ratio (PER and NPR) are routinely carried out in many laboratories and these ratios are derived from the weight gain of the test animals. When the growth method is used, PER and NPR are considered as the basic animal experiments. But precise information can be drawn only from nitrogen balance studies such as the estimation of relative net protein ratio. For animal experiments strict standardization of the experimental procedure is required to detect small difference in protein quality. Room temperature between 22 and 24°C and a relative humidity between 50 and 65% are recommended for rats. Healthy male rats of Wistar strains should be selected. Feed consumption measurement should be accurate.

Principle

After 4 weeks of feeding with the test protein the ratio of weight gain to protein consumed is calculated as PER. The ratio of the weight gain in test animal plus the weight loss in control animal to the protein consumption by test animal gives the NPR.

Materials

1. Weanling male rats of Wistar strain, 20–23 days of age. Ten animals for each diet with body weight 64–68 g.
2. Basal diet on an air-dried basis:

   Corn starch – 80%
   Corn oil or cotton seed oil – 10%
   Non-nutritive cellulose – 5%
   * Salts – 4%
   *Vitamin mixture – 1%
   Incorporate the protein food under test into the diet at the expense of corn starch to give 10% (9.7–10.3) protein.
3. Individual cages provided with feeders

Procedure

1. Randomize 10 rats for each diet and feed them for 4 weeks with diet and water ad libitum.
2. Record the food consumption and body weight at weekly and 10-day intervals.
3. *For PER determination*: In addition to the test group, maintain a reference group of rats on a diet consisting of the *basal ratio with casein* at the level of 10% protein.
4. For NPR determination: In addition to the test group, maintain a control group of rats, matched with the test animals with respect to weights, on a diet consisting of the unmodified basal ratio.

Calculation

1. PER: At 4 weeks, calculate the PER for each food and reference casein as:

$$\text{PER} = \frac{\text{Wt. gain of test animal}}{\text{protein consumed}}$$

* Composition mentioned subsequently.

Table 7.3 Salt mixture

Salt	Weight (g)
Calcium carbonate ($CaCO_3$)	78.6
Calcium citrate ($Ca_3C_{12}H_{10}O_{14}{\cdot}4H_2O$)	308.3
Calcium hydrogen phosphate ($CaHPO_4{\cdot}2H_2O$)	112.8
Dipotassium hydrogen phosphate (K_2HPO_4)	218.8
Potassium chloride (KCl)	124.7
Sodium chloride (NaCl)	77.1
Magnesium sulphate ($MgSO_4$)	38.3
Magnesium carbonate ($MgCO_3$)	35.2
Ammonium ferric citrate (Brown, 20.5–22.5% Fe)	15.3
Manganese sulphate ($MnSO_4{\cdot}H_2O$)	0.201
Cooper sulphate ($CuSO_4{\cdot}5H_2O$)	0.078
Potassium iodide (KI)	0.041
Aluminium ammonium sulphate ($AlNH_4(SO_4)_2{\cdot}12H_2O$)	0.507

2. For corrected PER, proceed as follows:

$$\text{Corrected PER} = \text{PER} \times \frac{2.5}{\text{determined PER for reference casein}}$$

3. NPR determination: At 10 days calculate the NPR for each food as follows:

$$\text{NPR} = \frac{\text{Weight gain of test animal} + \text{Weight loss of control animal}}{\text{Protein consumed by test animal}}$$

Composition of vitamin mixture
Prepare a vitamin mixture containing the following amounts in 1 g. Vitamin A, 1,000 IU; Vitamin D, 100 IU; Vitamin E, 10 IU; Vitamin K (menadione), 0.5 mg; thiamine, 0.5 mg; riboflavin 1 mg; pyridoxine, 0.4 mg; pantothenic acid, 4 mg; niacin, 4 mg; choline, 200 mg; inositol, 25 mg; *para*-aminobenzoic acid, 10 mg; Vitamin B_{12}, 2 μg; biotin, 0.02 mg; and folic acid, 0.2 mg. Add sufficient cellulose to make 1 g (Table 7.3).

7.12 Determination of Net Protein Utilization, Digestibility and Biological Value (Pellet and Young 1980)

For more precise information on the quality of a protein, the extent of utilization of protein in the animal system is required. The amount of protein absorbed and retained in the body should be determined. Since protein after digestion and absorption gets incorporated into nitrogenous compounds – both proteinaceous and non-proteinaceous of the body – a measure of nitrogen intake absorption and retention would be an ideal practical approach.

Based on nitrogen balance studies, net protein utilization (NPU), digestibility (D) and biological value (BV) are calculated.

Principle

The nitrogen content of the test diet is determined. The feed consumption over a 5-days-period is measured, leading to a calculation of the total N-intake. The faecal and urinary nitrogens on these 5 days are also determined. From these values calculations can be made, to know the "N retained" and "N absorbed". NPU, D and BV are then calculated as follows:

$$\text{NPU} = \frac{\text{N retained}}{\text{N intake}}$$

$$\text{D} = \frac{\text{N absorbed}}{\text{N intake}}$$

$$\text{BV} = \frac{\text{N retained}}{\text{N absorbed}}$$

Materials

1. Wistar male rats (weighing ~60 g)
2. Metabolic cages
3. Balance for weighing rats
4. Diet
5. Plastic container with tight lids
6. Sieves
7. N-free mixture

   Sucrose – 9.0%
   Cellulose powder – 5.2%
   Soybean oil – 5.2%
   N-free starch – 80.6%

Procedure

Preparation of diets: Five hundred gram diet (dry weight basis) is prepared for each treatment. Initially, determine the protein and dry matter content in the test sample, then calculate the amount of sample required to give 7.5 g N on dry weight basis. To this add 20 g mineral and 5 g vitamin mixture to make up 500 g diet (dry wt. basis) and rest of the amount from N-free diet after determining the moisture in N-free diet.

1. Weigh 40.0 g diet on dry weights basis (actual weight will vary depending upon the moisture in the diet) into plastic boxes for each rat sufficient for preliminary period for 4 days, and tightly close with the lid. Thus each animal receives 150 mg N and 10 g dry matter per day throughout the test period.
2. Put cages on the rack without funnel.

3. Weigh the rats in the beginning of the experiment, divide the rate into groups of five, such that average weight of the group differs by no more than 0.5 g and record the weight (lesser the difference better the standardization).
4. Transfer diet equivalent to 10 g dry weight to the feed box of respective cage. In all the experiments, group 1 is always allotted for casein diet.
5. Press the feed with suitable flat surface.
6. Place plastic bowls below each rat cage for the collection of urine and faeces.
7. Feed every rat once a day in the morning and check for the water in bottles.
8. On the last day of the preliminary period, when all the diet from diet box had been transferred to the rat cage feed box, weigh 50 g equivalent of dry weight diet and transfer to respective plastic boxes. Apply Vaseline grease to nylon net as well as Perspex funnel.
9. At the end of the preliminary period of 4 days, again weigh the rats and clean the rat cage feed box as well. Clean inside of the rat cage with lukewarm water.
10. Put both the greased Perspex funnel and nylon net in proper position.
11. Transfer 10.0 g dry weight equivalent diet from respective boxes to cage feed box and follow as in preliminary period for the 5 days of N-balance period.
12. Put flask containing 35 mL 5% H_2SO_4 and funnel with glass wool below the Perspex funnel to collect the urine and beaker containing 50 mL 5% H_2SO_4 below nylon net to collect faeces.
13. During the 5 days of the N-balance period daily weighing of the feed from diet box and its transfer to cage feed boxes is followed. Any of the feaces remaining, at the neck of the nylon netting is also transferred to beaker with the help of forceps. If by chance faeces have fallen on the funnel, these are also removed to respective beakers.
14. Every morning spray the net in situ with small quantity of 20% citric acid from a plastic wash bottle to prevent N losses and then finally wash the glass wool with small quantity of 5% H_2SO_4 for the same reason.
15. Thus following the procedure, during the N-balance period of 5 days, urine and faeces are collected, respectively, in flasks and beakers.
16. At the end of N-balance period, remove all the water bottles and prevent the access of rats to feed box 3 h before the termination of the experiment.
17. Weigh the rats and transfer to bigger cages.
18. Transfer any remaining feed in the feed boxes and spill tray to the respective diet boxes.
19. Wash bottom lids and lower portion of the case, funnel and nylon net with approximately 75 mL of lukewarm water, using a soft brush through a large glass funnel down the urine flask with funnel and glass wool. Further, wash funnel with glass wool *3 times* with water to ensure that all N has been washed out.
20. Transfer urine plus washings quantitatively to a graduated 500 mL flask and make up the volume. Mix well.
21. To beaker containing faeces, add 4 times 25 mL conc. H_2SO_4 at hourly intervals. After each addition stir and mix thoroughly with *spatula* and allow it to cool.

22. This process if followed 4 times, the resultant faeces solution would become homogenous. Transfer this mixture to 500 mL volumetric flask and make up the volume.
23. Weigh the remaining feed in diet boxes and record in the notebook.
24. Determine N in urine and faeces by taking out 25 mL sample of urine and 50 mL sample of faces by micro-Kjeldahl method. Samples are analysed in duplicate.
25. Calculate the total amount of N excreted in urine and in faeces by each rat during balance period.

Determination of metabolic and endogenous nitrogen: Feed a separate group with 4% freeze-dried, ether-extracted egg protein as the protein source. Since egg protein at this level is completely utilized by rats, nitrogen in the urine and faeces must be of endogenous origin.

Collect the determined nitrogen in the urine and faeces as described, and incorporate in the calculations.

Calculations

$$\mathrm{NPU} = \frac{\text{N retained}}{\text{N intake}} = \frac{\mathrm{I} - (\mathrm{F} - \mathrm{F_k}) - (\mathrm{U} - \mathrm{V_k})}{\mathrm{I}}$$

$$D = \frac{\text{N absorbed}}{\text{N intake}} = \frac{\mathrm{I} - (\mathrm{F} - \mathrm{F_k})}{\mathrm{I}}$$

and

$$\mathrm{BV} = \frac{\text{N retained}}{\text{N absorbed}} = \frac{\mathrm{I} - (\mathrm{F_k}) - (\mathrm{U} - \mathrm{U_k})}{\mathrm{I} - (\mathrm{F} - \mathrm{F_k})}$$

where I is the intake nitrogen (nitrogen in the diet), F = faecal nitrogen, F_k = endogenous faecal nitrogen, U = urinary nitrogen and U_k = endogenous urinary nitrogen.

7.13 Protein Estimation by Lowry's Method (Lowry et al. 1951)

Protein can be estimated by method described by Lowry and colleagues. The method is sensitive enough to give a moderately constant value and hence largely followed. Protein content of enzyme extracts is usually determined by this method.

Principle

The blue colour developed by the reduction of the phosphomolybdic-phosphotungstic components in the Folin-Ciocalteau reagent by the amino acids tyrosine and tryptophan present in the protein plus the colour developed by the biuret reaction of the protein with the alkaline cupric tartrate are measured in the Lowry's method.

Reagents

1. *Reagent A*: 2% sodium carbonate in 0.1 N sodium hydroxide.
2. *Reagent B*: 0.5% copper sulphate ($CuSO_4{\cdot}5H_2O$) in 1% potassium sodium tartrate.
3. *Reagent C*: alkaline copper solution: Mix 50 mL of A and 1 mL of B prior to use.
4. *Reagent D*: Folin-Ciocalteau reagent – reflux gently for 10 h a mixture consisting of 100 g sodium tungstate ($Na_2WoO_4{\cdot}2H_2O$), 25 g sodium molybdate ($Na_2MoO_4{\cdot}2H_2O$), 700 mL water, 50 mL of 85% phosphoric acid, and 100 mL of concentrated hydrochloric acid in a 1.5-L flask. Add 150 g lithium sulphate, 50 mL water and a few drops of bromine water. Boil the mixture for 15 min without condenser to remove excess bromine. Cool, dilute to 1 L and filter. The reagent should have no greenish tint. (Determine the acid concentration of the reagent by titration with 1 N NaOH to a phenolphthalein end-point). Folin-Ciocalteau reagent can be purchased commercially. Store refrigerated in amber bottles. A good quality reagent is straw yellow in colour.
5. *Protein solution* (*stock standard*): weigh accurately 50 mg of bovine serum albumin and dissolve in distilled water and make up to 50 mL in a standard flask.
6. *Working standard*: dilute 10 mL of the stock solution to 50 mL with distilled water in a standard flask. One millilitre of the solution contains 200 μg protein.

Procedure

Extraction of protein from sample: Extraction is usually carried out with buffers used for the enzyme assay. Weigh 500 mg of the sample and grind well with a pestle and mortar in 5–10 mL of the buffer. Centrifuge and use the supernatant for protein estimation.

Estimation of protein:

1. Pipette out 0.2, 0.4, 0.6, 0.8 and 1 mL of the working standard into a series of test tubes.
2. Pipette out 0.1 and 0.2 mL of the sample extract in two other test tubes.
3. Make up the volume to 1 mL in all the test tubes. A tube with 1 mL of water serves as the blank.
4. Add 5 mL of reagent C to each tube including the blank. Mix well and allow to stand for 10 min.
5. Then add 0.5 mL of reagent D, mix well and incubate at room temp in the dark for 30 min. Blue colour is developed.
6. Take the readings at 660 nm.
7. Draw a standard graph and calculate the amount of protein in the sample.

Calculation

Express the amount of protein mg/g or 100 g sample.

- If protein estimation is desired in a sample with high phenolic or pigment content, extract should be prepared with a reducing agent preferably cysteine and NaCl. Precipitate the protein with TCA, separate the protein and dissolve in 2N NaOH and proceed.
- If the protein concentration of the sample is high (above 50 μg/mL) measure the colour intensity at 550 nm.
- For complete enzyme extraction, sometimes the chemicals like ethylenediamine tetraacetic acid (EDTA), magnesium salts and mercaptoethanol are included. This method of protein (mercaptoethanol) compounds as they interfere with this procedure. When these chemicals are present in the extract, precipitate the protein by adding 10% TCA, centrifuge and dissolve the precipitate in 2 N NaOH and proceed for protein estimation.
- Rapid mixing as the Folin reagent is added is important for reproducibility.

7.14 Polycrylamide-Sodium Dodecyl Sulphate Slab Gel Electrophoresis (SDS-PAGE) of Proteins (Laemmli 1970)

Electrophoresis of proteins in polyacrylamide gels is carried out in buffer gels (non-denaturing) as well as in sodium dodecyl sulphate (SDS) containing (denaturing) gels. Polyacrylamide gel is more convenient than in any other medium such as paper and starch gel. Electrophoretic procedures are rapid and relatively sensitive requiring only micro-weights of proteins. Separation in buffer gels relies on both the charge and size of the protein, whereas it depends only upon the size in the SDS-gels. Analysis and comparison of proteins in a large number of samples is easily made on polyacrylamide gel slabs.

Polyacrylamide gels are formed by polymerising acrylamide with cross-linking agent (bisacrylamide) in the presence of a catalyst (persulphate ion) and chain initiator (TEMED; *N,N,N,N*-tetramethylethylene diamine). Solutions are normally degassed by evacuation prior to polymerization since oxygen inhibits polymerization. The porosity of the gel is determined by the oxygen inhibited polymerization. The porosity of the gel is determined by the relative proportion of acrylamide monomer to bisacrylamide. Gels are usually referred to in terms of the total percentage of acrylamide and bis present, and most protein separations are performed using gels in the range 7–15%. A low percentage gel (with large pore size) is used to separate high molecular weight proteins and vice versa. At high concentrations of persulphate and TEMED the rate of polymerization is also high. Among a number of methods commonly used, the sodium dodecyl sulphate-polyacrylamide gel electrophoresis (SDS-PAGE) that facilitates characterization of polypeptides and determination of their molecular weight is described.

Principle

SDS is an anionic detergent which binds strongly to, and denatures, proteins. The number of SDS molecules bound to a polypeptide chain is approximately half the number of amino acid residues in that chain. The protein–SDS complex carries net negative charges, hence moves towards the anode and the separation is based on the size of the protein (charge to mass ratio).

Reagents

1. Stock acrylamide solution

   Acrylamide 30% – 30 g
   Bisacrylamide 0.8% – 0.8 g
   DD Water to – 100 mL

2. Separating gel buffer

   1.875 M Tris–HCl – 22.7 g (pH 8.8)
   Water to – 100 mL

   Stacking gel buffer
   0.6 M Tris–HCl – 7.26 g (pH 6.8)
   Water to – 100 mL

3. Polymerising agents

   (a) Ammonium – 0.5 g/10 mL, prepare fresh before use persulphate 5%
   (b) TEMED – fresh from the refrigerator

4. Electrode buffer (may be used 2–3 times)

   0.05 M Tris – 12 g
   0.192 M Glycine – 28.8 g (pH 8.2–8.4)
   0.1% SDS – 2 g
   Water to – 2 L

5. Sample buffer (5× concentration)

   Tris–HCl Buffer, pH 6.8 – 5 mL
   SDS – 0.5 g
   Sucrose – 5 g
   Mercaptoethanol – 0.25 mL
   Bromophenol Blue – 1 mL
   (0.5% W/V solution in water)
   Water to – 10 mL
   Dilute to 1× concentration and use. Store frozen in small aliquots.

6. Sodium dodecyl sulphate 10% solution – store at room temperature and use

7. Standard marker proteins

Protein	MW (Daltons)
α-Lactalbumin	14,200
Trypsin inhibitor soybean	20,100
Trypsinogen	24,000
Carbonic anhydrase	29,000
Glyceraldehydes-3-phosphate dehydrogenase, rabbit	36,000
Albumin, egg	45,000
Albumin bovine	66,000

Dissolve the above proteins in single strength sample buffer at a concentration each to 1 mg/mL. Load the well with 25–50 μL.

8. Protein stain solution

   Coomassie Brilliant Blue R250 – 0.1 g
   Methanol – 40 mL
   Acetic acid – 10 mL
   Water – 50 mL
   Dissolve the dye in methanol first and use fresh preparation every time.

9. Destainer
   As for staining solution but without dye.

Procedure

1. Thoroughly clean and dry the glass plates and spacers, then assemble them properly. Hold the assembly together with clips. Clamp in an upright position. White petroleum jelly or 2% agar (melted in a boiling water bath) is then applied around the edges of the spacers to hold them in place and seal the chamber between the glass plates.
2. Prepare a sufficient volume of separating gel mixture (30 mL) for a chamber of about 18 × 9 × 0.1 cm by mixing the following:

	For 15% gel	For 10% gel
Stock acrylamide solution (mL)	20	13.3
Tris–HCl, pH 8.9 (mL)	8	8
Water (mL)	11.4	18.1
Degas the solution for 3–5 min and then add		
Ammonium persulphate solution (mL)	0.2	0.2
10% SDS (mL)	0.4	0.4
TEMED (μL)	20	20

Mix gently and pour the gel solution in the chamber between the glass plates. Layer distilled water on top of the gel and leave to set for 30–60 min.

3. Prepare stacking gel (4%) by mixing the following solutions (total volume 10 mL).

   Stock acrylamide solution = 1.35 mL
   Tris–HCl (pH 6.8) = 1 mL
   Water = 7.5 mL
   Degas as above, and then add:
   Ammonium persulphate solution (5%) = 50 μL
   10% SDS = 0.1 mL
   TEMED = 10 μL

   Remove the water from the top of the gel and wash with a little stacking gel solution. Pour the stacking gel mixture, place the comb in the stacking gel and allow the gel to set (30–60 min).
4. After the stacking gel has polymerized, remove the comb carefully without distorting the shapes of the well. Install the gel after removing the clips, agar, etc. in the electrophoresis apparatus. Fill it with electrode buffer and remove any trapped air bubbles at the bottom of the gel. Connect the cathode at the top and turn on the DC-power briefly to check the electrical circuit. The electrode buffer and the plates can be kept cooled using cooling facility so that heat generated during the run is dissipated and does not affect the gel and resolution.
5. Prepare samples for electrophoresis, following suitable extraction procedure. Adjust the protein concentration in each sample using the 5-strength sample buffer and water in such a way that the same amount of protein is present per unit volume. Again the concentration should be such as to give a sufficient amount of protein (50–200 μg) in a volume (25–50 μL) not greater than the size of the sample well. Generally, the sample solution is heated in boiling water for 2–3 min to ensure complete interaction between proteins and SDS.
6. Cool the sample solutions and the required volume is filled in a microsyringe and injected into sample well through the electrode buffer. Marking the position of wells on the glass plate with a marker pen and the presence of bromophenol blue in the samples, also load in to a well a standard marker proteins in the sample buffer.
7. Turn on the current to 10–15 mA for initial 10–15 min until the samples travel through the stacking gel. The stacking gel helps concentration of the samples. Then continue the run at 30 mA until the bromophenol blue reaches the bottom of the gel (about 3 h). However, the gel may be run at a high current (60–70 mA) for short period (1 h) with proper cooling.
8. After the run is compete, carefully remove the gel from between the plates and immerse in staining solution for at least 3–4 h or overnight with uniform shaking. The proteins absorb the Coomassie Brilliant Blue.
9. Transfer the gel to a suitable container with at least 200–300 mL destaining solution and shake gently and continuously. Dye that is not bound to proteins is removed. Change the destainer frequently, particularly during initial periods, until the background of the gel is colourless. The proteins fractionated into band are seen coloured blue. As the proteins of minute quantities are stained faintly,

destaining process should be stopped at appropriate stage to visualize as many bonds as possible. The gel can now be photographed or dried in vacuo for a permanent record.

Precautions

1. Acrylamide as a monomer is highly neurotoxic; handling should be with extreme care.
2. All solutions should be prepared fresh, old solutions result in poor resolution of proteins.
3. All chemicals and distilled water should be of high quality. The solution prepared should be filtered before use. The solution can be stored refrigerated for 1–2 weeks.
4. Prefer to use the gel immediately following polymerization although the separation gel after setting can be stored overnight by wetting with fourfold diluted separation gel buffer or with stacking gel and comb placed over it to avoid drying.
5. Degassing of gel mix should be adequate for easy polymerization.
6. The water layered over the separation gel should be completely removed for quick polymerization of the stacking gel.
7. Handle the polyacrylamide gel carefully to avoid any breakage.
8. The slab gel along the glass plates is placed vertically in the electrophoresis tank and run. It is therefore called "vertical slab gel electrophoresis".
9. In 10% polyacrylamide gels, the low molecular weight (~10,000 daltons) polypeptides will migrate diffused; for fine resolution of these polypeptides use gels of higher (15%) acrylamide concentration.
10. Any band of 0.1 μg protein is visualized by Coomassie Brilliant Blue staining in SDS-PAGE; for visualizing proteins of lower concentration below 0.1 μg high sensitive (silver staining) method is recommended (Table 7.4).

7.15 Fluorography of Polyacrylamide Gels (Bonner and Laskey 1974)

Fluorography is an improved version of autoradiography in the presence of a fluorescing compound. It is one of the technique to determine the radioactivity in gels and other media by a combination of fluorescence and photography. When radioactivity labelled macromolecules such as proteins are separated by electrophoresis much of the radiation is absorbed by the gel and the autoradiography of such gel gives poor results. For low-energy emitters such as tritium and ^{14}C the sensitivity of fluorography is many-fold higher than that of autoradiography. The fluorography may be photographed for qualitative information or analysed by scanning for quantitative results.

Table 7.4 Frequent troubleshooting and remedies during electrophoretic procedure

Trouble	Cause	Remedy
Failure or slow polymerization of the gel	Presence of oxygen,	Degas the solution efficiently
	Absence of catalysts	Check if all solutions mixed
	Stock solution aged	Use fresh solutions
	Glass plates	Degrease the plates with ethanol
Poor sample wells	Stacking gel and comb	Fit and or remove the comb carefully
Long duration of the run	Interference of air-bubbles	Flush air-bubbles
Staining is poor	The dye absorption not efficient	The dye may be old, use a strong dye solution
The staining is patchy	Solid dye	Dissolve the dye completely or filter
The stained bands are decolourised	The dye is removed excessively	Restain the gel and stop destaining appropriately
Protein bands are in adequately resolved	Insufficient electrophoresis separation gel	Run for longer time. Change the percentage of the gel
Protein bands wavy	Excess persulphate	Use optimum concentration of persulphate
Bands have become streaked	Proteins remain aggregated, denatured or insoluble	Use fresh sample buffer or extra SDS or centrifuge the sample extract sufficiently
Protein dye migration not even	Gel is partly insulated by air bubbles+ insufficient cooling	Remove air bubbles before electrophoresis improve the cooling or run at a low current
The protein bandlane broadens at the bottom of separation gel	Sample density	Load equal volume of sample in each well, equal strength sample buffer, leave no empty well in the middle
Sample diffuses while loading the wells	Low density of sample	Increase the concentration of sucrose/ glycerol in the sample buffer

Principle

In fluorographic technique, a fluro/scintillator impregnated into the gel absorbs the radiation from the isotope and re-emits light that passes through the gel to the film producing a photographic image analogous to an autoradiograph.

Materials

1. Dimethyl sulphoxide (DMSO)
2. PPO (*2,5 dipheynl oxazole*) *solution*: Dissolve 22 g PPO and make up to 100 mL in DMSO
3. *Fixing solution*: 7% acetic acid, 20% methanol in distilled water.
4. Film developer
5. Photographic flash unit (with red screen on the flash window)
6. X-ray film
7. Gel dryer

Procedure

1. Fractionate radioactively-label polypeptides by SDS-PAGE. Stain and destain the gel, otherwise, the proteins as fixed by immersing the gel in the fixing solution for an hour.
2. Immerse the destained/fixed gel in 20 volumes of DMSO for 30 min to dehydrate the gel. *Repeat this step for complete dehydration.*
3. Immerse the gel in the PPO solution to impregnate PPO into the gel for 4 h. The gel is shaken slowly in large Petri-dish throughout the steps 1–3. The gel shall appear shrunk due to dehydration.
4. Transfer the gel gently and evenly to a dish containing a large volume of water for 30–60 min. In water, PPO precipitates so that the gel turns opaque. Clean the gel in water.
5. Carefully dry the gel under vacuum at 70°C on a sheet of 3 mm filter paper.
6. In a photographic dark room, take out an X-ray film and lay a sheet of transparent paper over it. Pre-sensitize the film using a battery-operated small flash unit (camera flash) from 2 m height in the dark. Assemble the film and dried gel, with the presensitised side of the film facing the dried gel, between a fold of thick black sheet. Sandwich the assembly between two plywood plates slightly larger than the X-ray film. Wrap around the plywood plates 3–4 layers of aluminium foil and hold them firm using bulldog clips. A cassette may be used, if available, instead of the above assembly.
7. Place the assembly/cassette at −20°C in a freezer for 2–4 days to expose the film to the gel. In the case of gel which received low amounts of radioactivity, the exposure is done at −70°C to get the film sufficiently exposed. Take out the assembly/cassette from the freezer after the appropriate exposure time and allow to warm up for 1–1.5 h at room temperature.
8. In the dark room, open the assembly and develop the film.

Precautions

1. DMSO diffuses into the skin very easily proving harmful. Wear disposable gloves while handling.
2. Labelling of polypeptides is done using radioactive (^{3}H,^{14}C or ^{35}S) amino acid either in in vivo or in vitro in a cell-free protein-synthesizing system programmed with mRNAs. Use high specific activity precursor. Load each lane of the gel with equal counts of radioactivity.
3. The times given in steps 1–3 are sufficient for 15% acrylamide gel, thickness up to 1.5 mm. Thicker or more concentrated gels will require longer periods in all solutions.
4. *Drying the gels should be carefully done.* A commercially available gel dryer is preferred. Where such facility is unavailable, the procedure described below is useful. Place the PPO impregnated washed, opaque gel onto a sheet of 3-mm filter paper slightly larger than the gel itself. Do not trap air bubbles between them. Place them, gel uppermost, on top of a porous rigid polyethene pad. Introduce the whole assembly into a thick gauge polyethene bag(s). turn upside

down such that the porous pad is on top of the gel inside the polyethene bag. Place at the centre of the pad a glass funnel stem facing upward. Fasten the mouth of polyethene bag around the funnel stem using elastic bands or thread so that the bag is leak-proof. Connect the stem to a water pump vacuum line. Water in the gel is removed as droplets by suction. Place the gel assembly after 5–10 min over a hot plate at 70°C to hasten the drying. Many factors such as the gel thickness, efficiency of vacuum pump, the presence of DMSO in the gel etc., influence the gel drying period. Gel cracking may occur as an acute problem if the vacuum is turned off when the gel is partly dry. Continue until the gel is completely dried which may require 3–4 h or even longer. Finally, the dried gel on the filter paper is removed and used for fluorography.

5. Pre-sensitising of the film is done to improve the sensitivity and linearity of response of the film. The flash window of the unit should be covered with two layers of red filter (transparent sheets). While pre-sensitising neither the aluminum foil nor the gel should be at the vicinity of the film. Do not use the first flash after switching on the unit as it shall be stronger and different from other flashes. Pre-sensitising conditions such as the height of the film, the number of transparent sheets over the film, the red-filter over the flask window etc. *should be standardized* to get satisfactory results on the fluorography.
6. Do not place the gel-film sandwich near any radiation source during exposure as otherwise the fluorography plate will be fogged.
7. The exact exposure time is dependent on the amount of radioactivity in the gel. The recommended exposure times are of the order of 24 h for 1,000–10,000 dpm ^{3}H and correspondingly longer for lower amounts of radioactivity. Exposure at lower temperatures (say −70°C) is beneficial than at high temperatures.
8. The unused PPO in the DMSO solution can be recovered and recycled. Mix one volume of PPO solution with 10 volumes of distilled water and stir continuously. The PPO crystallizes out. Filter, collect the crystals and dry at 25°C for 2–3 days. Dissolve the PPO in a minimum volume of ethanol and precipitate, filter and dry again. Finally, dry the PPO in a vacuum oven for a few days and reuse.
9. View the developed plate for polypeptides which appear as black bands. Photograph the plate and make prints for qualitative information.
10. Quantitative information on the polypeptides can be obtained by scanning the X-ray plate (see Chap. 8).

7.16 Quantification of Protein in Polyacrylamide Gels (Smith et al. 1980)

The SDS-polyacrylamide gel electrophoresis (SDS-PAGE) is an easy and quick method to quantify a particular protein at microgram level from a mixture. This is done by scanning the gel by densitometry of the stained bands on it. Similarly, radioactivity-labelled polypeptides electrophoresed on SDS-PAGE can also be quantified by densitometric scanning of the fluorographic plate.

Principle

The percent absorption of incident light is directly proportional to the colour intensity of the protein–dye complex on the gel and is directly related to the protein concentration. Also, the intensity of darkening of the X-ray plate is directly proportional to the radioactivity in the protein in fluorographic plates.

Materials

For gel scan

1. A spectrophotometer with suitable scanning facility and chart.
2. *Protein stain (quantitative)*: 0.2% Proceion Navy MXRB dye in methanol: acetic acid: water 5:14. Dissolve the dye first in methanol and prepare fresh every time.
3. *Destaining solution*: methanol: acetic acid: water (1:1:8)

For fluorograph scan

Fluorograph plate

Procedure

1. After the electrophoresis immerse the gel in Proceion Navy solution and shake gently until the proteins are completely stained (for a fixed period say 2 h).
2. Destain the gel until the background is colourless.
3. Scan the gel at 580 nm to measure the degree of dye bound by each bond of protein. Depending upon the type of equipment available for scanning, the whole gel is used or each lane is cut out and scanned individually. The total absorption by the dye in each bond is proportional to the area of the peak in the scan profile. Each peak in the scan profile is traced using a planimeter to determine the area under it. Otherwise, each peak in the chart may be cut out and weighed. When an integrator is interposed, the area under each peak is automatically calculated.
4. A curve is obtained by plotting A_{580} vs. amount of protein used as standard. Bovine serum albumin (Fraction V) at different known concentrations co-electrophoresed in different lanes in the same gel is also used to construct the standard curve. The proteins both in the standard and under examination to have equal dye-binding property.

Scanning fluorographic plate: Scan the individual lane strip or the whole fluorographic plate at 620 nm as described above. The standard curve is obtained using a radioactivity labelled standard protein whose concentration and radioactivity are known.

Note

The following conditions need to be satisfied to quantify proteins on the gels:

- The protein bands should be well resolved.
- Dye should bind to the protein of interest, and the binding should be uniform to all proteins.
- If the peaks are not well resolved, use of a narrower beam of light will improve the situation but at the cost of baseline.
- Coomassie Brilliant Blue R250 staining is not suitable for quantitative analysis of proteins although it is a highly sensitive stain.
- Proceion Navy dye binds to the proteins stochiometrically and covalently. Destaining of this dye from the gel requires longer time.
- Sampling errors are inevitable but their effect can be reduced by repetition and averaging the results.
- Electrophoresis with a fixed sample volume, voltage and duration of run is necessary between runs to obtain satisfactory results.

7.17 Protein Electrophoresis Using Starch Gel (Smithies 1955)

Starch gel is another supporting medium used for electrophoresis, particularly as a horizontal gel for the fractionation of proteins. Partially hydrolysed starch dissolved in any of a variety of buffer solutions is used for casting gel. The gel may be prepared at any concentration between 2 and 15% (w/v) although a 10% gel is ideal for most separations. The starch gel electrophoresis is used for analytical as well as preparative purposes by changing the thickness of the gel.

Principle

Since starch gels exhibit molecular sieving property, the separation is not only on the basis of differences in chare but also of differences in the molecular size and shape of the proteins.

Materials

1. Glass plates (22 × 12 × 0.4 cm dimension). Glass edge strips 22- and 14-cm long, 5-mm wide and 1- and 3-mm thick.
2. Gel buffer (pH 8.6)
   - Tris–HCl – 9.2 g
   - Citric acid – 1.05 g
   - Water to – 1 L
3. Electrode buffer (pH 7.9)
   - Boric acid – 18.6 g
   - Sodium hydroxide – 2 g
   - Water to – 1 L
4. Hydrolysed starch (electrophoresis grade)

Procedure

1. Prepare the gel mold by sticking the 1-mm thick glass edge strips to one of the glass plates to give a shallow try 21 cm long, 14 cm wide and 1 mm deep. Place the mold on a horizontal surface.
2. To prepare a 10% gel for analytical purpose, mix 4 g of the dry starch with 40 mL gel buffer in a 200 mL conical flask to a fine suspension.
3. Heat the suspension while gently swirling the flask, and bring the starch solution to a boil. The solution attains a clear transparent from.
4. Degas the solution by applying vacuum to the flask for 5–10 s. Release the vacuum carefully to avoid any splashing of the solution.
5. Pour the hot starch solution into the centre and allow it to spread over the plate to form an even layer.
6. Leave the plate for 10 min or longer to allow the starch solution to cool and gel. Transfer the gel plate and maintain in a moisture cabinet at 4°C, preferably overnight, before use.
7. Make slots in the gel across the width of the plate approximately 6 cm from the cathode end for sample application. Press the comb into the gel and remove it carefully.
8. Soak pieces of 1-cm long cotton thread in the sample solution and insert into the slot in such a way that the thread touches the base glass plate and not above the gel level. By this way, approximately 2–3 μL of sample solution in a piece of 3-mm filter paper (1 × 0.5 cm) is used instead of cotton thread.
9. Place the gel plate in an electrophoresis tank and apply wicks to both ends of the gel. The wicks (several thickness of filter paper, cut to the width of the gel) should be soaked in electrode buffer, and placed overlapping the gel surface by 1–2 cm. Place a glass plate over the wicks and gel to hold them as well as to reduce evaporation from the gel.
10. Run the gel 5–10 V/cm. The running time may be between 2 and 15 h depending upon the voltage applied. The electrophoresis should however be conducted in a cold room (4–8°C).
11. For protein pattern, the gel may be stained with Coomassie Brilliant Blue R250 as described under SDS-PAGE. However, the starch gel electrophoresis of proteins is mainly used for studying a group of related proteins and therefore specifically stained as for isoenzymes.
12. After the run, remove the top plate and wicks, lift the gel plate from electrophoresis apparatus and apply a suitable stain.

Note

- Hydrolysed starch suitable for electrophoresis may be prepared by warming potato starch in acidified acetone.
- The buffer system should be carefully chosen. There are a number of buffer systems used for starch gel electrophoresis. As the separation of a particular group of proteins depends upon the pH, ionic strength, buffer constituents, etc.

- Prior to use, check that the gel does not contain any undissolved starch, bubbles, bacterial growth, etc.
- The sensitivity of the method depends upon the staining procedure used, therefore the samples that have a total protein concentration of between 1 and 100 mg/mL are generally used.

7.18 Isoelectric Focusing of Proteins (Wrigley 1968)

Isoelectric focusing (IEF) is an electrophoretic method for the separation of proteins according to their isoelectric points (pI). It is reproducible, sensitive and highly useful to resolve closely related proteins which may not be so well separated by other techniques.

Principle

IEF is carried out in a thin layer of polyacrylamide gel containing a large series of carrier ampholytes. When a potential difference is applied across the gel, the carrier ampholytes align themselves in order of increasing pI from the anode to the cathode, thus producing a uniform pH gradient across the gel. Under the influence of the electric field each protein migrates to the region in the gradient where the pH corresponds to its pI. The protein is electrically neutral at its pI and where it gets focused in the gel. After focusing, the separated components are detected by staining.

Materials and reagents

1. Acrylamide, bisacrylamide and sucrose.
2. Riboflavin solution, 1 mg/10 mL. The solution should be refrigerated in a brown bottle.
3. Carrier ampholytes of the suitable pH range (pH 3–10, 5–7 or 4–8), which should be stored at 4°C.
4. Glass plates of appropriate dimension.
5. *Fixing solution* : 5 g sulphosalicylic acid and 10 g trichloroacetic acid (TCA) in 90 mL distilled water.
6. *Destaining solution*: Methanol: acetic acid: water in the ratio 3:1:6 (v/v).
7. *Staining solution*: 0.2% Coomassie Brilliant Blue R250 in the destaining solution. *Filter before use*.
8. *Wick solutions*: 1 M phosphoric acid for anode and 1 M NaOH for cathode.
9. Electrofocusing system.

Procedure

1. Dissolve the following components

   Acrylamide – 1.94 g
   Bisacrylamide – 0.06 g

Sucrose – 5.0 g
Riboflavin solution – 0.25 mL
Water – 40 mL

2. Stick strips of insulating tape 1 cm wide and approximately 0.20 mm thick around the edge of a clean glass plate. This shall give a very shallow tray of 18 × 13 × 0.015 cm in which a thin polyacrylamide gel is cast.
3. Add 2 mL of carrier ampholyte solution of the appropriate pH range to the above mix. Ensure that all the components are uniformly mixed by gently swirling the flask until poured. The entire solution will be sufficient for six plates.
4. Place the glass mold in a large tray with the taped surface uppermost. Wipe clean the surface with an alcohol moistened tissue to remove any traces of grease.
5. Transfer approximately 7 mL of the above solution to one end of the glass mold.
6. Place a clean plain glass plate (20 × 15 cm) one of the short edges along the taped edge of the mold adjacent to the acrylamide-ampholyte solution. Gradually lower the top plate and allow the solution to spread over the mold without entrapping any air bubbles. Press the top plate into firm contact with the taped edge of the bottom plate.
7. Lift the complete mold and top plate out of the large tray remove any gel material at the edges and bottom of the mold plate.
8. Photopolymerize the gel for at least 3 h under white light or bright direct sunlight to provide sufficient UV light.
9. After polymerization wipe the outside surfaces with a wet tissue to remove any solid material, otherwise it may affect cooling during the run.
10. The plates may be stored for a month in the dark at 4°C. The removal of the top plate is easier when cooled at 4°C for at least overnight.
11. Prior to sample application, remove the top plate carefully, inserting a scalpel blade between the two glass plates at a corner. The whole gel should stick either on the bottom or top plate for use. Occasionally, part of the gel will stick to the top plate and part to the bottom plate. In such cases, the gel has to be discarded.
12. Absorb sample solutions (5–8 μL) on 5 × 5 mm pieces of Whatman No. 1 filter paper and lay these on the gel surface along the length of the plate at about 5 cm from the cathode edge. The protein concentration between 0.05 and 0.15 mg is generally sufficient.
13. Place the gel plat on the cooling plate of the electrofocusing apparatus through which water at 4–8°C is circulated.
14. Apply electrode wicks (strips of Whatman 17 filter paper) to the anode and cathode edges of the gel. The anode and cathode wicks are uniformly soaked in 1 M phosphoric acid and 1% ethanolamine may also be used as wick solutions.
15. Apply a potential difference between 130 and 160 V/cm across the plate. Focusing takes 2–4 h and is complete when the same components of the same sample placed on the cathode and anode sides in parallel lanes reached similar zones.

16. When focusing is complete, disconnect the power supply, remove the electrodes and wicks and lift the gel plate from the tank carefully.
17. Place the gel plate in the fixing solution for 15–20 min. Transfer the plate to destaining solution for 15 min. Then transfer the gel to staining solution for 30 min; and finally to destaining solution for about 20 min until the protein bands are clearly visible. The entire process is done at room temperature.
18. To preserve the gel, first immerse the destained gel in destaining solution containing 10% glycerol for 30–60 min. Soak a cellophane sheet in the same solution for a few minutes and wrap it around the gel and supporting glass plate to avoid curling of the gel. Avoid trapping air. Let the wrapped gel dry in a well-ventilated oven at 50°C. The gel is photographed for record.

Note

1. Preformed gels for focusing are available commercially.
2. Agarose can also be used as a supporting medium for analytical IEF.
3. Prefocusing of the plates prior to sample application for establishing pH gradient is desirable.
4. The sample may be applied at the cathode where the denaturation of most proteins does not take place.

7.19 Western Blotting (Towbin et al. 1979)

This method is an anology of the "Southern blot" used to transfer DNA from gels to nitrocellulose filter and thereby the transfer of protein bands from an acrylamide gel onto a more stable and immobilizing support is called as protein blotting or more precisely "Western blotting". During electrophoresis the separated proteins are buried in polyacrylamide gels and therefore further recovery of proteins is cumbersome. A number of supporting matrices such as nitrocellulose, nylon filters are used for the purpose. However, the proteins can be effectively transferred from the gel to the supporting medium by blotting. The transfer of proteins is usually carried out by electrophoresis (electroblotting) or by the capillary action of buffer (capillary blotting). A number of analysis involving immunoblotting, DNA-binding proteins, and glycoproteins could then be performed on the proteins blotted onto the filters. The benefits of proteins blot include rapid staining/destaining detection of proteins at low concentration and rapid localization of the protein in preparative gels. The blots can be preserved as replica of the original gels.

Principle

The separated proteins are transported out of the gel either by the capillary action of buffer or in an electric field. The presence of SDS increases the solubility of proteins and thus, facilitates the migration of proteins. Once out of the gel, the protein comes in contact with the nitrocellulose membrane which binds the protein very strongly onto the surface as a thin band thus producing a replica of original gel.

Materials and reagents

1. Nitrocellulose paper (pore size 0.20–0.45 μm)
2. Blotting buffer (Ph 8.3)
   - 0.02 M Tris–HCl – 2.42 g
   - 0.15 M Glycine – 10.25 g
   - 20% Methanol – 200 mL
   - Water to – 1 L
     (can be stored at 4°C for 2–3 weeks).
3. *Protein stain*: 0.01% Amido black 10B in methanol: acetic acid: water (5:15)
4. *Destaining solution*: methanol: acetic acid: water (5:1:5)

Procedure

1. Arrange a platform by placing a 25 × 20 cm glass plate at a suitable height on the work bench.
2. Assemble six layers of Whatman No. 1 filter paper and place over the platform. Dip the short ends of the papers in two glass trays containing the blotting buffer placed on either side of the platform. The papers should wet thoroughly and should be wide enough to accommodate the gel to be blotted.
3. After the separation of proteins in slab gels by SDS-PAGE, discard the stacking gel and carefully lay the separation gel to be blotted on the wetted filter paper. Cut the required portion of the gel with a scalpel blade if the whole gel is not to be blotted, and lay carefully.
4. Take a piece of nitrocellulose paper exactly the same size of the gel to be blotted. Wet it thoroughly by floating on the blotting buffer. Carefully place this wetted nitrocellulose paper over the gel without trapping any air bubbles. This is conveniently done by lowering first the middle of the paper on the gel and then laying towards the ends. It is again preferable to wet the top of the gel with blotting buffer before layering the nitrocellulose paper.
5. Cut out at the middle of a cling film to the size of the gel, and lay it over the nitrocellulose paper such that the film does not cover the nitrocellulose paper. This shall prevent by-pass of buffer from the bottom layer to top layers of filter papers directly. This is to ensure on any account the buffer should pass only through the gel.
6. On top of the film, lay six layers of Whatman 3 mm filter paper cut to the same size as the gel followed by a way of absorbent material (tissue paper or disposable nappy) also of the same size of the gel.
7. Place a heavy weight over the sandwich and leave the set up for 1–2 days. There should be ample blotting buffering 44e tanks during blotting.
8. The buffer from the bottom layer of filter papers moves upward by absorption via the gel and nitrocellulose paper. During this process, the proteins are

transferred by capillary action from the gel to the nitrocellulose which has more affinity for the proteins.

9. After blotting for required period, recover the nitrocellulose paper disassembling the set-up. The nitrocellulose paper can be stored pressed between a fold of filter papers until required for further analysis or can be stained directly.
10. Immerse the protein both in the amido black dye solution for 10–15 min with gentle shaking. Subsequently, destain the blot in the destaining solution with repeated changes. The transferred proteins are visualized as black bands. The blot is then dried between several sheets of filter paper held flat with a heavy weight.

Note

- The transfer of protein by electroblotting is mush faster and efficient than the capillary blotting method. Suitable cassettes to assemble the sandwich for electro-blotting are commercially available.
- Equilibration of the gel prior to blotting for renaturing of the separated proteins is suggested depending upon the further analysis of the blot. The gel is equilibrated in the following buffer with constant shaking for 30–60 min prior to capillary blotting for renaturing the separated proteins.

  1 M Tris–HCl, pH 7.0 – 5 mL
  5 M NaCl – 5 mL
  0.1 M $EDTANa_2$ – 10 mL
  0.1 M Dithiothreitol – 0.5 mL
  Urea – 120.12 g
  Water to – 350 mL

- The nitrocellulose sheets should be stored air tight at 4°C to prevent contamination.
- Nitrocellulose paper should be thoroughly wetted before being used. Any air bubbles between the gel and nitrocellulose paper will result in a poor transfer of proteins.
- While assembling the sandwich for blotting care must be taken to ensure that there is not direct contact of nitrocellulose paper or the above layers with the bottom layers of filter paper in order to avoid the flow of buffer directly. It should only flow through the gel to the nitrocellulose paper in an ideal blot assembly.
- Capillary blotting is a passive process and requires longer period (2–4 days). The duration of blotting largely depends on porosity of nitrocellulose, percentage of acrylamide and thickness of the gel, the ionic strength of the blotting buffer, the solubility of protein, etc. The low MW proteins are easily transferred than the high MW proteins.

7.20 Production of Antiserum (Hurn and Chantler 1980)

Antiserum production is an important aspect in immunological experiments. Different types of animals are used for the production of antiserum. Rabbit is general choice in most of the experiments. This usually depends upon many factors such as the amount of antigen available, the degree of immune response, the quantity of antiserum to be produced etc.

An immunogen (antigen) is a substance that when injected into a suitable animal gives rise to an immune response. This property depends on many factors of the immunogen molecule – its size, shape, chemical composition and structural difference from any related molecular species indigenous to the injected animal. Proteins of molecular weight above 5,000 daltons and certain large polysaccharides are effective immunogens. Low-molecular weight substance (hapten) coupled to a protein with immunogenicity can also be an immunogen. The immunogen should be used in a highly purified form so that specific antibodies against the antigen are produced. Antiserum so produced in the animal contains antibodies in response and to counter the immunogen. The immune response occurs in two phases. First administration of the immunogen induces the primary response during which only small amounts of the antibody molecules are produced. Subsequent administration of the immunogen results in the secondary phase during which large amounts of antibody molecules are produced by the lymphocytes.

Principle

A molecule which possesses an arrangement of atoms at its surface that differs from the surface configuration of normal host component is recognized as an antigen. Antibody synthesis is a defense reaction of higher vertebrates evolved to combat any foreign material that has entered the body. Antibodies are glycoproteins commonly called immunoglobulins (IgG) and are present in blood and other local secretions of the organs where antigen is present. The cellular events triggered by antigen lead to the differentiation and proliferation of specific antibody producing lymphocytes against a particular antigen. There are five major immunoglobulin classes. An antigen may stimulate the production of one or more classes of immunoglobulins. All the immunoglobulins have two identical light chains and two identical heavy chains. The antibody combining site is located in the amino terminal portion of the molecule which is also called the "V" (variable)-region. The primary structure of V-region varies with the type of antibody produced against an antigen.

Materials

1. White rabbits (approximately 2.5–3.0 kg body weight)
2. Purified immunogen in an appropriate buffer
3. Infrared lamp
4. Incomplete Freund's adjuvant

Procedure

1. Dilute the purified antigen (150 μg) to 0.2 mL in phosphate-buffered saline (PBS: 100 mg $CaCl_2$, 120 mg $MgSO_4$; 200 mg each of KCl and KH_2PO_4; 8.0 g NaCl; 1.15 g Na_2HPO_4 per litre) and emulsified with 0.8 mL of incomplete Freund's adjuvant by thorough vortexing. Draw the mixture inside a syringe. The syringe is emptied and refilled with the mixture 3–5 times to produce a uniform, stable emulsion which can then be administered to a rabbit.
2. Prepare the rabbit for immunization by shaving away the hair at one or two places on the hind thighs. Inject the emulsion in these places subcutaneously using a 21-gauge needle.
3. Upto three booster shots at 10-day intervals following the initial shot could be delivered. Each time, 50 μg protein (antigen) emulsified in a total volume of 0.5 mL of incomplete Freund's adjuvant is used.
4. Eight days after the last booster shot, prepare the animal for bleeding. Rub the surface of back of an ear with alcohol-moistened tissue to expand the veins.
5. Nick the marginal vein with a scalpel blade and collect the blood in a glass vessel (about 15 mL can be collected in 10 min). Every minute or so clean the clot at the puncture wound with alcohol-moistened tissue for continuous bleeding. Alternately, bleeding is carried out under infrared light. At the end, thoroughly clean the stained surface of the ear to avoid any infection. A test-bleeding may be carried out even before the third booster shot to examine the antiserum for the immune response and titre.
6. The blood is allowed to clot standing for 3–4 h at room temperature. The serum is separated from the clot by low speed centrifugation and stored at 4°C in the presence of 0.1% sodium azide as antibacterial agent. The serum is usually straw yellow-coloured. If RBCs are partly lysed then the serum is coloured red.
7. Bleeding up to 3 times can be made on successive days. The rabbit is allowed to rest for 4–6 weeks before further bleeding.
8. The animal can be used for about 6 months to collect blood and then abandoned. If after boostering, the antiserum is not of the desired quality, it is better to disregard the particular rabbit and look for the other rabbits for the antiserum. If the whole lot is unsatisfactory possibly due to the weak immunogens, either repeat the immunization or use other animals.
9. The antiserum is then tested by immunodiffusion or immunoelectrophoresis for its ability to form immunoprecipitate.
10. The antiserum may be stored at −20°C in small aliquots in the presence of 0.1% sodium azide for longer duration.
11. The antiserum may be further processed to isolate IgG or for other experimentation.

Note

- A good antiserum should possess three important qualities: avidity (measure of the strength of the interactions of its antibodies with antigen), specificity (ability of the antibody to recognize its antigen from related molecules) and titre (the concentration of antibodies present, and on their affinities for the antigen).

- When large amounts of pure immunogen are available, a high initial immunization dose can be used. However, lower the does of antigen, the greater is the avidity of the antiserum.
- Immunization can be done on any part of the body-skinfold of the neck and of back, footpad, etc.
- A long gap of 3–10 weeks is also normally used between the initial and booster immunizations in order to a state of tolerance to the antiserum by the animal.
- Since the antisera produced by conventional methods consist of mixtures of different antibody molecules, the properties of antisera collected during the prolonged period of immunization may change. Hence, each bleeding should be tested for the antiserum qualities before any use.
- Complete Freund's adjuvant induces antibody production greatly than the incomplete Freund's adjuvant. CFA contains killed *Mycobacterium* cells over the IFA, which are commercially available.

7.21 Immunodiffusion in Agarose Gels (Ouchterlony 1968)

Immunodiffusion in gels is often classified as single or double diffusion. In single diffusion technique, the antigen (Ag) is usually allowed to diffuse into a gel containing the gel antiserum/antibody (Ab). On the other hand, in the double diffusion technique, both the antigen and antibody are allowed to diffuse into the gel and meet each other. A lot of qualitative and quantitative information on the antigen can be derived from the diffusion techniques. This is a simple technique to test for the antibody titre during immunization and to study the antigen–antibody reactions in gels. The antigen–antibody complex forms an insoluble precipitate when the reaction is carried out both in solution and gels. Performing such reactions in agarose gel is advantageous because of higher sensitivity and resolution. The antigen and antibody react to produce the antigen–antibody complex. At equilibrium, the complex produced is immobile and forms a thin band of (protein) precipitin. The precipitin is visualized either directly or after protein staining for interpretation.

Materials

1. Agarose
2. 0.05 M Borate buffer (pH 8.0): (a) Dissolve 1.90 g borax ($Na_2B_4O_7{\cdot}10H_2O$) in 100 mL water (b) Dissolve 1.24 g boric acid in 100 mL water. Mix 30 mL of (a) and 70 mL of (b) dilute to 400 mL by adding 300 mL water.
3. Solutions of antigen and antiserum
4. Glass Petri dishes or rectangular plates or microscope slides.
5. Gel punch and template – various sizes

6. Humidity chamber
7. *Physiological saline solution*: Dissolve 0.9 g sodium chloride in 100 mL water.
8. *Staining solution*: 0.1% Coomassie Brilliant Blue R250 in methanol: acetic acid: water (4:1:5).
9. *Destaining solution*: methanol: acetic acid: water (4:1:5).

Procedure

1. Ouchterlony double immunodiffusion
   The double diffusion of the antigen and antibody was first described by Ouchterlony. This technique is most widely used in the qualitative analysis of antigens and antisera.

   (i) Prepare 100 mL of 0.05 M borate buffer (pH 8.0) containing 0.9% NaCl and 3% polyethylene glycol.
   (ii) Dissolve 0.9 g of agarose in the above buffer solution by heating to 90°C on a water bath with constant stirring or by autoclaving.
   (iii) Pour the agarose solution to a depth of 2–3 mm into the sterile Petri dishes or onto the rectangular plates or on a number of slides placed on a horizontal level surface. Allow the gel to set for 10–15 min and store in a humid chamber.
   (iv) Place a template of a suitable pattern depending upon the number of samples to be analysed on top of the gel. The gel punch that is connected to a water vacuum pump is carefully inserted in to the gel through each hole of the template to suck out the agarose plugs thus to form wells.
   (v) Fill the centre or inner well with the antiserum and the outer or radial wells with the antigen(s) solutions. Use serially diluted solutions of the same antigen or the same volume of different antigens. The concentration of antigen solutions and the dilution of the antiserum to be used have to be established largely by trial and error by running pilot experiments.
   (vi) Keep the gel plate in a humidity chamber at a constant temperature (between 16 and 20°C) for 1–3 days and examine for the precipitin line formation.
   (vii) The precipitin line is visualized directly. The presence of polyethylene glycol in the gel enhances the visibility of the precipitin. Alternatively, the gel can be stained for proteins, especially if the precipitin line is very faint.
   (viii) Prior to staining remove excess moisture from the gel plate by placing some weight to a wad of filter paper placed on top of the gel for 15–30 min. Wash out the unreacted antigen and antiserum with several changes in the physiological saline for 2–3 h. Dry the gel after further pressing, in a stream of cold air from a hair-dryer.
   (ix) Stain the dried gel for 10–15 min with Coomassie Brilliant Blue dye-Destain the plate for appropriate period to visualize the precipitin lines clearly. The gel may be photographed and dried subsequently for a permanent record.

2. Single radial immunodiffusion
   This method is widely used for the quantitative analysis of antigens.
   The method followed is essentially as described for the Ouchterlong double diffusion with the following modifications:

   (i) Double amount of agarose (1.8/100 mL) is used.
   (ii) The agarose solution is allowed to cool to 50–55°C, and is mixed with an equal volume of a suitable dilution of the monospecific antiserum also at 55°C; the gel is then poured on the gel plates. The plates are used within a day; may be stored for a week or so at 4°C.
   (iii) The wells of 2 mm diameter are cut in a row on the plate using a gel known concentration. The other wells are filled with the solution containing the antigen at unknown concentrations. The volume of antigens filled should be precisely known and equal in all wells.
   (iv) A few wells in the row are filled with standard solutions of antigen of known concentration. The other wells are filled with the solution containing the antigen at unknown concentrations. The volume of antigens filled should be precisely known concentrations. The volume of antigens filled should be precisely known and equal in all wells.
   (v) A disc of precipitin is formed around the antigen well. The disc size enlarges with progress of period of diffusion.
   (vi) The area of each disc is measured in terms of its diameter at every 12 h until no more increase is noticed. A magnifying lens on suitable oblique illumination such as colony counter may be used for measurement.
   (vii) At the end of diffusion, processed gel may be stained and preserved as described in the previous section.
   (viii) A standard graph is constructed by plotting the diameters of the disc against the logarithm of the antigen solutions of known concentration. The concentration of the antigen in the test sample can then be determined using the standard graph.

Note

Double immunodiffusion

- It is important to run a few pilot experiments to standardize the antigen concentration and the dilution of antiserum for satisfactory results.
- It is preferable to precoat with 0.4% agarose gel prior to formation of the main gel to prevent the latter from being detached during the staining procedure when Petri dishes or glass plates are used.
- Microscope slides are considered ideal for rapid microanalysis.
- When the antigen and antiserum wells are closer the sensitivity is greater and the time required for diffusion is shorter.
- If the antigen used is not highly purified, the antiserum will contain antibodies to the protein impurities a well. Consequently, there will be many precipitin lines in the diffusion patterns.

7.22 Enzyme-Linked Immunosorbant Assay (Wim Gaastra 1984)

The enzyme-linked immunosorbant assay (ELISA) technique is used for a semi-quantitative determination of the concentration of certain antigens/antibodies. It is vividly used in medicine to detect the antigen or antibodies in serum samples. At present, it has versatile applications in the immunodiagnosis of several infectious diseases.

Principle

The ELISA technique was first introduced in early 1970s by Engvall and Perlmann. The principle underlying the double antibody sandwich technique of ELISA is described below.

The antibodies against the antigen to be measured are adsorbed to a solid support, in most cases a polystyrene microtiter plate. The support after coating with antibody is washed. The antigen is now added and binds to the adsorbed antibodies. Then, an enzyme-linked antibody molecules called the conjugate is added which also binds to the antigen. A chromogenic substrate for the enzyme is added and the coloured product generated is measured. The intensity of the colour is proportional to the bound enzyme and thus to the amount of the bound antigen. Hence, the intensity of the colour produced by a series of standard antigens allows the calculation of the amount of antigen in an unknown sample.

Materials

1. Flat-bottomed polystyrene microtitre plate with 96 wells.
2. Micropipettes 0–250 μL.
3. Multi-channel Pipette 0–250 μL for pipetting of all reagents.
4. ELISA reader (Multiscanphotometer).
5. 0.1 M Carbonate buffer (pH 9.6): Prepare 0.1 M Na_2CO_3 solution and adjust to pH 9.6 with NaOH. The chemical should be of the highest quality and water double distilled.
6. *Wash solution*: Mix 90 mL Tween 80 with 910 mL water.
7. *BST*: 0.2% (w/v) Bovine serum albumin, 0.01% Tween 80 and 0.9% (w/v) sodium chloride in distilled water.
8. *Substrate solutions*: It depends upon the enzyme that is coupled to the conjugate. Two widely used enzymes in ELISA technique are horseradish peroxides (HRP) and alkaline phosphatase. For HRP, there are two substrate solutions and are prepared as below:
9. *Solution 1*: Dissolve 80 mg 5 amino-salicylic acid (purple-red brown colour) in 100 mL 0.05 M potassium phosphate buffer (pH 6.0) containing 0.001 M EDTA. Add 20 mL H_2O_2 (30%) and mix.
10. *Solution 2*: Mix 24.3 mL 0.1 M citric acid

    - 25.7 mL 0.2 M $NaHPO_4$
    - 50 mL H_2O

- 40 mg *ortho*-phenylenediamine (yellow colour)
- 40 μL H_2O_2 (30%)

11. *Stop solution*
    0.3 M NaOH (in the case of solution 1) or 1 M H_2SO_4 (in the case of solution 2) is used.
    (When alkaline phosphatase is the enzyme coupled, the substrate is then *p*-nitrophenyl phosphate and is released as yellow coloured *p*-nitrophenol).
12. A diluted solution of IgG against the antigen to be measured. The dilution is usually 1,500–2,500 fold depending on the titre of IgG. Dilutions are made in 0.1 M carbonate buffer.
13. Antigen solutions to be tested and standard antigen solutions.
14. Enzyme (HRP) labelled diluted IgG solution. As a rule the conjugate solution has to be diluted 500–2,000 times in BST.

Procedure

The double antibody sandwich technique

1. Pipette 150 μL of the diluted IgG solution to each of the wells of a microtitre plate manually or using multichannel pipette. Cover the plate and incubate overnight at room temperature.
2. Wash the plates with wash solution. The wells can be emptied by tapping the plate over a sink and then beating the plate upside down against a filter paper. The plates can be stored several months, covered and cooled. An appropriate volume of wash solution is pipetted into the wells and left for a couple of minute. The wells are then emptied as described above. Washing is repeated a few times to ensure good results.
3. After washing add 100 μL BST to each well.
4. Add 100 μL of an antigen solution to be tested to the first well of each row. Avoid air bubbles or splashing of small drops. Mix carefully and thoroughly.
5. Take 100 μL from the first wells and transfer to the second wells in each row. Repeat the mixing procedure. Take 100 μL from the second wells and add to the third wells and so on. By this way a twofold dilution series from wells 1–12 is created. Finally, remove the 100 μL excess from the last wells.
6. Incubate the plate for 2 h at 37°C, to allow the antigen bind to the coated antiserum, and then wash thoroughly.
7. Add 100 μL of the diluted conjugate solution to each well and incubate for 2 h at 37°C, then wash the plate thoroughly.
8. Add 100 μL of substrate solution to each well and incubate for 1–2 h at 37°C in the dark.
9. Stop the reaction by adding 100 μL of stop solution.
10. Read the titre of the antigen solutions by either using an ELISA Reader or visually by observing the last well that still gives some colour and could be observed with the naked eyes.

Preparation of conjugate

A conjugate is the covalent complex of IgG and an enzyme. The coupling of HRP is described below:

1. Dissolve 5 mg of HRP in 1 mL 0.3 M Na_2CO_3 (pH 8.1). This solution should be prepared fresh.
2. Add 0.1 mL of 1% fluorodinitrobenzene in pure ethanol. If the HRP used is not pure, a precipitate may be formed that must be removed by centrifugation (10 min, 18,000 rpm).
3. Mix thoroughly and incubate for 1 h at room temperature.
4. Add 1 mL of 0.16 M ethylene glycol, mix, and incubate for another hour at room temperature. The total volume is now 2.1 mL.
5. Dialyze the mixture against 0.01 M sodium carbonate buffer (pH 9.5) for 25 h. The buffer should be changed at least 3 times.
6. Add IgG dissolved in 0.01 M sodium carbonate buffer (pH 9.5) to the peroxidase aldehyde solution in the ratio of one volume of IgG solution to one volume activated peroxidasealdehyde or 5 mg purified IgG protein to 3 mL peroxidase solution.
7. Mix well and incubate 2–3 h but not longer at room temperature. If any precipitate is formed, clarify by centrifugation (10 min, 10,000 rpm).
8. Dialyze extensively against 0.01 M phosphate buffer (pH 7.2) containing 0.9% NaCl at 4°C. Store the conjugate in a refrigerator or freezer in small aliquots and use once only.

Precautions

1. Purification of IgG fraction from whole serum:

   - Mix 100 mL serum with 200 mL of 0.06 M sodium acetate (pH 4.6). The final pH of the mixture should be 4.8.
   - Add 8.2 mL (for rabbit serum) of caprylic acid dropwise at room temperature. The volume of caprylic acid (6.8–8.2 mL) needed to precipitate IgG varies from sera to sera depending upon the source.
   - Stir for 30 min and remove the precipitate (10,000 rpm, 10 min).
   - Dialyze the IgG fraction against 0.9% NaCl solution and store after lyophilization.

The experiment is wrong when many or all wells develop the same amount of colour. In such case, the problem can be overcome by using freshly prepared solutions.

7.23 Preparations of S-30 Extract for Protein Synthesis In Vitro (Roberts and Paterson 1973)

Protein synthesis is a complex biosynthesis reaction involving a large number of cell components and molecules. The process takes place on ribosomes and involves polymerization of amino acids at the expense of energy as directed by messenger ribonucleic acid. The protein synthesis as takes place in vivo could be conducted in vitro using cell-free extracts. These extracts are prepared from a variety of sources such as reticulocyte, wheat embryo, wheat germ, etc. These extracts are very useful tools to study the synthesis of protein products programmed with exogenous mRNAs. This procedure describes a method to prepare an efficient cell-free extract from wheat embryos/germs.

Principle

The starting material is extracted in a suitable buffer to release that cell content, centrifuges to get rid off fat and mitochondria and the post-mitochondrial supernatant containing ribosomes and soluble components is used for protein synthesis in vitro.

Materials

- Wheat Embryo/Wheat Germ
- Standard HEPES buffer (SHB)
- 20 mM HEPES-KOH (pH 7.6)
- 120 mM KCl
- 2 mM Magnesium acetate
- 6 mM 2-Mercaptoethanol
- SHB (pH 6.25)

Procedure

1. Prechill a mortar and pestle by adding a few millilitre of liquid nitrogen. Add 3 g of wheat embryo/wheat germ when liquid nitrogen is still there. Grind to a fine powder before thawing. Add 20 mL of SHB (pH 6.25) in increments and continue grinding to gel a fine homogenate.
2. Centrifuge the homogenate at 17,000 rpm (30,000 g) for 10 min in a refrigerated centrifuge.
3. Remove the supernatant carefully using a Pasteur pipette avoiding both the precipitate and the top lipid layer.
4. Repeat the steps 2 and 3.
5. Load the supernatant onto a Sephadex G-25 column (25 × 1 cm) pre-equilibrated with SHB (pH 7.6) and elute with the same buffer.

6. Collect the eluant in fraction. Combine the most turbid fractions. Immediately, dilute 20 μL of the combined fraction to 2 μL with water and measure the absorbance. The A_{260}/A_{280} should be above 1.6.
7. Freeze immediately and store the combined fractions in small aliquots (0.2–0.5 mL) in microfuge tubes under liquid nitrogen (−190°C) or at −70°C.

Note

1. Embryos can be collected from freshly harvested wheat grains. Final purification of embryos is carried out briefly by floating in a mixture of organic solvents – cyclohexane: carbon tetrachloride (1:4) – change the solvent ratio slightly to float the embryos on the surface. Collect the floating embryos quickly and dry on filter paper. Embryos can be stored in sealed vials in freezers for a few weeks before extraction.
2. The translation efficiency of the extract will be lower if A_{260}/A_{280} is below 1.6. Carefully combine only peak turbid fractions.
3. Storing of the extract at −20°C leads up to 70% activity loss in the 3 weeks.
4. The thawed extract should be used once.

7.24 In Vitro Translation Assay (Marcus et al. 1974)

In vitro translation study is an excellent procedure to study protein synthesis and to characterize the product encoded by mRNAs. The cell-free protein synthesizing system efficiently translates mRNAs from exogenous source under optimal conditions. The assay is usually carried out in a small volume of 20–25 μL, which could be scaled-up for preparative purposes.

Principle

The cell-free (S-30) extract containing the necessary protein synthesis machinery components translates the genetic message in the mRNAs into protein when provided with energy source under proper ionic conditions. The hot TCA perceptible radioactivity due to the labelled amino acid incorporation is measured.

Materials

1. Cell-free extract (see preceding experiment)
2. Salt mix (10×)

   200 mM HEPES-KOH (pH 7.6)
   750 mM Potassium chloride
   25 mM Magnesium acetate
   20 mM Dithiothreitol
   6 mM Spermidine (optional)
   Store in aliquots at −20°C

3. Energy mix (5×)

   2.5 mM ATP (pH 7.0)
   25 μL 10 mM ATP (dipotassium salt)
   1.5 mM GTP(pH 7.0)
   15 μL 10 mM GTP (trisodium salt)
   100 mM creatine phosphate (pH 7.0)
   10 μL 1 M creatine phosphate (dipotassium salt)
   250 μg/mL creatine phosphokinase
   25 μL creatine phosphokinase 1 mg/mL
   25 μL H_2O
   Prepare the 5× mix afresh from stock solution kept at −20°C.

4. Amino acid mix

   2.4 mM each amino acid except the labelled amino acid stored at −20°C.

Procedure

1. In an Eppendorf tube mix successively:

   10 μL of 10× salt mix.
   20 μL of 5× energy mix.
   10 μL of amino acid mix.
   4 μL (>40 Ci) of ^{35}S methionine (>1,000 Ci/immol) or ^{3}H Leucine 36 μL of S-30 extract (freshly thawn)

2. Pipette out into three numbered 0.5-mL Eppendorf tubes three different volumes (1, 2 and 4 μL) of mRNAa (1 mg/mL) solution. Normalize the volume to 5 μL in each tube by adding sterile distilled water.
3. Transfer 20 μL of the assay mix (step 1) to each tube containing mRNA.
4. To the remaining assay mixture add 5 μL of sterile water (control).
5. Mix and incubate all the tubes at 25°C for 1 h.
6. Meanwhile warm 5 mM square filter papers (Whatman 3 mm) over a hot plate at 70°C.
7. After incubation transfer 2 or 5 μL of assay in triplicate onto the filters papers, and dry.
8. Precipitate the proteins onto the filters by transferring them to ice-cold 10% trichloro acetic acid (TCA) containing excess unlabelled amino acid for 10 min.
9. Boil the filters in 5% TCA for 10 min in a water bath to deacylate the charged tRNAs.
10. Wash the filters successively in 5% TCA, ethanol, ethanol-ether mixture and finally ether each step proceeding for 2–3 min.
11. Dry the filters at 70°C. Transfer each filter to a scintillation vial, add 2 mL scintillation fluid (4 g PPO/L toluene) or suitable mixture and count the radioactivity.

12. Include three filters with no solution pipetted onto them during the above processing to subtract the background noise.
13. Arrest the reaction after incubation (step 5) by adding 5 μL of sample buffer (5×) for SDS-PAGE analysis.

7.25 Ammonium Sulphate Fractionation of Proteins

The solubility of proteins is markedly affected by the ionic strength of the medium. As the ionic strength is raised, protein solubility at first increases, which is referred to as "salting in". However, beyond a certain point the solubility begins to decrease and this is known as "salting out".

As low ionic strengths the activity coefficients of the ionizable groups of the proteins are decreased so that their effective concentration is decreased. This is because surrounded by counter ions which prevent interaction between the ionizable groups. Thus protein–protein interactions are decreased and the solubility is increased.

At high ionic strengths water becomes bound by the added ions which is not enough to properly hydrate the proteins. As a result, protein–protein interactions exceed protein–water interactions and the solubility decreases. Because of differences in structure and amino acid sequence, proteins differ in their slating in and salting out behaviour. This forms the basis for the fractional precipitation by application of salt.

Ammonium sulphate is preferably a useful salt for the fractional precipitation of proteins. It is available in highly purified form and has great solubility allowing significant changes in the ionic strength. Moreover, it is not so expensive. Changes in the ammonium sulphate concentration of a solution can be brought about either by adding solid substance or by adding a solution of known saturation, generally, a fully saturated (100%) solution (Table 7.5).

7.26 Methods for Determining Amino Acid Sequences of Protein

The sequence of amino acids in proteins can be determined by means of three basic analytical procedures:

(a) Identification of the NH_2 – terminal amino acid in the protein.
(b) Identification of the COOH – terminal amino acid.
(c) Partial cleavage of the original polypeptide into smaller polypeptides whose sequence can be determined.

In the last procedure, cleavage of the original protein must be carried out in at least two different ways so that the smaller polypeptides produced in one procedure "overlap" those produced in the second procedure and provide on opportunity for

Table 7.5 The amount of solid ammonium sulphate to be added to a solution to give the desired final saturation at 0°C

Final concentration of ammonium sulphate, % saturation 0°C																	
	20	25	30	35	40	45	50	55	60	65	70	75	80	85	90	95	100
Initial concentration of ammonium sulphate																	
	g solid ammonium sulphate to add to 100 mL of solution																
0	10.7	13.6	16.6	19.7	22.9	26.2	29.5	33.1	36.6	40.4	44.2	48.3	52.3	56.7	61.1	65.9	70.7
5	8.0	10.9	13.9	16.8	20.0	23.2	26.6	30.0	33.6	37.3	41.1	45.0	49.1	53.3	57.8	62.4	67.1
10	5.4	8.2	11.1	14.1	17.1	20.3	23.6	27.0	30.5	34.0	37.9	41.8	45.8	50.0	54.5	58.9	63.6
15	2.6	5.5	8.3	11.3	14.3	17.4	20.7	24.0	27.5	31.0	64.8	38.6	42.6	46.6	51.0	55.5	60.0
20	0	2.7	5.6	8.4	11.5	14.5	17.7	21.0	24.4	28.0	31.6	35.4	39.2	43.3	47.6	51.9	56.5
25		0	2.7	5.7	8.5	11.7	14.8	18.2	251.4	24.8	28.4	32.1	36.0	40.1	44.2	48.5	52.9
30			0	2.8	5.7	8.7	11.9	15.0	18.4	21.7	25.3	28.9	32.8	36.7	40.8	45.1	49.5
35				0	2.8	5.8	8.8	12.0	15.3	18.7	22.1	25.8	29.6	33.4	37.4	41.6	45.9
40					0	2.9	5.9	9.0	12.2	15.5	19.0	22.5	26.2	30.0	34.0	38.1	42.4
45						0	2.7	3.0	9.1	12.5	15.8	19.3	22.9	26.7	30.6	36.7	38.8
50							0	3.0	6.1	9.3	12.7	16.1	19.7	23.3	27.2	31.2	35.3
55								0	3.0	6.2	9.4	12.9	16.3	20.0	23.8	27.7	31.7
60									0	3.1	6.3	9.6	13.1	16.6	20.4	24.2	28.3
65										0	3.1	6.4	9.8	13.4	17.0	20.8	24.7
70											0	3.2	6.6	10.0	13.6	17.3	21.2
75												0	3.2	6.7	10.2	13.9	17.6
80													0	3.3	6.8	10.4	14.4
85														0	3.4	6.9	10.6
90															0	3.4	7.1
95																0	3.5
100																	

identifying the sequence of amino acids in the area of the original chain where the cleavage occurs. The protein whose structure is to be determined must obviously be free of any contaminating amino acids or peptides. Knowing its molecular weight and amino acid composition, the number of times each residue occurs in the protein can be determined. The determination of sequence can proceed in following steps.

7.26.1 Identification of the NH_2-Terminal Amino Acid

When a polypeptide is reacted with 2,4-dinitrofluorobenzene the NH_2-terminal group (and the e-amino group of any lysine that is present in peptide linkage) reacts to form the intensely yellow 2,4-dinitrophenyl derivative of the polypeptide. Subsequent hydrolysis of the peptide with 6 N HCl hydrolyzes all the peptide bonds, and the yellow derivative of the NH_2-terminal residue (and that of lysine) can be separated by paper chromatography from the free amino acids, compared with know derivatives of the amino acids, and identified. The NH_2-terminal residue can also be identified with the dansyl reagent.

The reaction of polypeptides with phenylisothiocyanate in dilute alkali is the basis for a sequential degradation of a polypeptide that has been devised by P. Edman. In this procedure, the NH_2-terminal group reacts to form a phenylthicarbamyl derivate. Next, treatment with mild acid causes cyclization and cleavage of the NH_2-terminal amino acid as its phenylthiohydatoin. This compound can be separated and compared with the same derivative of known amino acids and thereby identified. The acid conditions utilized to cleave off the phenylthiohydantoin are not sufficiently drastic as to break any other peptide linkages. As a consequence, this method results in the removal and identification of the NH_2-terminal amino acid together with the production of a polypeptide containing one less amino acid than the original. This new polypeptide can now be treated with more phenylisothiocyanate in alkali in the same manner and the process repeated many times to degrade the original polypeptide in a stepwise manner (Fig. 7.1).

7.26.2 Identification of the COOH-Terminal Amino Acid

The carboxyl-terminal group of a polypeptide (and the distal carboxyl groups of aspartic and glutamic acid residues in the peptide) can be reduced to the corresponding alcohol with lithium borohydride, $LiBH_4$. It is first necessary to protect the free amino groups by acetylation and to esterify the carboxyl groups. The polypeptide can then be hydrolyzed with acid to produce its constituent amino acids and the amino alcohol corresponding to the COOH-terminal residue. The alcohol can be separated, compared with reference compounds, and identified.

The action of the enzyme carboxypeptidase on polypeptides can also be used to identify the COOH-terminal amino acid, since its action is to hydrolyze that amino

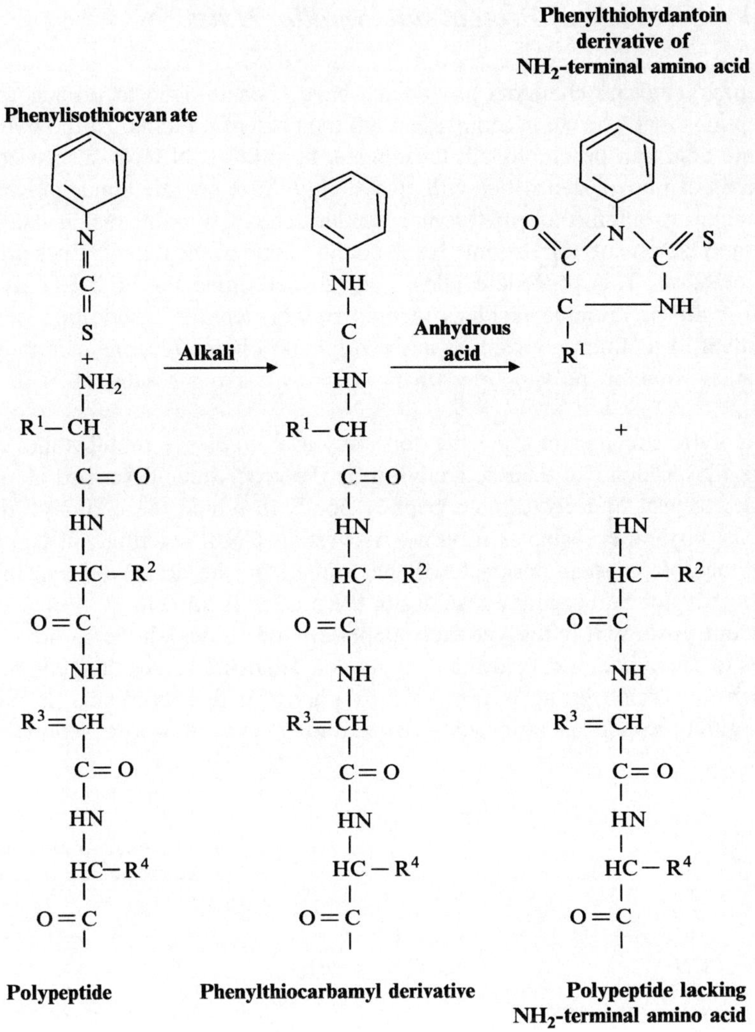

Fig. 7.1 Edman degradation for N-terminal amino acid

acid off the polypeptide. The major disadvantage is that the enzyme does not act exclusively on the original polypeptide but will also hydrolyze the new COOH-terminal peptide bond as soon as it is formed. Therefore, the investigator must follow the rate of formation of free amino acids to learn which residue represents the terminal in the original polypeptide.

7.26.3 Cleavage of Protein into Smaller Units

Both enzymatic and chemical procedures have been utilized to produce smaller polypeptides that overlap in sequences with the present proteins. Partial hydrolysis by dilute acid can be employed. Cyanogens bromide (CNBr) is also used since conditions can be chosen which will cleave only those peptide bonds in which the carbonyl group belongs to a methionine residue. The methionine residue becomes a substituted lactone of homoserine that is bound to one of the two peptides produced in the reaction. This procedure allows one to determine the amino acids in the region of the methionine residues in the original peptide. In addition, knowing the number of methionine residues in the original polypeptide, one can predict the number of smaller polypeptides that will result from treatment with CNBr (Fig. 7.2).

Proteolytic enzymes have been extensively used to cleave proteins into smaller polypeptides which can then be analysed by the procedures described above. For example, trypsin hydrolyze those peptide bonds in which the carbonyl group is contributed by either lysine or arginine. As with the CNBr reaction, one can predict the number of polypeptides that will be formed by the action of trypsin if the number of lysine and arginine residues in the protein is known.

Chymotrypsin will hydrolyze those peptide bonds in which the carboxyl group belongs to phenylalanine, tyrosine or typtophan. Pepsin cleaves the peptide bonds in which the amino group is furnished by phenylalanine, typrosine, tryptophan, lysine, glutamic and aspartic acids. By utilizing trypsin, whose action is quite

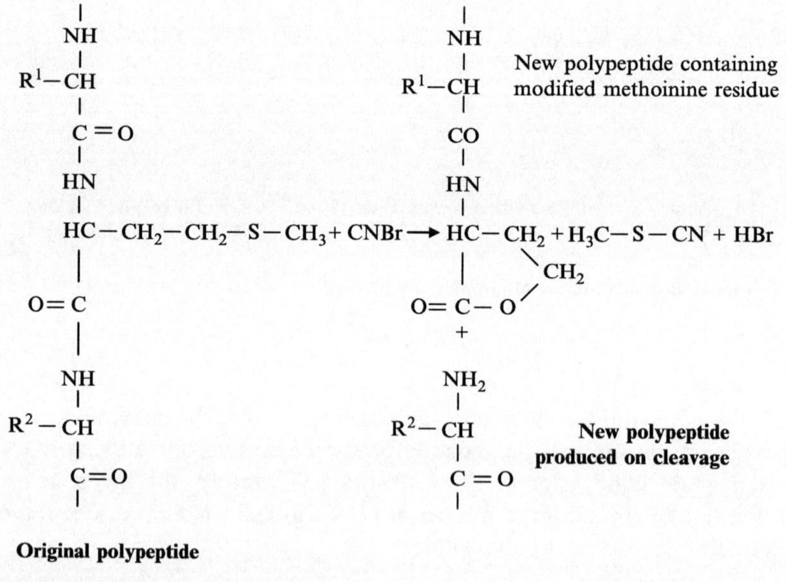

Fig. 7.2 C-terminal sequencing of protein

specific, and either chymotrypsin or pepsin, the investigator can obtain fragments of the original protein or polypeptide that overlap in sequence. Once the sequence of amino acids in these fragments is known, the process of fitting together the individual fragments can proceed. If, as in the case of insulin, the original protein can be easily separated into two parts by simple reduction of disulfide bonds, the sequence determination of the two separate chains can proceed.

Chapter 8
Protein Purification Techniques

A wide variety of protein purification methods that can be combined to generate a suitable purification scheme are available. To understand the nature of a biomolecule like protein, it must be purified to near homogeneity. Purified protein may be used for a cloning or may be used to learn about its catalytic activities and its responsiveness to regulatory molecules that raise or lower the activity or interactions with other proteins. For attaining the goal of a pure protein, the cardinal rule is that the ratio of the target protein (in terms of its activity) to the total protein is increased to the limit. Usually, one executes a series of purification steps. Combining early ones of high capacity and low resolution (when large amounts of protein is present) with lower capacity and higher resolution ones (when less protein is present) at later stages of purification scheme. The widely used purification methods involve the following steps:

1. Protein extraction, involving
   - Extract preparation
   - Sub-cellular fractionation
   - Solubilization of bound proteins
2. Bulk techniques of protein separation
   - Salting out
   - Precipitation with organic solvents
   - Precipitation with decreased pH or heat
3. Chromatographic techniques
   - Ion exchange chromatography
   - Adsorption chromatography
   - Gel filtration
   - Affinity chromatography

R. Katoch, *Analytical Techniques in Biochemistry and Molecular Biology*,
DOI 10.1007/978-1-4419-9785-2_8, © Springer Science+Business Media, LLC 2011

4. Electrophoretic procedures
   - Native (PAGE) or SDS-PAGE
   - Isoelectric focusing
   - Two-dimensional gel electrophoresis (2D-electrophoresis)

8.1 Protein Extraction Procedure

The first steps of a typical protein isolation procedure usually consist of washing the tissue and applying the lysis method. The method should be efficient to disrupt the cells to release the protein in soluble form of intracellular compartments into a solution of well-defined composition. The selection of an appropriate buffer is important in order to maintain the protein at the desired pH and to ensure reproducible experimental results. A 50–100 mM buffer is generally suitable for extraction. Along with the buffer, other components like chelators, reducing agents, detergents can be added depending on the nature and location of the protein. Grinding of tissue with abrasive materials is an effective means of lysis which is achieved by the abrasive action of grinding the thick paste of sample by hand with alumina or sand. Then, centrifugation separates the soluble proteins from the membrane fraction and insoluble debris. Finally, the protein sample may be analysed, further purified or stored for further use.

1. Materials, equipment and solutions
   (a) Prechilled pestle and mortar
   (b) Quartz sand or alumina
   (c) Extraction buffer: 0.1 M sodium acetate buffer (pH 5.2) containing 12 mM β-mercaptoethanol
   (d) Polyvinyl polypyrrolidone
   (e) Muslin cloth
   (f) Refrigerated centrifuge
   (g) Centrifuge tubes
   (h) Beaker and funnel
2. Protocol
   (a) Weigh about 40 g of fresh leaf tissue or 25 g of dry leaf tissue and homogenize with 150 mL of extraction buffer (1:4 or 1:6 w/v) by addition in increments in a chilled pestle and mortar. Add 12 mg of polyvinyl polypyrrolidone during grinding to remove the phenolics. A pinch of quartz sand can be added to facilitate thorough grinding.
   (b) Squeeze the homogenate through four layers of muslin cloth and clarify the extract by centrifugation at 12,000 × *g* for 10 min.

(c) Transfer the more or less opalescent brown supernatant to a clean 250-mL beaker and discard the pellet. This supernatant can be termed as crude extract.
(d) A small portion of the supernatant can be used to test the presence of desired protein.

8.2 Ammonium Sulphate Fractionation

Presence of high concentration of salt in the protein solution makes the proteins to aggregate and precipitate out of solution. This technique is referred to as "salting out". Since different proteins precipitate at different salt concentration, salting out is often used during protein purification by fractionation. Factors such as pH, temperature and protein purity play important roles in determining the salting out point of a particular protein. Ammonium sulphate is the salt of choice because of its salting out effectiveness, pH versatility, high solubility, low heat of solution and low price. To ensure maximal precipitation, it is best to start with a protein concentration of at least 1 mg/mL. ammonium sulphate concentrations are generally expressed in percent saturation, and a simple equation for calculation of grams of ammonium sulphate needed to make an $X\%$ saturation solution starting from $X\%$ saturation is

$$G = \frac{533(X - X_O)}{100 - 0.3X} \quad \text{(for a liter solution at 20°C)},$$

where, G is grams of ammonium sulphate needed (Table 8.1).

1. Materials, equipment and solutions

   (a) Beaker
   (b) Magnetic stirrer and stir bar
   (c) Centrifuge tubes
   (d) Refrigerated centrifuge
   (e) Ammonium sulphate
   (f) 20 mM sodium phosphate buffer (pH 6.2) containing 10 mM NaCl

2. Protocol

   (a) Place beaker containing the protein solution in a cooling bath on top of a magnetic stirrer. This can be accomplished by placing the breaker within another beaker containing water-ice slurry.
   (b) While agitating gently on a magnetic stirrer, slowly add pre-weighed quantity of ammonium sulphate so as to reach a saturation of 30%. Add salt more slowly as final saturation is approached. This step should be completed in 5–10 min.
   (c) Continue stirring for 10–30 min after all salt has been added. Put the beaker in cold for 15–30 min.

Table 8.1 Ammonium sulphate table

		Final concentration of ammonium sulphate percent saturation																			
		10	15	20	25	30	33	35	40	45	50	55	60	65	70	75	80	85	90	95	100
	0	56	84	114	144	176	196	209	243	277	313	351	390	430	472	516	561	610	662	713	767
	10		28	57	86	118	137	150	183	216	251	288	326	365	406	449	494	540	592	640	694
	15			28	57	88	107	120	153	185	220	256	294	333	373	415	459	506	556	605	657
	20				29	59	78	91	123	155	189	225	262	300	340	382	424	471	520	569	619
	25					30	49	61	93	125	158	193	230	267	307	348	390	436	485	533	583
	30						19	30	62	94	127	162	198	235	273	314	356	401	449	496	546
	35							12	43	74	107	142	177	214	252	292	333	378	426	472	522
	40								31	63	97	132	168	200	238	278	319	364	411	457	506
	45									31	63	97	132	168	205	245	285	328	375	420	469
Initial concentration of ammonium sulphate percent saturation	50										32	65	99	134	171	210	250	293	339	383	431
	55											33	66	101	137	176	214	256	302	345	392
	60												33	67	103	141	179	220	264	307	353
	65													34	69	105	143	183	227	269	314
	70														35	70	107	147	190	232	275
	75															35	72	111	153	194	237
	80																36	74	115	155	198
	85																	38	77	117	157
	90																		39	77	118
	95																			38	77
																				39	

This table indicates the grams of solid ammonium sulphate (at 25°C) to be added to 1 L of solution to produce a desired change in the percent saturation of ammonium sulphate. Saturated ammonium sulphate at 25°C is 4.1 M and requires 767 g of salt per litre. "Percent saturation" in this table is percent of 4.1 M. The listed values were calculated from tables of percent salt, specific gravity and grams per litre at various concentrations. This table is an extension of the widely used one published in Green and Hughes (1955)

(d) Spin at 10,000 × g for 10 min.
(e) Decant supernatant into a beaker and resuspend the pellet in 1–2 pellet volumes of buffer. Any insoluble material remaining is probably denatured protein and should be removed by centrifugation.
(f) Add pre-weighed ammonium sulphate to the decanted supernatant so as to reach a saturation of 60% to salt out the proteins that precipitate between 30 and 60%, and repeat the cycle.
(g) Subsequently, make the saturation of ammonium sulphate of the above supernatant to 80% to salt out the proteins that precipitate between 60 and 80% and repeat the cycle.
(h) Finally, make the saturation of ammonium sulphate of the supernatant 100% to salt out the proteins that precipitate between 80 and 100%.
(i) Suspend each pellet in 1–2 pellet volumes of the buffer.
(j) Remove the ammonium sulphate from the protein solution by dialysis or ultrafiltration or by using a desalting column.
(k) Dialyse the protein suspension using a processed dialysis bag against the same buffer in which the protein was suspended.
(l) Make a minimum of three changes of the surrounding buffer at 3 h interval.
(m) Remove any undissolved material from the protein solution by centrifugation.

8.3 Ion-Exchange Chromatography

It is the most commonly used chromatographic method for protein purification. The ion exchange principle permits the protein to bind even when a large buffer volume is applied, making this method especially useful for an initial purification step from a crude extract. Ion-exchange chromatography requires that a protein contains a net ion charge under experimental conditions.

At a given pH, a protein's net charge will depend on its relative number of positive and negative charges. At a lower pH, the net charge will be more positive, and at a higher pH, the net charge will be more negative. As a result, the protein will displace a low molecular weight ion from an ion exchange matrix (an "exchange of ions" hence the term ion exchange) and becomes bound. A change in experimental conditions (such as an increase in the counter-ion concentration or a decrease in the protein net ion charge) will cause another exchange of ions, this time releasing the protein from the ion exchange matrix in favour of the counter-ion. This process of successive exchanges of ions allows the separation of proteins with different charge properties.

8.3.1 Selection of Ion-Exchange Matrix

Selection of the matrix depends on the binding abilities of the target protein to a given matrix under experimental conditions. If the protein to be purified is negatively charged under experimental conditions, an anion exchange matrix must be used and conversely, a positively charged protein will adsorb to a cation exchange matrix. The two main decisions for choosing an adsorbent are: (1) the charge, plus or minus, and the nature of the group responsible for that charge, and (2) the nature of the matrix particles, in terms of bead size, flow rate required under pressure, capacity and often, cost per unit volume (Table 8.2).

The pH at which the positive charges on a protein equals the negative charges (in other words, the pH at which the net charge of the protein is zero) defines that protein's isoelectric point. The isoelectric point indicates how strongly charged a protein will be at a given pH. When this is known for a given protein, it is desirable to select a working pH (buffer pH), which is 1 unit away from the pI of the protein. At this pH, the protein will possess a high enough net charge to bind with the ion exchange column without being so highly charged so that harsh elution conditions are required (such as extremely high ionic strength or significant change in pH). Thus, for a protein with a pI of 6.9, buffers of pH 5.9 (and a cation exchange matrix such as CM-Sepharose CL-6B) or 7.9 (with an anion exchanger such as DEAE-Sepharose CL-6B) are recommended. Ion exchangers can be strong or weak exchangers. Most protein purifications rely on a weak exchanger. For most introductory experiments, an agarose-based ion exchanger such as DEAE- or CM-Sepharose is appropriate. Generally salt elution is simple to perform and is easily reproducible, besides being effective and inexpensive.

8.3.1.1 CM-Sepharose Chromatography

CM-Sepharose is a weak cation exchanger (negatively charged matrix). It has a carboxymethyl substituent attached to the agarose. Carboxymethy-adsorbents are widely used, and are fully charged above pH 4.5. The carboxymethyl groups progressively protonate and become uncharged as the pH falls below 4.5. This can be used to fractionate proteins, which are positively charged at physiological pH. In other words, the proteins rich in basic amino acids (basic proteins) would bind to this matrix at the pH above 5.0.

1. Materials, equipment and solutions

   (a) CM-Sepharose
   (b) Glass column
   (c) Tubes for fractions' collection
   (d) Fraction collector system which includes peristaltic pump, UV monitor, fraction collector and recorder
   (e) UV-visible Spectrophotometer
   (f) Column buffer: 20 mM sodium acetate (pH 5.2) containing 10 mM NaCl and 0.02% sodium azide
   (g) 0.05, 0.075, 0.125 and 0.15 M NaCl in column buffer

Table 8.2 Types of commercially available ion exchangers

Supplier	Name	Type	Matrix	Loading capacity (mg/mL)	Flow rate (cm/min)[a]	pH stability
GE	DEAE Sepharose Fast Flow	Weak anion	X-linked agarose	3–110	12.5	1–14
GE	DEAE Sepharose CL-6B	-do-	-do-	2.170	1.7	2–14
GE	DEAE Sephacel	-do-	Beaded cellulose	10–160	0.17	2.12
GE	DEAE Sephadex A-50	-do-	X-linked dextran	2–110		
BioSepra	DEAE Trisacryl M	-do-	Synth. polymer	80–90	3	1–11
Bio-Rad	DEAE Bio-Gel A	-do-	X-linked agarose	45	>0.3	2–9.5
GE	CM Sepharose Fast Flow	Weak cation	-do-	15–50	12.5	2–14
GE	CM Sepharose CI-6B	-do-	-do-	10–120	2	2–14
BioSepra	CM Trisacryl M	-do-	Synthetic polymer	90–100	3	1–11
Bio-Rad	Bio-Rex 70	-do-	-do-		0.4–15	5–14
Bio-Rad	CM Bio-Gel A	-do-	X-linked agarose	45	>0.3	4.5–10
GE	CM Sephadex C-50	-do-	X-linked dextran	7–140		6–10
GE	Q Sepharose Fast Flow	Strong anion	X-linked agarose	3–120	6.7–11.7	2–12
GE	QAE Sephadex A-50	-do-	X-linked dextran	1.2–80		2–10
GE	SP Sepharose Fast Flow	Strong cation	X-linked agarose	60	12.5	3–14
GE	SP Sephadex C-50	-do-	X-linked dextran	8–110		2–10
BioSepra	SP Trisacryl M	-do-	Synthetic polymer	100	6	1–11

[a] cm/min = mL/min cm^2 column cross-sectional area

2. Column packing
   (a) Equilibrate the CM-Sepharose matrix with the column buffer prior to packing. This is done by mixing the matrix in a beaker with the column buffer by swirling, test the buffer pH. If the pH changes, decant most of the buffer, add new column buffer, and repeat the slurry swirling and pH test.
   (b) To pack the column, the matrix slurry should contain one volume of matrix to one volume of buffer.
   (c) For high resolution, the column dimension should have a height:width ratio of 4:1 or 5:1.
   (d) Degas the buffer and matrix slurry solutions.
   (e) Add a small amount of buffer to the column.
   (f) Open the column outlet to allow some of the buffer to pass, then close the outlet. This will remove air from the dead space at the bottom of the column.
   (g) Swirl the matrix solution, then pour into the column down a glass rod, making sure that air bubbles do not become trapped in the matrix.
   (h) Open the column outlet and add more buffer as the matrix packs.
   (i) After column packing, wash the matrix further with buffer to complete packing process and to bring the matrix into final equilibrium with the buffer pH and ionic conditions.
3. Sample application
   A turbid protein solution should be clarified by ultracentrifugation or by filtration through a 0.45-μm pore size filter. The sample should be in a low ionic strength buffer (preferably below 50 mM).
   (a) Drain the column until the buffer reaches the surface of the matrix bed and close the column outlet.
   (b) With a pipette, apply the sample gently to the bed surface.
   (c) Open the column outlet until the sample has entered the matrix, close the column outlet.
   (d) Gently apply some buffer to the bed surface.
   (e) Open the column outlet, so that the buffer enters the matrix (carrying with it any remaining sample solution), then close the outlet when the liquid reaches the bed surface.
   (f) Again add buffer gently to the bed surface, then hook up buffer reservoir.
   (g) The column is now ready for washing and protein elution.
4. Washing and eluting the protein of interest
   Washing the column removes unbound protein before eluting any bound protein. In general, 3–10 column bed volumes of buffer are adequate for column washing. It is useful to monitor the column eluant for protein concentration or optical density during the wash and elution steps. Elution is carried out by step elution. In this, discrete increases in ionic strength (e.g. 0.05 M NaCl, the 0.075 M NaCl, so on) causes protein elution.

(a) Wash the column with 3–10 bed volumes of first elution buffer (e.g. 0.05 M NaCl in column buffer).
(b) Allow buffer to reach the bed surface before applying a new elution buffer. This will ensure that the salt concentration increases in a well-defined manner.
(c) Apply 3–10 bed volumes of the second elution buffer (e.g. 0.075 M NaCl in column buffer), followed by any further elution buffers.
(d) Measure the absorbance of the peak fractions at 280 nm using spectrophotometer.
(e) Assay each of the pooled protein fractions for the presence of protein of interest.
(f) Pooled peak fractions of interest may be concentrated by lyophilization or dialysis on a sucrose pad at 4°C.

5. Regenerating and storing the matrix

Column regeneration removes tightly bound contaminants and prepares the matrix for future separations. Regeneration is possible in the column if the matrix is stable to swelling caused by ionic strength fluctuations (Sephadex ion exchangers should be removed from the column before regenerating) most ion exchange matrices are regenerated with high concentrations of salt (e.g., 1–2 M NaCl). More tenacious contaminants such as lipids are often removed with 0.1 M NaOH or with a detergent. Anion exchangers can be stored in a buffered solution containing 0.002% hibitane (Chlorohexidine) and cation exchangers can be stored in buffers containing 0.02% sodium azide or 0.005% merthiolate.

8.4 Gel Filtration

This is also referred to as size exclusion, gel exclusion, molecular sieve or gel permeation chromatography. This method is unique in fractionating without requiring protein binding, thus significantly reducing the risk of protein loss through irreversible binding or protein inactivation. This method separates proteins according to their size. The gel filtration matrix contains pores which permit the buffer and smaller proteins to enter but which exclude larger proteins and protein complexes. Therefore, larger proteins migrate around the matrix particles and elute from the column before the smaller proteins. The largest proteins emerge from the column first since they have the smallest volume to pass through before reaching the end of the column. Medium-sized proteins can enter the larger size pores in the matrix, and so they reach the end of the column later. Small proteins are able to enter all the pores, and they have the largest volumes to pass through before emerging from the column as the last component (Tables 8.3–8.5).

Table 8.3 Characteristics of gel filtration matrices

Gel type	Fractionation range (mol. wt) Peptides	Dextrans	Bed volume (mg/g dry material)	Dry bead diameter (μm)
Sephadex G-10	−700	−700	2–3	40–120
Sephadex G-5	−1,500	−1,500	2.5–3.5	40–120
Sephadex G-25 (Fine)	1,000–5,000	100–5,000	4–6	20–80
Sephadex G-50 (Fine)	1,500–30,000	500–10,000	9–11	20–80
Sephadex G-75	3,000–80,000	1,000–50,000	12–15	40–120
Sephadex G-100	4,000–150,000	1,000–100,000	15–20	40–120
Sephadex G-150	5,000–150,000	1,000–150,000	20–30	40–120
Sephadex G-200 (Super fine)	5,000–250,000	1,000–200,000		10–40
Sephacryl S-200	$5{,}000–2.5 \times 10^5$	$1 \times 10^3–8 \times 10^4$		40–150[a]
Sephacryl S-300	$1 \times 10^4–1.5 \times 10^6$	$1 \times 10^3–7.5 \times 10^5$		40–150[a]
Sepharose 2B	$7 \times 10^4–40 \times 10^6$	$1 \times 10^5–20 \times 10^6$		60–120[a]
Sepharose 4B	$6 \times 10^4–20 \times 10^6$	$3 \times 10^4–5 \times 10^4$		60–140
Sepharose 6B	$1 \times 10^4–4 \times 10^6$	$1 \times 10^4–1 \times 10^6$		45–165[a]

[a]Diameter of wet beads

Table 8.4 Characteristics of gel filtration matrices

Type	Exclusion limit	Water regain (g/g gel), mL	Bed volume per g dry gel	Particles size (μm)
Polyacrylamide types				
Bio-Gel P-2	1,600	1.5	3.8	50–400
Bio-Gel P-4	3,600	2.4	5.8	50–400
Bio-Gel P-6	4,600	3.7	8.8	50–400
Bio-Gel P-10	10,000	4.5	12.4	50–400
Bio-Gel P-30	30,000	5.7	14.8	50–400
Bio-Gel P-60	60,000	7.2	19.0	50–400
Bio-Gel P-100	100,000	7.5	19.0	50–400
Bio-Gel P-150	150,000	9.2	24.0	50–400
Bio-Gel P-200	200,000	14.7	34.0	50–400
Bio-Gel P-300	300,000	18.0	40.0	50–400
Agarose types				
Bio-Gel A-0.5 m	500,000			50–400
Bio-Gel A-1.5 m	1.5×10^6			50–400
Bio-Gel A-0.5 m	5.0×10^6			50–400
Bio-Gel A-1.5 m	15×10^6			50–400
Bio-Gel A-50 m	50×10^6			50–400
Bio-Gel A-150 m	150×10^6			50–400

For high-resolution, column must be long and narrow (column height is generally 20–40 times the column diameter). This method can also be used to determine molecular weight, to remove the low-molecular weight impurities, separate monomers from dimmers and other oligomers, change the buffer of a protein, etc.

Table 8.5 Characteristics of affinity chromatography matrices

Type of material	Type of ligands attached
Cyanogens bromide activated Sepharose 4B	All types of proteins and ligands containing primary amino group
AH-Sephorse 4B	Has 6-C long spacer arm (1.6-diaminohexane) and has free amino group. Any ligand having amino group can be attached
Epoxy activated Sepharose 6B	Has a hydrophilic spacer and oxirane ring. Ligands having hydroxyl, amino or thiol group can be attached
Activated CH-Sepharose 4B	Activated form of CH sepharose 4B formed by esterification of carboxylic group using *N*-hydroxysuccinimde compounds having free primary amino groups
Con A sepharose	Concanavalin A attached to Sepharose 4B. A group specific adsorbent for polysaccharides and glycoproteins
Poly(U) Sepharose 4B	Used for isolation of mRNA having poly (A) sequences
5′-AMP Sepharose 4B	For adsorption of kinases and dehydrogenases which require a cofactor having an adenylate residue

1. Materials, equipment and solutions

   (a) Fraction collector system
   (b) Gel filtration matrix (Superose – 12)
   (c) Tubes for fractions' collection
   (d) UV-visible spectrophotometer
   (e) Beaker
   (f) Column buffer: 20 mM sodium acetate (pH 5.2) containing 200 mM NaCl and 0.2% sodium azide

   Choose the matrix based on the molecular weight of the protein to be separated. The molecular weight should be in the middle of the matrix fractionation range for the optimal separation.

2. Preparing the matrix and pacing the column

   (a) Preparation of matrix

   - Add approximately ten parts buffer to one part matrix material.
   - Agitate the slurry on a rotary shaker or swirl the mixture by hand occasionally. Agitation permits uniform hydration of the gel filtration matrix.
   - Swelling may take overnight depending on the matrix. Do not agitate the slurry with a magnetic stirrer.

   (b) Packing the column

   - Make the matrix into a thick slurry with the matrix material taking up ¾th of the volume and the buffer taking up the remaining ¼th.
   - Pour the matrix slurry into the column, either along the side of the column or down a glass rod.

- After the matrix has settled, wash the column with several column bed volumes to allow stabilization and equilibration.
- Do not allow the column matrix to run dry at any time.

(c) Checking the column

- Inspect the column packing by applying a 2 mg/mL solution of blue dextran. The applied volume should be 1% of the column bed volume, and the dye should elute in no more than twice the applied volume.
- Blue dextran can also be used to determine the void volume, which is the earliest point at which a protein could possibly emerge from the column. The volume required for eluting other proteins can be measured relative to this standard.

3. Sample application and elution

The protein sample should be highly concentrated (10–20 mg/mL) and the sample volume should be small (typically 1–5% of the column bed volume). The protein sample should be clarified by filtration (0.2-μm pore size filter) or centrifugation (5 min at 10,000 × *g*) to remove debris which may interfere with the chromatography. The chosen buffer should preserve protein activity and prevent nonspecific protein–protein or protein–matrix interactions. In general, a low ionic strength salt (20–200 mM) is sufficient.

(a) Elute the buffer until the meniscus reaches the top of the matrix bed, then close the column outlet.
(b) Gently apply the sample solution onto the top of the matrix.
(c) Open the column outlet and allow the sample to enter the matrix, then close the outlet.
(d) Add a small amount of buffer to the top of the matrix.
(e) Open the column outlet until the buffer enters the matrix, then shut the outlet.
(f) Apply more buffer of the matrix.
(g) The reservoir and pump can be attached now and elution can start.
(h) When using a peristaltic pump do not exceed the maximum operating pressure of the matrix. Peak resolution generally improves with a slower flow rate.
(i) To estimate the molecular weight of a protein (keeping in mind that size estimations are relative to globular proteins except under denaturing conditions), a standard curve can be generated by running protein standards over the gel filtration column. The elution volumes of standard proteins plotted as a function of the logarithm of their molecular weights should generate a straight line. Then, the elution volume of an unknown protein can be plotted onto this standard curve to find out molecular weight.
(j) Regeneration of the column can be done by washing with dilute sodium hydroxide (0.1–0.2 M NaOH).

8.5 Characterization of Proteins by SDS-PAGE (Laemmli, UK 1970)

Polyacrylamide gel electrophoresis (PAGE) is widely used to separate and characterize the biological molecules such as peptides, proteins, nucleotides and nucleic acids, which possess ionizable groups. These molecules at any given pH, exist in solution as electrically charged species either as cation (+) or anion (−). Under the influence of an electric field these charged particles migrate either to cathode or to anode depending on the nature of charged particles. In the case of sodium dodecyl sulphate-polyacrylamide gel electrophoresis (SDS-PAGE) the separation is based only on Mass. SDS is an anionic detergent, which binds to and denatures proteins. The protein–SDS complex carries net negative charges hence it moves towards the anode, and the separation is based on the size of the proteins.

Electrophoretic methods provide some of the simplest, least expensive and often most sensitive approaches to determine the number of protein components in a sample. If the expected contaminants differ in molecular weight from that of the desired protein, SDS gel electrophoresis would be the method of choice. On the other hand, molecules of similar molecular weight but different amino acid composition would appear to be identical in SDS gel electrophoresis, but could have different electrophoretic mobilities in native gel electrophoresis. Proteins of almost any molecular weight might be separable using isoelectric focusing technique. Finally, the use of 2D-PAGE with silver staining provides one of the best methods to estimate protein purity. This analysis, in conjunction with one-dimensional analysis of proteins visualized by silver stain (to detect proteins whose pI is outside the pH range of the ampholites), will provide a rigorous estimate of protein purity.

Reagent and solutions

1. Tris–glycine–SDS electrode buffer, pH 8.3
   Tris: 15.15 g
   Glycine: 72 g
   SDS: 5 g
   Dissolve in water and make up the volume to 5,000 mL.
2. 0.5 M Tris–HCl, pH 6.8
   Tris–HCl: 7.8 g
   Tris-base: 0.45 g
   Dissolve in 90 mL distilled water. Adjust pH to 6.8 with HCl and make up the volume to 100 mL.
3. 1.5 M Tris–HCl, pH 8.8
   Tris–HCl: 3.69 g
   Tris-base: 15.39 g
   Dissolve in 90 mL distilled water. Adjust pH to 8.8 with HCl and make up the volume to 100 mL.

4. 10% SDS
   Dissolve 1 g of SDS in 10 mL of double distilled water.
5. Sample buffer
   0.5 M Tris–HCl: 0.12 mL
   10% SDS: 0.20 mL
   Glycerol: 0.10 mL
   0.5% Bromophenol blue: 0.05 mL
   Distilled water: 0.48 mL
   2-mercaptoethanol: 0.05 mL
   Add 2-mercaptoethanol just before use. Store in a tightly sealed bottle.
6. Acrylamide stock
   Acrylamide: 30 g
   N′N′-Methylene bisacrylamide: 0.8 g
   Dissolve in water and make up the volume to 100 mL. Store in dark bottle at ~4°C for upto 2 weeks.
7. Ammonium persulphate solution
   Ammonium persulphate: 0.1 g
   Distilled water: 1.0 g
   Make fresh solution each time.
8. Resolving gel solution (for 1 nm (1.0) gel 12%)
   30% Acrylamide stock solution: 8.0 mL
   1.5 M Tris–HCl stock solution (pH 8.8): 5.0 mL
   10% SDS: 0.2 mL
   Distilled water: 6.6 mL
   TEMED: 10 μL
   10% Ammonium persulphate solution: 0.2 mL
9. Stacking gel solution (for 1 mm (1.0) gel, 5%)
   30% Acrylamide stock solution: 1.02 mL
   0.5 M Tris–HCl stock solution (pH 6.8): 1.50 mL
   10% SDS: 0.06 mL
   Distilled water: 3.33 mL
   TEMED: 6 μL
   10% Ammonium persulphate solution: 0.06 mL
10. Sample preparation
    Dissolve lyophilized proteins directly in the sample buffer. If the sample solution is with high ionic strength, dialyze it against 1:10 dilution in water of Tris–HCl stock solution, pH 6.8. Sample solutions are then mixed in 1:1 ratio with twice the normal sample buffer concentration before boiling.
11. Staining solution
    Commassie brilliant blue R-250: 1.15 g
    Methanol: 227 mL
    Glacial acetic acid: 46 mL
    Distilled water: 500 mL
12. Destaining solution
    Methanol: 50 mL

Glacial acetic acid: 50 mL
Distilled water: to 1,000 mL

Procedure

1. Vertical slab gel

   (a) Clean and dry the glass plates and spacers, then assemble them properly.
   (b) Degas the resolving gel solution (without ammonium persulphate) using a vaccum pump for 3–5 min or by using a stir bar on a magnetic stirrer, and then add ammonium persulphate solution.
   (c) Mix gently and carefully, pour the gel solution in the chamber between the glass plates, leaving the space for stacking gel. Layer distilled water or isobutanol on the top of the gel and leave to set for ~60 min.
   (d) Degas stacking gel solution (without ammonium persulphate) and then add ammonium persulphate solution.
   (e) Remove the water from the top of the gel. Pour the stacking gel mixture, place the comb in the stacking gel and allow the gel to set.
   (f) After the stacking gel has polymerized, remove the comb without distorting the shapes of the well. Wash the wells with distilled water using a syringe.
   (g) Prepare samples for electrophoresis. Adjust the protein concentration in each sample so that the same amount is present per unit volume. Again the concentration should be in the range of 20–200 μg protein in a volume (about 40 μL) not greater than the size of the sample well.
   (h) Fill the upper and lower tank with electrode buffer and connect upper through to cathode (−) and lower trough to anode (+). Turn on the voltage to 50–75 V until the sample travels through the stacking gel. Then continue the run at 100 V until the bromophenol blue reaches the bottle of the gel (about 3 h).
   (i) After the run is complete, carefully remove the gel from the plate and immerse in staining solution for at least 2–3 h.
   (j) Transfer the gel to destaining solution. Dye that is not bound to the proteins is thus removed. Change the destainer many times until the background of the gel is clear. The gel can now be photographed or stored in polythene bags or dried in vacuum for record.

Determination of the molecular weight of the purified protein: For determination of molecular weight of purified protein by SDS-PAGE, the protein has to be co-electrophoresed with at least three to four standard proteins with molecular weights spread above and below the unknown. The relative mobilities of standard proteins are plotted against their molecular weights on a semilog paper. Since the mobilities of these proteins are a linear function of the logarithms of their molecular weights, a linear plot would be obtained. The molecular weight of the unknown (purified) protein can be known by positioning its relative mobility on the linear plot (Fig. 8.1).

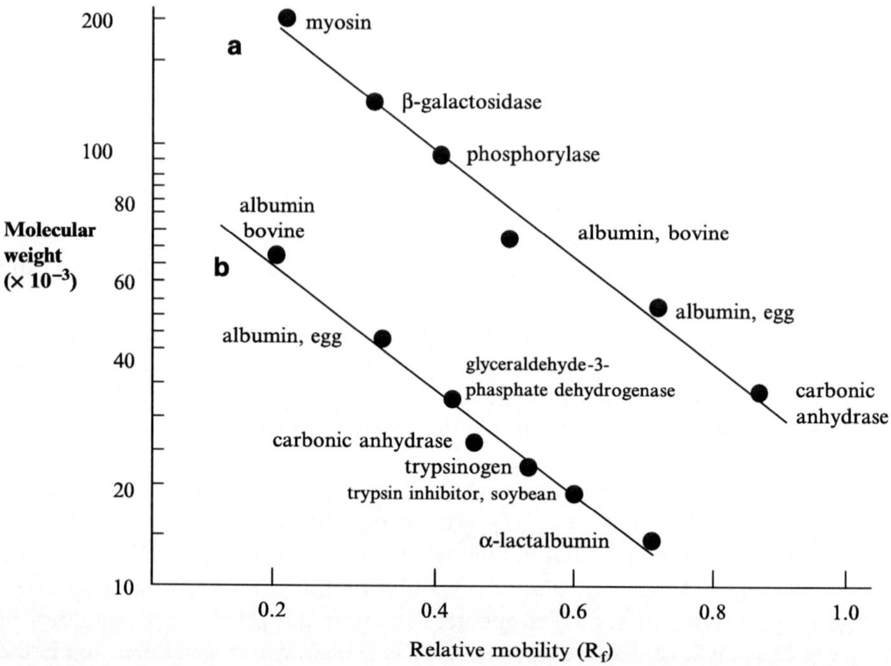

Fig. 8.1 Typical calibration curves obtained with standard proteins separated by denaturing (SDS) discontinuous gel electrophoresis based on the method of Laemmli (1970). **(a)** 7% Polyacrylamide gel. **(b)** 11% Polyacrylamide gel

Cautions

1. Availability of suitable assay method for measuring the presence of protein of interest in the sample is essential for successful purification.
2. Knowledge of appropriate conditions like temperature, extraction medium (buffer), solubility, etc. for preservation of protein activity and stability is essential.

   – Ideally, different buffers with a similar pKa should be tested to determine the existence of any undesired interactions between a certain buffer and the protein under investigation.
   – All the chemicals should be of reagent grade or higher.

3. Once the cells have been disrupted, care must be taken to counteract effects due to increased contact with oxygen and dilution of naturally occurring reducing agents.
4. Detergents (e.g. nonionic detergents <0.1%) may be used for extraction and purification of membrane proteins, which otherwise are usually insoluble in aqueous solution.
5. It is advisable to test various storage methods before committing a large amount of protein to any one them.

6. Stirring should be regular and gentle during salting out process.
7. Precipitation should be carried out in a buffer of at least 50 mM in order to compensate for a slight acidification upon dissolving ammonium sulphate.
8. Allow space for volume increases during dialysis (when dialyse against a buffer with low salt) to avoid the risk of the membrane bursting.
9. Degas buffers and avoid introducing air bubbles in the chromatographic column system.
10. Filter the sample before loading on to the column to avoid blocking at the top of the matrix.
11. Use appropriate antimicrobial agents in the column buffer to avoid microbial growth in the matrix.
12. Ensure proper cleaning and equilibration of the column to enable binding of the protein to the matrix in ion exchange chromatography.
13. Avoid pH of column buffer that is near to the pI of the target protein, else the desired protein may get precipitated.
14. The elution buffer chosen for chromatographic techniques should preserve protein activity and prevent nonspecific protein–protein or protein–matrix interactions. In general, a low ionic strength salt (20–100 mM) is sufficient to block non-specific ionic interactions.
15. The choice of a gel filtration matrix should be based primarily on the molecular weight of the protein to be separated (for details, please see Appendix). The molecular weight should be in the middle of the matrix fractionation range for optimal separation.
16. Some trial and error experiments may be required to determine which flow rate can be tolerated by the given matrix without compromising the resolution of protein peaks.
17. An application volume (sample volume) greater than 5% of the bed volume may reduce resolution.
18. If peaks obtained upon gel filtration have a long tail, the protein might be adsorbed to the matrix, include a higher ionic strength salt or some detergent in the buffer. If hydrophobic interaction is involved, ionic strength of the salt should be reduced.
19. Avoid large dead space at bottom of the column.
20. Choose appropriate column length, because resolution increases in proportion to the square root of column length.

8.6 Protein Analysis by Reversed Phase HPLC (Dong and Gant 1989)

High pressure liquid chromatography (HPLC) is based on fundamental principle of separation due to differences in the equilibrium distribution of the sample components between stationary phase (solid, column) and liquid phase (solvent) (Fig. 8.2).

The sample components move through the column, towards the detector, only when they are in the mobile phase, hence, velocity of migration of the component is a function of equilibrium distribution between the mobile and stationary phase.

Traditionally, the separation of proteins have been carried out by a wide range of well-established techniques, such as precipitation procedures, gel chromatography, ion-exchange chromatography, affinity chromatography and electrophoretic separations. Although these methods are still frequently and successfully employed, each has certain disadvantages, e.g. poor resolution, low recovery and long separation time. In recent years, HPLC has been used successfully for the separation of macromolecules, ionic species, labile natural products and wide variety of other high-molecular weight and less stable products. HPLC has been found superior to older methods in separation power, yield, speed and lower level of detectability. It acts as a combination of two or more classical purification methods, as the HPLC separations, more obviously, are governed simultaneously by several different properties of proteins, e.g. both charge and polarity. The unprecedented degree of resolution which may be obtained by HPLC makes it particularly suitable for purity control, peptide mapping or search for genetical abnormalities in a protein. HPLC has, for several reasons, proved to be of special importance when handling minute amounts of protein material. Thus, fewer purification steps are needed and losses are accordingly reduced.

In reversed-phase high pressure liquid chromatography (RPHPLC), the packing is nonpolar and the solvent is polar with respect to the sample. Retention is the result of the interaction of the nonpolar components of the solutes and the nonpolar stationary phase. Typical stationary phases are nonpolar hydrocarbons, waxy liquids or bonded hydrocarbons (such as C_{18}, C_8, C_4, etc.) and the solvents are polar aqueous-organic mixtures such as methanol-water or acetonitrile-water.

Instrumentation

HPLC consists essentially of a high pressure pumping system, relatively narrow bore column packed with small particle size stationary phase and on line highly sensitive detectors (Fig. 8.2).

The solvent reservoir: Generally made of glass. Before use it is necessary to degas the solvent to remove dissolved gases (particularly oxygen), which may interact either with mobile or stationary phase.

Pumping system: Different types of pumps are used for solvent delivery system, e.g. constant volume type and constant pressure type.

Flow controller: These consists of flow through pressure transducer, which measures the flow rate by measuring the pressure drop across restrictor of fixed value placed at the pump outlet. The flow rate signal is fed back to a control unit which compares the actual and pre-set flow rate.

Sample injectors: The sample is introduced by syringe through the septum of an injection port into the system.

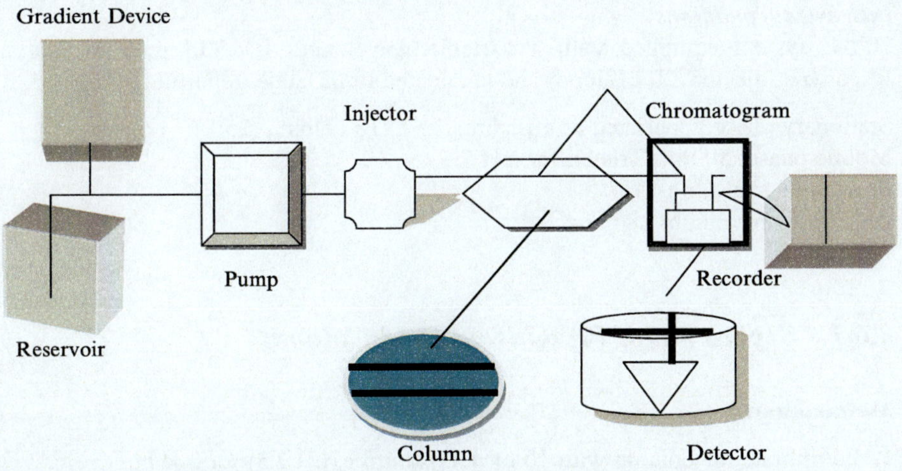

Fig. 8.2 Schematic diagram of a HPLC

Detector system: An ideal detector should have a high sensitivity (be able to detect less than 1 ppm concentration), non-destructive. Most commonly used detector is UV-visible detector (170–700 nm).

Materials

Working solutions

1. 50 mM Tris buffer, pH 8.0
2. 50 mM Tris buffer, pH 8.0 containing 0.5 M NaCl
3. 10 mM Tris buffer, pH 7.5
4. Acetonitrile

Preparation of sample
Extract 10 g of seedlings in 50 mL of 50 mM Tris buffer, pH 8.0, by grinding in a mortar and pestle. Centrifuge it at 15,000 rpm at 4°C for 15 min. Use the supernatant for soluble protein separation. Store at 4°C.

Solvent
Prior to use, filter the solvent/eluent through 0.5-µm membrane. This can be done using solvent clarification kit, which removes the particulates very quickly and easily. Degas the solvent for detector stability by passing through nylon membranes as they can be used for both hydrophilic and hydrophobic solvents.

Sample filtration
Pass the sample through sample filter (0.5-µm nylon membrane) to assure complete particulate removal before injecting.

Instrument/conditions

HPLC system equipped with a variable wavelength UV-VIS detector and a Rheodyne injector (20 μL loop) and connected to a Datajet reporting integrator.

Stationary phase: Consisted of a Lichrosorb C-18 column (250 × 4.6 mm i.d.)
Mobile phase: 50 mM Tris buffer, pH 7.5
Flow rate: 1 mL/min
Detector wavelength: 280 nm

8.6.1 Separation of Proteins on C-18 Column

Methodology

1. Equilibrate the column with 10 mM Tris buffer (pH 7.5) for 3–4 h.
2. Inject the sample using a syringe with blunt tipped needles.
3. Wash the column with the same buffer till the O.D. at 280 nm comes to zero. Elute the proteins using a linear gradient of 50 mM Tris buffer, pH 7.5 and acetonitrile.
4. Collect the fractions of the separated proteins.
5. Calculate the concentration of the protein using the peak area of that protein.

Precautions

- Degas solvent before use to remove dissolved oxygen.
- Use only blunt tipped needle for sample injection.
- Column should be washed thoroughly after the analysis is over and should be preserved either in methanol or 0.1% sodium azide to prevent from microbial growth.

Chapter 9
Cell Disruption and Fractionation

The living cell contains a number of subcellular fractions. The fractionation of cells involves two distinct phases: disruption of the tissue or cells in a suitable medium and the subsequent separation of the subcellular particles, by differential centrifugation which exploits differences in their size and density. The procedure results in rather crude subcellular fractions which are enriched with one particular component and are by no means pure. These fractions are then purified by different techniques.

9.1 Homogenization Media

The ideal homogenization medium should be capable of maintaining the morphological and functional integrity of the organelles.

Homogenization media usually has the following composition:

- 0.3 M mannitol so as to be isotonic with the cytosol.
- A buffer of pH 7–8 (often Tris) at about 50 mM concentration to neutralize the acidic vacuolar contents.
- A sulphydryl compound (dithiothreitol or mercapto-ethanol) at about 10 mM concentration to minimize the inactivation of enzymes.
- Mg^{2+} at about 10 mM concentration to keep ribosomes intact.
- Ca^{2+} at about 1 mM concentration to prevent the clumping of nuclei.
- Polyvinylpyrrolidone and bovine serum albumin (0.1–0.2%) to precipitate out the tannins and phenolics.

9.2 Cell Disruption

Since the plant cell wall is tough, disruption is done by mechanical means. The most common method employs a blender whose high-speed blades exert large shearing forces on the cells. The tissue is immersed in an equal weight of

R. Katoch, *Analytical Techniques in Biochemistry and Molecular Biology*, 169
DOI 10.1007/978-1-4419-9785-2_9, © Springer Science+Business Media, LLC 2011

homogenizing medium and blended for 0.5–2 min at full speed. Unfortunately, this procedure is far severe for the delicate cellular components and may be partly damaged. Probably the gentlest procedure of all is hand-grinding of chopped pieces of sample with a pestle and mortar in an equal volume of homogenizing medium sometimes with a little acid-washed sand to act as an abrasive. All these cell disruption procedures are carried out rapidly at 2–4°C to minimize autolytic changes.

9.3 Fractionation of Tissue Homogenate

The cell homogenate is fractionated using differential centrifugation into at least five major fractions namely nuclei, chloroplasts, mitochondria, microsomes and supernatant. These fractions are however impure and can be purified by a number of techniques such as centrifugation, phase separation, electrophoresis and specific adsorption of these components. Centrifugation remains the most generally used procedure. Again, a number of centrifugation media such as sucrose, ficoll and metrizamide are used either on a linear or nonlinear density gradients. A generalized procedure for separating a tissue homogenate into crude fractions enriched in a particular cell component is given in the flow-sheet (Fig. 9.1).

The particular fraction is resuspended in homogenization medium and then carefully layered onto the top of a sucrose gradient in a centrifuge tube. The gradient is made by successively adding layers of sucrose solution of decreasing concentration to the centrifuge tube so gently that they do not mix to any great extent. The tube may be allowed to stand at 2–4°C for about 30 min to allow diffusion to smooth the steps in the gradient. After the resuspended cell fraction has been added, the tube is centrifuged in a swinging bucket rotor. Depending upon the nature of gradient and the length of the centrifugation period different particles separate and band at zones where their density equals that of sucrose solution. After centrifugation, these bands are carefully pipetted out separately as pure fractions.

9.4 Isolation of Mitochondria (Douce et al. 1972)

Due to their crucial role in the processes such as photorespiration and fatty acid oxidation, they often result in their close proximity to other cellular organelles such as peroxisomes, glyoxysomes and chloroplasts. Isolated mitochondria show marked changes following fractionation suggesting some degree of structural damage during homogenization or from the presence of disruptive enzymes. Most of the problems involved in the isolation of intact mitochondria occur during the initial homogenization because the cell wall of plant tissues is a rigid structure and the high shearing forces necessary to rupture cell walls often have a deleterious effect on sub-cellular organelles. The fresh tissue is gently homogenized to disrupt the

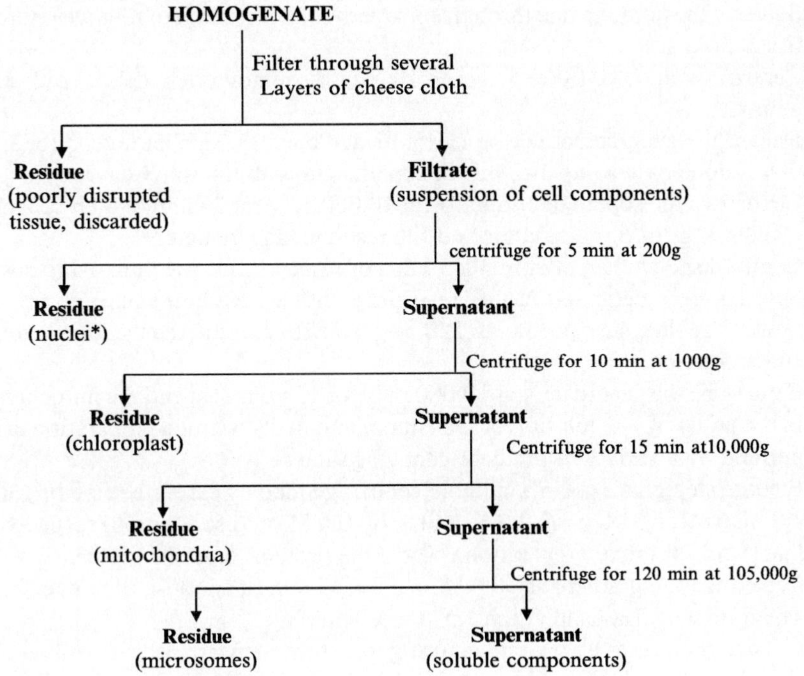

Fig. 9.1 Scheme for the separation of tissue homogenate

cells and release the contents and the mitchondria are pelleted by differential centrifugation. Further purification is carried out by sucrose gradient centrifugation.

Reagents

- Isolation medium (pH 7.8) containing 30 mM 3-(*N*-Morpholino) ethane sulfonic acid (MOPS), 0.3 M mannitol, 4 mM cysteine, 1 mM EDTA and 0.1% (w/v) defatted BSA adjusted to pH 7.8. For green leaf tissue, 0.6% (w/v) insoluble polyvinyl pyrrolidone (acid-washed) is included and the BSA concentration increased to 0.2% (w/v).
- Re-suspension medium: As above medium but without 4 mM cysteine.
- Non-linear sucrose gradient: Sucrose solution of 1.8 M (52.1% w/v), 1.5 M (43.1%), 1.2 m (35.6%) and 0.6 M (19.1%) separately prepared in 10 mM MOPS or phosphate buffer (pH 7.2), 0.1% (w/v) BSA.
- Tissue homogeniser

Procedure

1. Collect and chop 100–200 g of fresh tissue into two volumes of chilled isolation medium (4°C).
2. Disrupt the tissue using either a mixer for 20 s, a waring blender at low speed for 2–3 s.

3. Squeeze the homogenate through six layers of cheese cloth to remove unbroken tissue pieces.
4. Centrifuge at 700–1,000 × *g* for 10 min to remove cell debris and starch grains.
5. Decant the supernatant taking care to leave the starch pellet undisturbed, and this is done by leaving 1–2 mL of supernatant with the starch layer.
6. Centrifuge the supernatant fraction at 10,000 × *g* for 20 min or alternatively at 39,000 × *g* for 5 min and discard the resultant supernatant.
7. Gently disperse the pellet in 40–50 mL of resuspension medium using a wide-bore 10-mL pipette and further resuspend with a glass homogenizer.
8. Centrifuge the suspension at 250 × *g* for 10 min to reduce the levels of contamination.
9. Centrifuge the supernatant at 10,000 × *g* for 15 min. Suspend the mitochondria in the pellet in 1–2 mL of resuspension medium. This crude preparation can be purified by a variety of gradient centrifugation.
10. Prepare step gradients in a suitable centrifuge tube by carefully pipetting 6 mL 1.8 M, 6 mL 1.5 M, 6 mL 1.2 M and 3 mL 0.6 M sucrose solutions respectively, load 1 mL of crude preparation (40–50 mg protein) onto the gradient.
11. Centrifuge the gradient at 40,000 × *g* for 45 min in an ultracentrifuge.
12. The mitochondria band at the 1.5–1.2 M interface.
13. Collect the band by side-puncturing the tube using a hypodermic needle slightly below the band. Alternatively appropriate fractions in drops can be collected by injecting 2 M sucrose into the base of the tube.
14. Dilute the gradient fraction containing mitochondria to isotonic conditions (0.3 M) by slow, careful addition of buffer.
15. Pellet the mitochondria by centrifuging at 10,000 × *g* for 15 min and finally suspend in a small volume of relevant medium.

Note

- The most important step is that the disruption of the cells should be done gently to avoid damage to the organelle and at the same time ensuring maximum recovery.
- Mitochondria from green leaf tissue can be isolated in a similar way described above. The medium is identical to that used for etiolated tissues except for the addition of 0.6% (w/v) acid washed insoluble PVP and increasing the defatted BSA concentration 0.2% (w/v). After filtration through cheese cloth, chloroplasts sediment at 3,000 × *g* for 5 min and the mitochondria are collected from the supernatant by centrifugation at 12,000 × *g* for 20 min. The pellets are resuspended in approximately 50 mL of medium except for addition of 0.2% defatted BSA as described previously. Following a low speed centrifugation at 15,000 × *g* for 10 min, mitochondria are sedimented from the supernatant by centrifugation at 11,000 × *g* for 15 min.

9.5 Isolation of Chloroplasts (Walker 1980)

Isolated chloroplasts are required for different studies including the electron transport system of the photosynthetic apparatus. The separation is carried on the basis that the cell organelles, depending upon their size and weight, sediment at different centrifugal fields.

Reagents

1. Isolation medium: Weigh 2.42 g Tris (20 mM); 72.8 g sorbitol (0.4 M); 1.168 g NaCl (20 mM); 0.61 g $MgCl_2 \cdot 6H_2O$ (3 mM) and dissolve in 1 L of distilled water and adjust the pH to 7.8.

Procedure

1. Cut 5–10 g of leaf tissues into small bits. Add 20 mL of the prechilled isolation medium.
2. Homogenize with standard homogenizer.
3. Filter the debris through eight-layered cheese cloth.
4. Centrifuge at 3,000 × *g* for 2 min.
5. Discard the supernatant and suspend the pellets in the isolation medium.
6. Centrifuge again at 3,000 × *g* for 2 min.
7. Discard the supernatant and resuspend the pellet in a small volume of the grinding medium and store on ice.
8. Since any further advanced study on isolated chloroplasts is expressed on the basis of chlorophyll, estimate the chlorophyll by diluting 0.1–0.2 mL of chloroplast suspension to a total volume of 4 mL with 80% acetone.

   Calculate the chlorophyll content as follow:

$$(12.7 \times A_{663}) - (2.69 \times A_{645}) = \text{chl.a}\ (\mu g/mL),$$
$$(22.9 \times A_{645}) - (4.68 \times A_{663}) = \text{chl.a}\ (\mu g/mL),$$
$$(20.2 \times A_{645}) - (8.02 \times A_{663}) = \text{total chl.}(\mu g/mL).$$

   Calculate the chlorophyll concentration of the stock chloroplast suspension.

Considerations

Most leaves yield better chloroplasts if freshly harvested except spinach, which can be stored for 4 weeks in cold for better yield. If leaves are brightly illuminated for 20–30 min prior to grinding, the chloroplast yield is increased.

Chapter 10
Enzymes in Metabolism

Enzymes are the biocatalyst in metabolic processes. The study of the enzymes involve estimation of activities; isolation, purification and characterization of different enzymes

10.1 Main Steps During Enzyme Purification

During the first step of purification, the tissue is usually homogenized in buffer at 0–4°C, if mitochondria or particles are to be isolated, an isotonic or hypertonic solution is employed, namely 0.25–0.8 M sucrose, with a suitable buffer to control the pH.

1. *Tissue homogenization*: The under given scheme can be applied for the separation of particulate systems.
2. *Centrifugation*: Centrifugation of the homogenate is the next step during protein purification.
3. *Fractional precipitation with ammonium sulphate*: A homogenate, a soluble protein extract, or an acetone powder extract can be used for a series of standard purification procedures. By the addition of a saturated solution of ammonium sulphate, proteins will be salted out and separated by centrifugation. If conditions are kept constant, remarkable reproducibility can be obtained.
4. *Selective adsorption and elution on calcium phosphate gels*: Proteins are readily adsorbed on these gels and then are differentially eluted by increasing salt concentrations.
5. *Ion exchange and gel permeation chromatography*: The gels, such as carboxymethyl cellulose (CMC) or DEAE, are extremely useful in purification procedures. Sephadex columns – gel filtration techniques using unmodified gels or gels with DEAE or CM side chains on the polysaccharide molecule – are widely employed for enzyme purification. These methods are the general

R. Katoch, *Analytical Techniques in Biochemistry and Molecular Biology*,
DOI 10.1007/978-1-4419-9785-2_10, © Springer Science+Business Media, LLC 2011

approaches to enzyme purification. All steps must be checked for enzyme units, specific activities, yields, and recoveries.

6. *Purity of enzyme*: In order to examine detailed structures of complex proteins, it is mandatory to have proteins that are homogeneous entities. Over a period of years techniques have therefore been developed to analyze protein solutions for homogeneity.

Gel electrophoresis of proteins: Since proteins are polyelectrolytes with their charges dependent on the pH of the surrounding medium, electrophoresis techniques have been developed which can separate a mixture of proteins in an electric field. The mobility of a protein in an electric field depends on the number of charges on the protein, the sign of the net charges, the degree of dissociation which is a function of pH, and the magnitude of the electrical field potential. An opposing resistance against the mobility of the protein molecule relates to the size and shape of the ion, viscosity of the medium, concentration of the ion, solubility of the protein, and adsorptive properties of the support medium.

The most widespread and useful support medium presently employed for electrophoresis is a polymer of acrylamide cross-linked with *N,N*-dimethyl-bis-acrylamide.

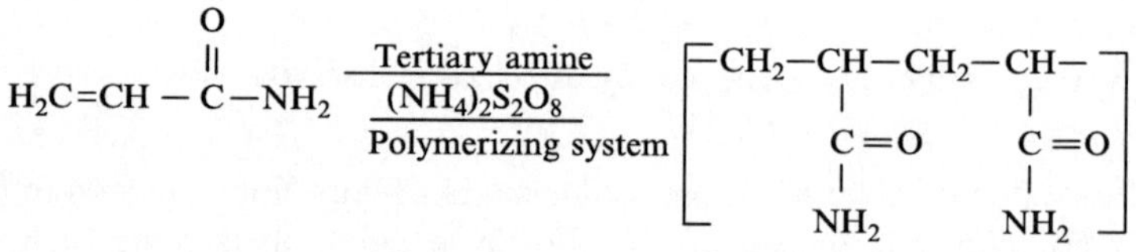

The advantage of the gel electrophoresis method is that the "pore size", that is, the sieving action of the gel, is directly related to the concentration of the gel. Thus, by increasing the range of gel concentration from 3 to about 9–10%, the pore size decreases and the proteins move more slowly. By this simple variant, one can alter the mobility of charged proteins and thus study a wide range of protein sizes.

The apparatus involves a direct-current power supply, and upper and lower reservoir buffer systems, the proteins sample, a stacking gel (2.5% gel), and the running gel (about 6–7%). The gel is prepared by mixing acrylamide with the cross-linking component, methylene-bis-acrylamide and the polymerizing initiator, and ammonium per sulphate. With the running gel in place, the stacking gel is prepared above it, the gel cassette is appropriately mounted in the apparatus, the protein solution is added above the stacking gel, and current is turned on. Frequently, a tracking dye is added with the protein mixture to serve as an indicator of the front of the moving zone as it descends down the tube. When the tracking dye moves to the bottom of the gel column, the current is turned off, and the gel tube is removed, stained with an appropriate dye, and inspected for the number of protein components.

By minor modifications, namely running an unknown protein and a known protein at different gel concentrations and plotting their log R_m against gel concentration and in turn the slopes of each curve against the molecular weight, an accurate molecular weight of the unknown protein can be determined. Thus, one can ascertain the purity as well as the molecular weight by gel electrophoresis employing microgram quantities of proteins. Modifications which involve incorporation of detergents or urea in the gel system allow an estimation of the number of subunits and their molecular weights in a given protein.

10.2 Preparation of Acetone Powder

On some occasions acetone powders of tissues are prepared for the study of enzymes. These powders, which are mostly stable, can be stored for long periods of time with little loss in activity. For the preparation of acetone powder, tissue (1 volume) is homogenized in a Waring blender in 5–10 volume of acetone at 0°C. The smooth slurry is filtered on a Buchner funnel and the cake resuspended in 5 volume of cold acetone and again filtered. This process is repeated until the powder appears thoroughly dehydrated and defatted. One volume of fresh, cold, diethyl ether is then poured over the cake on the Buchner funnel and the cake is sucked dry. Traces of acetone and ether are removed in a vacuum desiccator over paraffin strips. These acetone powders are excellent initial sources of enzymes for purification.

10.3 Estimation of Activities of Some Important Enzymes

10.3.1 Malate Dehydrogenase (L-Malate: NAD⁺ Oxidoreductase EC 1.1.1.37) (Gnanam and Francis 1976)

Malate dehydrogenase (MDH) is one of the enzymes involved in TCA cycle. It catalyzes the reversible conversion of oxaloacetic acid to malic acid. This enzyme is also involved in carbon dioxide assimilation in C_4 plants. It is coupled with PEPcase in CO_2 assimilation in the chloroplasts of mesophyll cells. Here, it uses NADPH as the coenzyme.

$$\text{Oxaloacetic acid} + \text{NADH} \underset{}{\overset{Mg^{2+}}{\rightleftarrows}} \text{Malic acid} + \text{NAD}^+$$

Principle

It is an oxidoreductase involving nicotinamide-adenine dinucleotide. Therefore, the decrease in absorbance due to the oxidation of NADH is followed during estimation.

Reagents

- Oxaloacetic acid (5 μmol) 66 mg/50 mL in distilled water
- Magnesium chloride (10 μmol) 203.5 mg/50 mL distilled water
- Tris–HCl buffer (0.1 M) pH 7.8
- NADH (0.4 μmol) 5.32 mg/10 mL

Crude Enzyme Extract

Grind the tissue thoroughly with acid-washed sand in a pre-chilled pestle and mortar in grinding medium (1 mL/1 g tissue) containing 50 mM Tris–HCl (pH 8.0), 50 mM $MgCl_2$, 5 mM 2-mercaptoethanol, and 1 mM EDTA. Pass the homogenate through four layers of cheese cloth, and centrifuge the filtrate at 3,000 × *g* for 20 min at 5°C. Use the supernatant as enzyme source.

Procedure

S. No	Reagents	Test (mL)	Blank (mL)
1.	Oxaloacetic acid	0.5	0.5
2.	Magnesium chloride	0.5	0.5
3.	Tris–HCl buffer, pH 7.8	1.3	1.8
4.	Enzyme extract	0.2	0.2
5.	NADH (refer step 3)	0.5	nil

Pipette out all reagents as shown in table.

1. Set the spectrophotometer to 0 absorbance at 340 nm without adding NADH in the test against the blank in the reference cuvette.
2. Add NADH as quickly as possible into the test, mix well and record the initial absorbance.
3. Record the absorbance every 30 s for at least 3 min.

Calculation

Calculate the enzyme activity as follows with decrease in absorbance for 1 min.

Micromoles NADH oxidized per minute per 0.2 mL enzyme extract

= Absorbance decrease/minute × 0.1613

× 3 (volume of the reaction mixture in millilitre)

Determine the protein content by the method of Lowry et al. (1951) in the enzyme extract. Calculate the value for milligram protein to calculate the specific activity.

10.3.2 Glutamate Dehydrogenase (L-Glutamate: NAD Oxidoreductase (Deaminating) EC 1.4.1.2) (L-Glutamate: NADP Oxidoreductase (Deaminating) EC 1.4.1.4) (Delma 1970)

Glutamate dehydrogenase (GDH) occurs in almost all living organisms. In higher plants, GDH activity has been found in most species. The existence of two distinct GDH enzymes in higher plants is well known. The enzyme has high Km for ammonia (10–80 mM). The molar extinction coefficient of NADH at 340 nm is 6.22×10^3. This indicates that 1 μmol NADH/mL will have an absorbance of 6.22.

(a) A mitochondrial enzyme which specifically requires coenzyme NAD.
(b) A chloroplast enzyme which specifically requires coenzyme NADP.

Both fungal and bacterial enzyme has a single coenzyme specificity. GDH activity has been found in both bacteriod and cytosol fractions of root nodules of a number of legumes.

The reaction catalyzed by this enzyme is presented below:

$$\text{L-Glutamate} + H_2O + NAD^+(P) \rightleftarrows \text{2-Oxoglutarate} + NH_4^+ + NAD(P)H + H$$

Principle

GDH is assayed by following the oxidation of the reduced coenzyme, NADH or NADPH. *These reduced coenzymes absorb light at 340 nm, which in most biological systems is uniquely uncluttered with interfering absorption by other compounds*. Thus, the absorption of NADH at 340 nm is easily detected.

Reagents

- Potassium phosphate buffers 1.0 M (pH 7.8 and 7.0)
- 2-Oxoglutarate 0.1 M: Dissolve 14.6 g in 1 L distilled water
- NH_4Cl (0.1 M): Dissolve 53.5 g in 1 L of distilled water
- NADH 10 mg/mL
- NADPH 10 mg/mL
- Enzyme extract: Follow the method given for the preparation of extract.

Procedure

Reagent	NADH assay (mL)	NADPH assay (mL)
Potassium phosphate buffer pH 7.0	1	–
Potassium phosphate buffer pH 7.8	–	–
2-Oxoglutarate	0.3	0.3
NH_4Cl	0.5	0.5
NADH	0.12	–
NADPH	–	0.12
Enzyme extract	0.2	0.2
Water	8	8

1. Proceed for the assay of NADH- and NADPH-dependent GDH as per the composition of assay given in table.
2. Add 0.3 mL of water in the blank instead of 2-Oxoglutarate.
3. Incubate the reaction mixture at 37°C for 15–30 min.
4. Record the change in absorbance at 340 nm.

Calculation

The amount of NADH or NADPH oxidized is calculated from the molar extinction coefficient. Activities are expressed as *n* mole NAD(P)H oxidized per minute per milligram protein.

Nanomole of NAD(P)H oxidized/minute/milligram protein

$$= \frac{\text{A340} \times \text{volume of assay solution} \times 1,000}{6.22 \times \text{Time of incubation (min)} \times \text{milligram protein in enzyme extract used}}$$

10.3.3 Glutamate Synthase (L-Glutamate: NADP⁺ Oxidoreductase (Transaminating) EC 1.4.1.13) (Tempest et al. 1970)

The original name given to this enzyme was glutamine (amide) 2-Oxoglutarate amino transferase (oxidoreductase $NADP^+$) from which the acronym GOGAT is derived. The bacterial enzyme is highly specific for the pyridine nucleotide electron donor (NADH or NADPH). In angiosperms ferredoxin glutamate synthase has been reported. Angiosperm pyridine nucleotide glutamate synthase appears to exhibit activity with both NADH and NADPH. The only exception to this is the NADH-specific enzyme from legume root nodules. The root enzyme has low affinity for 2-Oxoglutarate (Km 0.4–1.0 mM). The reaction catalyzed by this enzyme is as follows.

$$\text{Glutamine} + \text{2-Oxoglutarate} + \text{NADPH} + H^+ \longrightarrow 2\,\text{Glutamate} + \text{NADP}^+$$

Principle

Glutamate synthase is assayed spectrophotometrically by recording the rate of oxidation of NADPH or NADH, as indicated by a change in absorbance at 340 nm following the addition of enzyme extract.

Reagents

1. Tris–HCl buffer 50 mM pH 7.6
2. Prepare the following reagents in Tris–HCl buffer 50 mM pH 7.6

   (a) Glutamine, 5 mM (36.5 mg/10 mL)
   (b) 2-Oxoglutarate 5 mM (36.5 mg/10 mL)
   (c) NADPH 0.25 mM (10 mg/10 mL)

Enzyme Extract: 1 g of the plant material with 5 mL of 100 mM phosphate buffer (pH 7.5) containing 1 mM disodium EDTA, 1 mM dithioerythritol and 1%

polyvinyl pyrrolidone (PVP) and centrifuge at 10,000 × *g* for 30 min at 4°C. Collect the supernatant and use it for enzyme assay.

Procedure

S. No.	Reagent	mL to the added
1	Glutamine	1.0
2	2-Oxoglutarate	1.0
3	NADPH	1.0
4	Enzyme extract	0.2
5	Buffer	1.8

- Prepare reaction mixture as per the table.
- Omit 2-Oxoglutarate in the blank, instead add 1 mL buffer.
- Incubate for 15–30 min at 37°C.
- Record the change in absorbance at 340 nm.

Calculation

Activity is expressed as *n* mole of NAD(P)H oxidized per minute per milligram protein.

10.3.4 Ascorbic Acid Oxidase (L-Ascorbate: Oxygen Oxidoreductase EC 1.10.3.3) (Oberbacher and Vines 1963)

Ascorbic acid oxidase is widespread in plant tissues. The role of this enzyme is to regulate the levels of oxidized and reduced glutathione and NADPH. The activity of this enzyme is increased during infection.

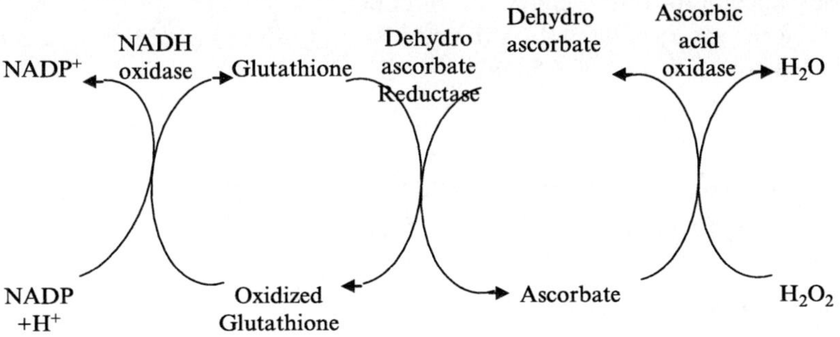

Principle

The method used for assay of ascorbic acid oxidase relies on the fact that within a certain range of enzyme concentrations, the rate of oxygen consumption during ascorbic acid oxidation is proportional to the amount of enzyme present. Ascorbic acid has an absorption maximum at 265 nm. Therefore, the decrease in the absorption peak of the acid due to oxidation by ascorbic acid oxidase is followed spectrophotometrically.

Reagents

- Phosphate buffers 0.1 M (pH 5.6 and 6.5 separately)
- Substrate solution: Dissolve 8.8 mg ascorbic acid in 300 mL phosphate buffer (pH 5.6)
- Enzyme extract: Grind one part of plant tissue with five parts (w/v) of 0.1 M phosphate buffer (pH 6.5) in a homogenizer. Centrifuge the homogenate at 3,000 × *g* for 15 min. Use the supernatant as enzyme source. All procedures are carried out at 0–5°C

Procedure

1. Pipette out 3 mL of substrate solution to each of sample and reference cuvettes of a spectrophotometer.
2. Add 0.1 mL of enzyme extract to the reference cuvette (enzyme is added so as get a positive increase in absorbance values).
3. Measure the absorbance change at 265 nm in 30 s intervals for 5 min.

Calculation

1. From the linear phase of reaction, compute the change in absorbance per minute.
2. Activity may also be expressed in enzyme units. One enzyme unit is equivalent to 0.405 μmol oxygen/min.
3. Ascorbic acid requires only 1 atom of oxygen/mol oxidized. Therefore, 0.81 μmol of ½ oxygen/min is utilized per enzyme unit. Ascorbic acid has $E_{1\ cm}^{1\%}$ of 760 at 265 nm. From this it is calculated that ascorbic acid has an absorbance of 4.4 /μmol in a 3 mL volume.
4. Therefore, one enzyme unit (0.81 μmol ½ oxygen/min) would be equivalent to an absorbance change of 3.58/min.

10.3.5 Catalase (Hydrogen Peroxide: Hydrogen Peroxide Oxidoreductase EC 1.11.1.6) (Luck 1974)

Enzyme catalase catalyzes the following reactions:

1. Decomposition of hydrogen peroxide to give water and oxygen

$$2H_2O_2 \xrightarrow{\text{Catalase}} 2H_2O + A$$

2. Oxidation of H donors, e.g. methanol, formic acid, phenol with the consumption of 1 mol of peroxide

$$ROOH + AH_2 \xrightarrow{\text{Catalase}} H_2O + ROH + A$$

Principle

The UV light absorption of hydrogen peroxide solution can be easily measured between 230 and 250 nm. On decomposition of hydrogen peroxide by catalase, the absorption decreases with time. The enzyme activity could be arrived at from this decrease. But this method is applicable only with enzyme solutions which do not absorb strongly at 230–250 nm.

Reagents

1. Phosphate buffer, 0.067 M (pH 7.0). Dissolve 3.522 g KH_2PO_4 and 7.268 g $Na_2HPO_4{\cdot}2H_2O$ in distilled water and make up the volume to 1 L.
2. Hydrogen Peroxide–Phosphate Buffer. Dilute 0.16 mL of H_2O_2 (10% w/v) to 100 mL with phosphate buffer. Prepare fresh. The absorbance of the solution should be about 0.5 at 240 nm with a light path.
3. Enzyme extract: Homogenize plant tissue in a blender with M/150 phosphate buffer (assay buffer diluted 10 times) at 1–4°C and centrifuge. Stir the sediment with cold phosphate buffer, allow to stand under cool conditions with occasional shaking and then repeat the extraction once or twice. The extraction should not take longer then 24 h. Use the combined supernatants for the assay. The catalase activity can change considerably on storage of the tissue. In comparative studies, always use the same conditions of extraction, storage and temperature.

Procedure

1. Maintain wavelength = 240 nm and volume = 3 mL, approximately at room temperature
2. Read against a control cuvette containing enzyme solution as in the experimental cuvette, but containing H_2O_2-free PO_4 buffer (M/15)
3. Pipette into the experimental cuvette
   3 mL H_2O_2–PO_4 buffer. Mix in 0.01–0.04 mL sample with a glass or plastic rod flatten rod flattened at one end. Note the time Δt required for a decrease in absorbance from 0.45 to 0.4. This value is used for calculations. If t is greater than 60 s repeat the measurements with a more concentrated solution of the sample

Calculation

If one gram tissue is homogenized in a total volume of 20 mL, diluted 1–10 volume with water and taken 0.01 mL for assay then the absorbance at 240 nm decreased from 0.45 to 0.4 in 13.9 s is calculated as:

$$\frac{17}{13.9} = 1.22 \text{ units in the assay mixture}$$

or

$$\frac{1.22 \times 10}{0.01} = 1,220 \text{ units/mL extract i.e., } 2.44 \times 10^4 \text{ units/g tissue}$$

Calculate the concentration of H_2O_2 using the extinction coefficient 0.036/μmol/mL.

10.3.6 *Nitrate Reductase (NADH: Nitrate Oxidoreductase EC 1.6.6.1 (NADH-Dependent)) (NAD(P)H: Nitrate Oxidoreductase EC 1.6.6.2 (NAD(P)H-dependent)) (NADPH: Nitrate Oxidoreductase EC 1.6.6.3 (NADPH-Dependent)) (Hageman and Reed 1980)*

The nitrate reducing system consists of nitrate reductase which catalyze stepwise reduction of nitrate to nitrite and then to ammonia. According to the specificity of electron donor, two major types of nitrate reductase occur:

(a) Ferredoxin-dependent nitrate reductase (blue-green algae)
(b) Pyridine nucleotide-dependent nitrate reductase (higher plants)

The assimilatory reduction of nitrate by plants is a fundamental biological process in which a highly oxidized form of inorganic nitrogen is reduced to nitrite and then to ammonia.

$$NO_3^- + AH_2 \rightarrow NO_2^- + A + H_2O$$

Principle

Nitrate reductase (NR) is capable of utilizing the reduced form of pyridine nucleotides, flavins or benzyl viologen as electron donors for reduction of nitrate to nitrite. NADH-dependent nitrate reductase is most prevalent in plants. Hence, NR activity in plants can be measured by following the oxidation of NAD(P)H at 340 nm. However, NR activity is commonly measured by colorimetric determination of nitrite produced.

Reagents

1. Potassium phosphate buffer 0.1 M (pH 7.5).
2. Potassium nitrate 0.1 M: Dissolve 1.01 g potassium nitrate in 100 mL water.
3. NADH 2 mM: Dissolve 14 mg NADH disodium salt in 10 mL water.
4. Sulphanilamide, 1% (w/v): Dissolve 1 g suphanilamide in 100 mL 2.4 N hydrochloric acid.
5. *N*-(1-naphthyl) ethylenediamine dihydrochloride, 0.02%: Dissolve 20 mg in 100 mM water.

6. Potassium nitrite standard solution (0.01 M):
   Dissolve 851 mg pure potassium nitrite in 100 mL water in a standard flask. Dilute 10 mL of this solution to 100 mL and use as working standard solution.
7. Enzyme extract
   Homogenize a weighed quantity of the plant material in a known volume of medium (6 mL for 1 g fresh tissue) containing 1 mM EDTA, 1–25 mM cysteine and 25 mM potassium phosphate adjusted to a final pH 8.8 with KOH. Filter through four layers of cheese cloth and centrifuge for 15 min at 30,000 × *g*. Decant the supernatant through glass wool and use for assays. Extract under ice-cold conditions.

Procedure

1. Pipette out 0.5 mL phosphate buffer (pH 7.5) in test tube.
2. Add 0.2 mL potassium nitrate, 0.4 mL NADH solution and 0.7 mL water.
3. Initiate the reaction by the addition of 0.2 mL enzyme extract. Set up a control in the same way but with water instead of enzyme extract.
4. Incubate at 30°C for 15 min.
5. Terminate the reaction by the rapid addition of 1 mL sulphanilamide followed by 1 mL naphthyl ethylenediamine reagent and wait for 30 min.
6. Measure the absorbance at 540 nm.
7. Prepare a standard graph with sodium nitrite. Pipette out different known aliquots of potassium nitrite standard solution into a series of test tubes and make up the volume in each tube to 2 mL by adding water. Proceed from step 5 to 7.

Calculation

Activity is expressed as micromole nitrite produced per minute per milligram protein (or per gram fresh tissue). The enzyme reaction rate is linear over a 30 min period. The pink colour produced by nitrite is stable for 2–3 h.

10.3.7 Nitrite Reductase (NAD(P)H: Nitrite Oxidoreductase EC 1.6.6.4) (Vega et al. 1980)

The enzyme present in both eukaryotic and prokaryotic cells accepts electrons from photosynthetically reduced ferredoxin but not from reduced pyridine nucleotides.

Nitrite is directly reduced to ammonia without the liberation of free intermediates by nitrite reductase

$$NO_2^- \xrightarrow[+6e]{\text{Nitrite reductase}} NH_4^+$$

Principle

The disappearance of nitrite is measured in reaction. Reduced methyl viologen is used as elected donor.

Reagents

- Tris–HCl buffer 0.5 M (pH 7.5).
- Sodium nitrite solution: Dissolve 43.2 mg $NaNO_2$ in 20 mL distilled water.
- Methyl Viologen solution: Dissolve 60.1 mg methyl viologen in 20 mL water.
- Sodium Dithionite-Bicarbonate solution: Dissolve 250 mg each of $Na_2S_2O_4$ and $NaHCO_3$ in 10 mL water.
- Enzyme extract: Homogenize the leaf tissue (10 g/100 mL) with Tris–HCl buffer (pH 7.5) in a waring blender at high speed for 3 min and force the homogenate to flow through eight layers of cheese cloth at 4°C. Use the filtrate as enzyme source.

Procedure

1. Prepare a reaction mixture by mixing 6.25 mL of Tris–HCl buffer, 2 mL of sodium nitrite solution, 2 mL methyl viologen solution and 14.75 mL water.
2. Pipette out 1.5 mL reaction mixture and 0.3 mL of enzyme preparation into a test tube.
3. Run a blank without the enzyme.
4. Start the reaction by adding 0.2 mL of recently prepared dithionite-sodium bicarbonate solution.
5. Incubate for 15 min at 30°C.
6. Stop the reaction by vigorous shaking (vortex mixer) until blue colour disappears.
7. Use a 20 μL aliquot for nitrite determination.
8. Follow steps 5–8 as under nitrate reductase.
9. Estimate the amount of nitrite disappeared using blank as reference.

Calculation

The enzyme activity is expressed as the amount of nitrite (micromolar) reduced per minute per milligram protein.

10.3.8 Peroxidase (Donor: H_2O_2 Oxidoreductase E.C. 1.11.1.17) (Malik and Singh 1980)

Peroxidase (POD) catalyzes dehydrogenation of large number of organic compounds such as phenols, aromatic amines, hydroquinones etc. POD occurs in animals, higher plants and other organisms. The best studied is horse radish POD. Guaiacol is used as substrate for the assay of POD.

Principle

$$\text{Guaiacol} + H_2O_2 \xrightarrow{\text{POD}} \text{Oxidized guiacol} + 2H_2O$$

The resulting oxidized (dehydrogenated) guaiacol is probably more than one compound and depends on the reaction conditions. The rate of formation of guaiacol dehydrogenation product is a measure of the POD activity and can be assayed spectrophotometrically at 436 nm.

Materials

- Phosphate Buffer 0.1 M (pH 7.0).
- *Guaiacol solution* 20 mM: Dissolve 240 mg guaiacol in water and make up to 100 mL. It can be stored frozen for many months.
- Hydrogen peroxide solution (0.042% = 12.3 mM). Dilute 0.14 mL of 30% H_2O_2 to 100 mL water. The extinction of this solution should be 0.485 at 240 nm. Prepare freshly.
- Enzyme extract: Grind 1 g of fresh plant tissue in 3 mL of 0.1 M phosphate buffer (pH 7) in pre-cooled mortar and pestle. Centrifuge the homogenate at 18,000 × *g* at 5°C for 15 min. Use the supernatant as enzyme source within 2–4 h. Store on ice till the assay is carried out.

Procedure

1. Pipette out 3 mL buffer solution, 0.05 mL guaiacol solution, 0.1 mL enzyme extract and 0.03 mL hydrogen peroxide solution in a cuvette (bring the buffer solution to 25°C before assay).
2. Mix well. Place the cuvette in the spectrophotometer.
3. Wait until the absorbance has increased by 0.05. Start a stopwatch and note time required in minutes (Δt) to increase the absorbance by 0.1.

Calculation

Since the extinction coefficient of guaiacol dehydrogenation product at 436 nm under the conditions specified is 6.39/μmol, the enzyme activity per litre of extract is calculated as follows:

$$\text{Enzyme activity units/litre} = \frac{3.18 \times 0.1 \times 1,000}{6.39 \times 1 \times \Delta t \times 0.1} = \frac{500}{\Delta t}$$

Considerations

- Most accurate values are obtained when Δt is between 1 and 3 min. The enzyme extract has therefore to be diluted appropriately.

- *O*-dianisidine (1 mg/mL method) may be used as an alternative substrate for the assay. The oxidized *O*-dianisidine (yellow/orange coloured compound) is measured at 430 nm. Take 3.5 mL phosphate buffer (pH 6.5) in a clean dry cuvette. Add 0.2 mL enzyme extract and 0.1 mL freshly prepared *O*-dianisidine solution. Bring the assay mixture to 28–30°C and then place the cuvette in the spectrophotometer set at 430 nm. Then, add 0.2 mL 0.2 M H_2O_2 and mix. Immediately start the stopwatch. Read the initial absorbance and then at every 30 s intervals up to 3 min. If the rate of increase is very high, repeat the assay with diluted extracts. Plot increase in absorbance against time. From the linear phase, read the change in absorbance per minute. Express enzyme activity in terms of rate of increased absorbance per unit time per milligram protein or tissue weight. A water blank is used in the assay.

10.3.9 Polyphenol Oxidase (Monophenol, Dihydroxyphenylalanine: Oxygen Oxidoreductase EC 1.14.18.1) (Esterbaner et al. 1977)

Phenol oxidases are copper protein of wide occurrence in nature which catalyzes the aerobic oxidation of certain phenolic substrates to quinones which are autoxidized to dark brown pigments generally known as melanins. These enzymes are assumed to be single enzymes with broad specificity although there is some evidence for the presence of more than one phenol oxidase in certain tissues. Each individual enzyme tends to catalyze the oxidation of one particular phenol or phenolic compound more readily than others. The polyphenol oxidase (PPO) comprises of catechol oxidase and laccase. The activities of these enzymes are important with regard to (a) plant defence mechanism against pests and diseases and (b) appearance palatability and use of plant products.

Principle

The intensely yellow 2-nitro-5-thiobenzoic acid (TNB) with an absorption maximum at 412 nm reacts with the quinones generated through enzymatic oxidation of 4-methylcatechol (catechol oxidase) and 1,4 dihydroxybenzene (laccase) to yield colourless adducts. The decrease in the absorbance of yellow colour due to enzyme activity is taken into consideration and measured accordingly.

Reagents

1. Citrate-phosphate buffer 0.2 M (pH 6.0).
2. TNB: Add 30 mg sodium borohydride to a suspension of Ellman's reagent, i.e. 5,5-dithiobis (2-nitrobenzoic acid) (19 mg) in 10 mL water. Within 1 h, the

disulphide is quantitatively reduced to the intensely yellow, water-soluble thiol. This solution is stable for at least 1 week when stored at 4°C.

3. Quinine solutions:
   Dissolve 4-methly-1,2-benzoquinone in double-distilled water in a 50 mL volumetric flask by bubbling nitrogen gas until the quinone is completely dissolved.
   Prepare *p*-benzoquinone solution in a similar manner. Both solutions are stable for 30 min, a time sufficient to carry out the spectrophotometric assay.
4. Substrate solution:
   Dissolve 4-Methylcatechol (2 mM) for catechol oxidase assay and Quinol (1,4-dihydroxybenzene, 2 mM) for laccase assay.

Enzyme extract

To get a crude enzyme preparation, mix 100 mg acetone powder with 2.5 mL of 0.2 M citrate-phosphate buffer (pH 6.0), 1 mL of 1% Triton X-100, 6.5 mL of water and 500 mg polyamide. Shake for 1 h and filter. Use the filtrate as enzyme source.

Procedure

1. Pipette out into clean 1 cm cuvette of 1.4 mL citrate 0.1 M phosphate buffer (pH 6.0), 0.5 mL of TNB and 1 mL of the substrate solution.
2. The reaction is initiated by the addition of 0.1 mL of enzyme preparation and the absorbance at 412 nm in a spectrophotometer already set is immediately noted down.
3. Follow the decrease in absorbance at 30 s intervals and record.

Calculation

Read the change in absorbance per minute from the linear part of the curve.

Calculate the enzyme units according to the following equation:

$$\text{Units in the test} = K \times (\Delta A/\min)$$

where, K is 0.272 for catechol oxidase and 0.242 for laccase.

One unit of either catechol oxidase or laccase is defined as the enzyme which transforms 1 μmol of dihydricphenol to 1 μmol of quinone/min under the assay conditions. One unit is equivalent to the consumption of 1 μmol of TNB.

Considerations

- The method described is rapid and sensitive than any other methods available.
- The enzyme concentration required to get satisfactory linearity of time vs. absorbance decrease has to be standardized.
- The relation between rate and enzyme concentration is linear for all catechol oxidases; however, with laccase it is observed only for low concentrations.

- A simpler method of assaying PPO is given below.

  Add 2.5 mL of 0.1 M phosphate buffer (pH 6.5), 0.3 mL of catechol solution (0.01 M) into cuvette and set the spectrophotometer at 495 nm. Now add 0.2 mL of enzyme extract and start recording the change in absorbance for every 30 s up to 5 min.

  For the above procedure, the enzyme extract may be prepared by grinding 5 g leaves with a mortar and pestle in about 20 mL medium containing 50 mM Tris–HCl, pH 7.2, 0.4 M Sorbitol and 10 mm NaCl. Centrifuge the homogenate at 20,000 × *g* for 10 min and use the supernatant for assay.

10.3.10 Indole Acetic Acid Oxidase (Byrant and Lane 1979)

Indole acetic acid (IAA) is the best known naturally occurring plant auxin. It participates in controlling many phase of plant growth and differentiation. Levels of free IAA are in turn regulated through synthesis binding, esterification and enzyme degradation. Indolyl acetic acid oxidase (IAAO) is the enzyme involved in the catabolic degradation of IAA to 3-methylene oxindole.

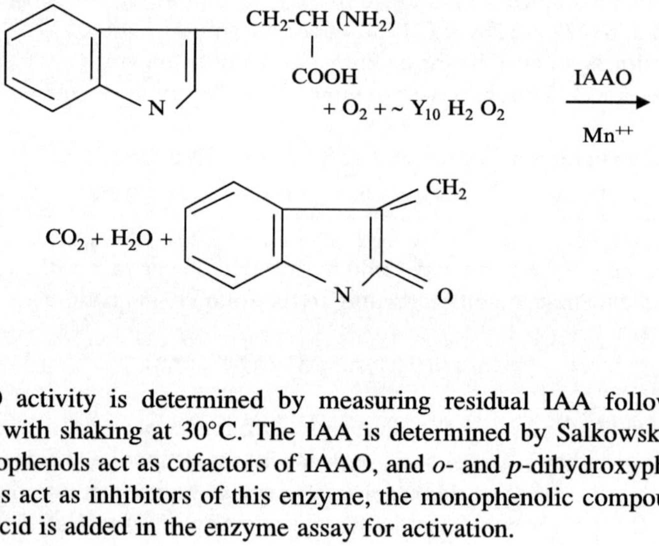

Principle

The IAAO activity is determined by measuring residual IAA following dark incubation with shaking at 30°C. The IAA is determined by Salkowski reaction. Since monophenols act as cofactors of IAAO, and *o*- and *p*-dihydroxyphenols and polyphenols act as inhibitors of this enzyme, the monophenolic compound, *para*-coumaric acid is added in the enzyme assay for activation.

Reagents

1. Phosphate buffer, 0.07 M, pH 6.2.
2. *Para*-Coumaric acid solution: Dissolve 25 mg *p*-coumaric acid in 50 mL water.
3. IAA solution: Dissolve 10 mg IAA in 40 mL water.
4. Manganese chloride solution: Dissolve 118 mg $MnCl_2{\cdot}4H_2O$ in 20 mL water.
5. Perchloric acid 5 M.

6. Ferric nitrate, 0.1 M: Dissolve 24.18 g of substance in 100 mL water.
7. Enzyme extract: Prepare acetone powder from the frozen tissue by homogenizing 25 g tissue in 100 mL aliquot of cold acetone. The homogenate is collected by Buchner filtration through Whatman No. 1 filter paper following both grindings. The homogenate is air-dried until free of acetone odour, the resulting dry powder was weighed and freezer-stored in cold containers.

   One gram of acetone of powder is ground in two successive 20 mL aliquots of 250 mM phosphate buffer (pH 6.2) in an ice bath. Collect the extract by Buchner filtration through Whatman No. 1 paper after each grinding. Combine the filtrates and dilute to 50 mL with phosphate buffer.

Procedure

1. Pipette out the following solutions in a test tube in the order given below:

Phosphate buffer (pH 6.2)	2 mL
Para-coumaric acid	1 mL
Manganese chloride	1 mL
Enzyme extract	2 mL

2. Start the reaction by the addition of 4 mL IAA solution.
3. Incubate the reaction mixture in the dark with shaking at 30°C.
4. Withdraw 2 mL of the mixture after 0 and 50 min of incubation and add 5.2 mL perchloric acid and 0.5 mL ferric nitrate solution.
5. Then dilute to 10 mL with water.
6. After incubating the reaction mixture in the dark for 60 min, measure the absorbance at 535 nm.
7. Estimate the protein content in the enzyme extract following the method of Lowry et al. (1951).

Calculation

Express the enzyme activity as micromole IAA oxidized per minute per milligram protein.

10.3.11 Lipase (Triacylglycerol Acylhydrolase EC 3.1.1.3)

During germination of oilseeds, lipases play an important role in hydrolyzing the stored oils so that the required energy for growth and carbon skeleton for synthesis of new compounds are produced. The germinating seeds of castor bean, sunflower, groundnut are good sources of lipases. Lipase hydrolyses triglycerides to release free fatty acids and glycerol.

$$\text{Triglycerides} + H_2O \rightarrow \text{Glycerol} + \text{Fatty acids}$$

Principle

The quantity of fatty acid released in unit time is measured by the quantity of NaOH required to maintain pH constant. The milliequivalent of alkali consumed is taken as a measure of the activity of the enzyme.

Reagents

1. Enzyme source: Grind a known quantity of sample with a mortar and pestle. Homogenize the tissue with twice the volume of ice-cold acetone. Filter and wash the powder successively with acetone, acetone:ether (1:1) and ether. Air-dry the powder. This acetone powder can be stored in a refrigerator. Extract 1 g of the powder in 20 mL ice-cold water or a suitable buffer. Centrifuge at 15,000 rpm for 10 min and use the supernatant as enzyme source.
2. Substrate: Take 2 mL of any clear vegetable oil, neutralize to pH 7.0, if necessary, and stir well with 25 mL of water in the presence of 100 mg bile salts (sodium taurocholate) till an emulsion is formed. Addition of 2 g gum Arabic hastens emulsification.
3. 0.1 N NaOH.
4. 20 mM phosphate buffer (pH 7.0).

Procedure

1. Take 20 mL of substrate in 500 mL beaker. Add 5 mL of phosphate buffer (pH 7.0).
2. Set the beaker on top of a magnetic stirrer cum hot plate and stir the contents slowly. Maintain the temperature at 35°C. Dip the electrodes of a pH meter in the reaction mixture and adjust the pH to 7.0.
3. Add enzyme extract (0.5 mL), immediately record the pH and set the timer on. Let it be pH at zero time.
4. At frequent intervals (say 10 min) or as the pH drops by about 0.2 unit, add 0.1 N NaOH to bring pH to the initial value. Continue the titration for 30–60 min period.
5. Note the volume of alkali consumed.

Calculation of Enzyme Activity

The enzyme activity is defined as the amount of enzyme which releases 1 meq of free fatty acid/min/g sample. Specific activity is expressed as milliequivalent per minute per milligram protein.

Activity milliequivalent/minute/gram sample

$$= \frac{\text{Volume of alkali consumed} \times \text{strength of alkali}}{\text{Weight of sample in gram} \times \text{time in minute}}$$

10.3.12 Acetylcholine Esterase (Acetylcholine Hydrolase EC 3.1.1.7) (Kasturi and Vasantharajan 1976)

Acetylcholine esterase enzyme is found in the free state mainly in brain, nerve cells, muscle, lung and erythrocytes plays an important role in the transmission of nerve impulses. The enzyme activity is strongly inhibited in cases of poisoning with organophosphorus compounds. The determination of AChE in whole blood is of importance when there is a question of possible poisoning with organophosphorus compounds. AChE is widely assayed in toxicological studies on insects.

Acetylcholine esterase has also been reported to be present in roots, stem and leaves of pea plant and its properties have been well studied. Acetylcholine is involved in phytochrome action and hence the involvement of AChE in plants for the purpose of regulating the levels of endogenous acetylcholine is suggested.

Principle

The hydrolysed acetylcholine per unit time is measured by comparison of the initial concentration in a reference tube with the final concentration in the experimental tube. Acetylcholine is made to react with hydroxylamine to form the corresponding acylhydroxamic acid which forms a strongly coloured ferric hydroxamate with ferric salts and the colour of hydroxamate is measured at 490 nm.

Reagents

1. Extraction of enzyme from plant tissues: Fresh sample material (root, leaf or any other part) is finely ground and extracted in 10 mM veronal buffer (pH 8.6) followed by centrifugation at 20,000 × *g* for 10 min; the pellet containing the enzyme is again ground and extracted with the earlier mentioned buffer containing 5% ammonium sulphate. The extract is centrifuged at 20,000 × *g* for 10 min and the supernatant is used as enzyme source.
2. Collect 0.2 mL blood from the test animal and transfer to 5 mL water. Use the haemolysate for the assay.
3. Veronal buffer, 0.1 M pH 8.6: Dissolve 4.92 g sodium veronal and 3.24 g sodium acetate in about 300 mL water, add 3 mL 1 N HCl and dilute to 500 mL with water. Check the pH.
4. Acetylcholine stock solution (200 mM): Put 1.82 g acetylcholine chloride (hygroscopic, cheek the purity) in a 50 mL volumetric flask, dissolve in water and make up to volume.
5. Substrate acetylcholine solution (1.33 mM): Mix 150 mL veronal buffer and 1 mL of acetylcholine stock solution thoroughly.
6. Sodium hydroxide 2.5 N: Dissolve 7 g hydroxylammonium chloride in water and make up to 100 mL. Store the solution in a well-stoppered polyethylene flask in a refrigerator.
7. Alkaline hydroxylamine solution: Mix equal volumes of sodium hydroxide (2.5 N) and 1 N hydroxylamine solutions.

8. Iron solution 1 M (0.7 M): Dissolve 33.75 g Fe $(NH_4)(SO_4)_2 \cdot 12\ H_2O$ in about 70 mL water with gentle warming. Add 2.5 g potassium nitrate (dissolved separately in water). Transfer to a 100 mL volumetric flask and dilute to the mark.
9. Citrate buffer 1 M (pH 1.4): Dissolve 2.10 g citric and 0.8 g NaOH in minimum quantity of water in a 100 mL volumetric flask, add 89 mL 1 N HCl and dilute to the mark with water. Dilute 10 mL of this solution in a volumetric flask to 100 mL; pH of this solution must be between 1.4 and 1.2. With the exception of substrate solution all other solutions are stable for several months.

Procedure

1. Pipette out into 50 mL volumetric flasks as given below:

	Reference (mL)	Test (mL)	Blank (mL)
Simple (enzyme source)	–	–	–
Substrate solution	25	25	–

Mix well and incubate at 37°C for 30 min

	Reference (mL)	Test (mL)	Blank (mL)
Alkaline hydroxylamine solution	5	5	5
Sample	2	–	–
Citrate buffer	5	5	5
Ferric solution	10	10	10

Allow the ferric solution to run slowly down the wall of the flask
2. Dilute with water to the mark and shake thoroughly.
3. Allow to stand for 20 min at room temperature.
4. Filter the solution through a double-folded filter paper and discard the first portion of filtrate.
5. Measure the absorbance of the filtrates at 490 nm against blank.

Calculation

The absorbance difference (ΔE) between reference (initial concentration of substrate) and test (final concentration of substrate) is used for calculation. For measurements at 490 nm the extinction of the dye is 0.961/μmol. Hence, the amount of dye formed from the non-hydrolysed acetylcholine in 50 mL.

$$C = \frac{E \times 50}{0.961 \times 1.0} (\mu\text{mol}/50\,\text{mL})$$

The AChE activity in whole blood is

$$\frac{E \times 50}{0.96 \times 1.0} \times \frac{1}{0.08 \times 30} \times 1,000$$

$$E \times 21,667 (\text{U/L})$$

Considerations

- For assay of acetylcholinesterase in insect tissues, homogenize the tissues in Veronal buffer solution, centrifuge and use aliquots of supernatant.
- In the case of plant extracts the enzyme activity is expressed as millimole acetylcholine hydrolysed per hour milligram protein.
- The enzyme can be assayed using thiocholine esters as substrate instead of acetylcholine.

10.3.13 Pectin Methylesterase (Pectin Pectyl Hydrolase EC 3.1.1.11) (Hobson 1964)

Pectin methylesterase (PME) (Pectase or pectinesterase) de-esterifies pectin into pectic acid and methanol and thus makes it susceptible to precipitation by calcium or other polyvalent cations. The roots, stem, leaves and fruits of most higher plants contain PME.

Principle

Since PME produces methanol and free carboxyl groups in pectic acid, either methanol or the carboxylic group may be measured as a function of the enzyme. The measurement of methanol is cumbersome. The increase in free carboxyl groups can be easily followed titrimetrically while a constant pH in the reaction mixture is maintained. The procedure given below is based on this principle.

$$\text{Pectin} + H_2O \xrightarrow{\text{PME}} \text{Pectic acid} + CH_3OH$$

Reagents

1. Solid salt mixture of NaCl and EDTA disodium salt 10:1 (w/w).
2. Salt mixture solution: Dissolve 7.5 g of the above mixture in 100 mL water.
3. Pectin: Prepare 2% (w/w) pectin solution in 1.5 M NaCl. From this stock solution, dilute accordingly to get 1% solution in 0.1 M NaCl.
4. 0.1 or 0.02 N NaOH. For low enzyme activities use dilute alkali.
5. Enzyme extract: To about 30 g macerated tissue, add 2.25 g NaCl-EDTA salt mixture and stir for 15 min and centrifuge. Collect the supernatant. Again stir the residue with 20 mL of salt mixture solution for 15 min, centrifuge and collect the supernatant. Repeat the extraction step. To the supernatant collected each time, add 100 mL cold ethanol the precipitate. Pool the precipitate, dissolve in 15 M NaCl and use as enzyme source.

Procedure

1. Take 100 mL of pectin (1%) solution in a 500 mL beaker and add 30 mL of 1 M NaCl and 60 mL distilled water.

2. Place it in a water bath at 30°C. The water bath should be maintained on a magnetic stirrer cum hot plate so that the solution can be stirred continuously.
3. Place a pH meter adjacent to the magnetic stirrer and dip the electrode in the reaction mixture.
4. Bring the reaction mixture to pH 7.5 by adding 0.1 N NaOH with continuous mixing.
5. Now add 10 mL of the enzyme solution and set the timer on.
6. Maintain a constant pH of 7.5 in the reaction mixture by continuous addition of 0.1 N NaOH for definite period of time.
7. Measure the alkali consumption as a function of time.
8. Run a blank using the inactivated enzyme. The enzyme is inactivated by placing it in boiling water for 10 min.

Calculation

One unit of pectin methylesterase activity is that amount of enzyme which releases 1 meq of carboxyl groups/min/g sample. Milliequivalent of carboxyl group is calculated by multiplying titre value with the normality of the alkali used.

10.3.14 Aspartate Aminotransferase (Glutamate: Oxaloacetate Aminotransferase EC 2.6.1.1)

Aminotransferases are enzymes catalyzing the transfer of an amino group plus a protein and an electron pair, from an amino-donor compound to the carbonyl position of an amino-accepter compound. Usually, an amino acid acts as the amino donor and a 2-oxo acid as the amino acceptor. In some cases aldehydes may serve as amino acceptors, and amines may act as donors. The enzymatic transfer of amino groups plays an important role in many metabolic processes where the interconversion of nitrogen containing molecules is involved. Nitrogen, following its initial assimilation into glutamine and glutamate can be distributed to many other compounds by the action of aminotransferases. Glutamate is often the amino-donor substrate in biosynthetic transamination reactions. This reaction regenerates 2-oxoglutarate for necessary ammonia assimilation through GDH and the glutamine synthetase/glutamate synthase routes.

Principle

Glutamate–oxaloacetate aminotransferase (GOT) catalyzes the reversible interconversions between glutamate and aspartate and their 2-oxo analogues. The oxaloacetic acid is measured colorimetrically by a reaction with

2,4-dinitrophenylthydrazine giving a brown coloured hydrazone after the addition of 0.4 N sodium hydroxide

$$\text{Glutamate} + \text{oxaloacetate} \underset{}{\overset{\text{GOT}}{\rightleftarrows}} \text{2-Oxoglutarate} + \text{asparatate}$$

Reagents

1. Phosphate buffer, pH 7.4. Add 11.3 g dry anhydrous disodium hydrogen phosphate and 2.7 g dry anhydrous potassium dihydrogen phosphate in 1 L volumetric flask and make up to the mark with water. Check the pH and store at 4°C.
2. *Substrate solution*: Dissolve 13.3 g DL-aspartic acid in minimum amount of 1 N sodium hydroxide and prepare a solution with pH 7.4 (about 90 mL is required). Add 0.146 g 2-oxoglutarate and dissolve it by adding a little more sodium hydroxide solution. Adjust to pH 7.4 and then make to 500 mL with phosphate buffer. Divide into 10 mL portions and store frozen at −15°C.
3. *Pyruvate standard*: Dissolve 22 mg sodium pyruvate in 100 mL water in a standard flask.
4. *2,4-Dinitrophenylhydrazine (DNPH)*: Dissolve 19.8 mg DNPH in 10 mL concentrated hydrochloric acid and make to 100 mL with water. Store it in an amber bottle at room temperature.
5. *Sodium hydroxide 0.4 N*: Dissolve 16 g sodium hydroxide in 1 L water.
6. *Enzyme extract*: Prepare the crude extract by grinding the plant tissue in 0.2 M potassium phosphate, pH 7.5 in a homogenizer for 2 min. Pass the slurry through eight layers of cheese cloth and then centrifuge at 25,000 × *g* for 15 min to get the enzyme fraction.

Procedure

1. Warm 0.5 mL of substrate solution in a water bath at 37°C for 3 min.
2. Add 0.2 mL enzyme extract and mix gently.
3. Incubate for 60 min at 37°C.
4. Remove the tubes from the bath and immediately add 0.5 mL DNPH solution and mix well.
5. Mix 0.5 mL substrate with 0.5 DNPH solution and then add 0.1 mL enzyme extract for control.
6. Allow the DNPH to react for 20 min at room temperature.
7. Add 5 mL of 0.4 N sodium hydroxide, mix well and leave for a further 10 min.
8. Record the absorbance at 510 nm.
9. Pipette out pyruvate standard 0.05–0.20 mL and make up to 0.2 mL. Add 0.5 mL substrate and 0.5 mL DNPH solution. For blank mix 0.5 mL substrate, 0.2 mL water and 0.5 mL DNPH solution.
10. Thereafter proceed with steps 6–8.

Calculation of enzyme activity

The pyruvate formed by enzyme is responsible for the absorbance difference between test and control. The pyruvate in standard produces the difference between standard and blank. Express the enzyme activity as micromole of pyruvate formed per minute per milligram protein.

10.3.15 Alanine Aminotransferase (Glutamate: Pyruvate Aminotransferase EC 2.6.1.2) (Bergmeyer and Bernt 1984)

Principle

Glutamate pyruvate aminotransferase catalyzes the reversible interconversions between glutamate and alanine and their 2-oxo analogues. The pyruvate formed after 30 min incubation period is measured colorimetrically by reaction with DNPH.

$$\text{2-Oxoglutrate} + \text{alanine} \rightleftarrows \text{Glutamate} + \text{pyruvate}$$

Reagents

1. Phosphate buffer, pyruvate standard, DNPH and sodium hydroxide, 0.4 N – as described in GOT.
2. Substrate solution: Dissolve 9.0 g alanine in 90 mL water with addition of about 2.5 mL sodium hydroxide, 1 N and adjust to pH 7.4. Then add 0.146 g 2-oxoglutarate and dissolve it by adding a little more sodium hydroxide solution and adjust to pH 7.4, make up to 500 mL with phosphate buffer. Divide into 10 mL portions and store frozen at −15°C.

Procedure

Similar procedure as in the enzyme GOT is followed here and in this case alanine is used as substrate and incubated for 30 min.

Calculation of Enzyme Activity

The enzyme activity is expressed as micromole of pyruvate formed per minute per milligram protein.

10.3.16 Phosphatases (Orthophosphoric-Monoester Phosphohydrolase, Alkaline Medium 3.1.3.1) (Orthophosphoric-Monoester Phosphohydrolase, Acid Medium 3.1.3.2)

Phosphatases liberate inorganic phosphate from organic phosphate esters. Acid phosphatase (3.1.3.2) hydrolyzes a number of phosphomonoesters and phosphoproteins. Alkaline phosphatase catalyzes the hydrolysis of numerous phosphate esters, such as esters of primary and secondary alcohols, sugar alcohols, phenols and amines. Phosphodiesters are not hydrolysed by either of them.

Principle

The enzyme phosphatase hydrolyzes *p*-nitrophenol phosphate. The released *p*-nitrophenol is yellow in colour in alkaline medium and is measured at 405 nm. The optimum pH for acid and alkaline phosphatases are 5.3 and 10.5, respectively.

10.3.16.1 Acid Phosphatase

Reagents

- Sodium hydroxide 0.085 N: Dissolve 0.85 g sodium hydroxide in 250 mL water.
- Substrate solution: Dissolve 1.49 g EDTA, 0.84 g citric acid and 0.03 g *p*-nitrophenylphosphate in 100 mL water and adjust to pH 5.3.
- Standard: Weigh 69.75 mg *p*-nitrophenol and dissolve in 5.0 mL distilled water (100 mM).
- Enzyme extract: Homogenize 1 g fresh tissue in 10 mL of ice-cold 50 mM citrate buffer (pH 5.3) in a pre-chilled pestle and mortar. Filter through four layers of cheese cloth. Centrifuge the filtrate at 10,000 × *g* for 10 min. Use the supernatant as enzyme source.

Procedure

1. Incubate 3 mL of substrate solution at 37°C for 5 min.
2. Add 0.5 mL enzyme extract and mix well.
3. Remove immediately 0.05 mL and mix it with 9.5 mL of sodium hydroxide 0.085 N. This corresponds to zero time assay (blank).
4. Incubate the remaining solution (substrate + enzyme) for 15 min at 37°C.
5. Draw 0.5 mL sample and mix it with 9.5 mL sodium hydroxide solution.
6. Measure the absorbance of blank and incubated tubes at 405 nm.
7. Take 0.2–1.0 mL (4–20 mM) of the standard, dilute to 10.0 mL with NaOH solution. Read the colour and draw the standard curve.

Calculation

Specific activity is expressed as *m* mole *p*-nitrophenol release per minute per milligram protein.

10.3.16.2 Alkaline Phosphatase

1. For alkaline phosphatase extract the enzyme in 50 mM glycine NaOH buffer (pH 10.4).
2. Alkaline phosphatase functions optimally at about pH 10.5. The assay procedure is similar to that for acid phosphatase, except for the substrate solution. Prepare the substrate solution as follows.

Dissolve 375 mg glycine, 10 mg magnesium chloride, 165 mg *p*-nitrophenylphosphate in 42 mL of 0.1 N sodium hydroxide and dilute to 100 mL. Adjust to pH 10.5.

10.3.16.3 Assay Using Glycerophosphate

Acid and alkaline phosphatases can also be assayed by measuring the amount of inorganic phosphorus released from sodium glycerophosphate. Phosphorus is estimated by Fiske–Subbarow method as described below.

Reagents

1. Substrate: 0.1 M solution of β-glycerophosphate 3.153 g in 100 mL water.
2. 10% TCA.
3. Ammonium molybdate (2.5%): Dissolve 25 g ammonium molybdate in 400 mLof water. Add 500 mL of 10 N H_2SO_4 and make up the volume to 1 L with water.
4. 1-amino 2-naphthol 4-sulphonic acid (ANSA) reagent: Dissolve 30 g sodium metabisulphite, 6 g sodium sulphite and 550 mg ANSA separately in small quantities of water. Combine all the solutions and make up to 250 mL with water. Allow to stand overnight and filter. Store refrigerated in an amber-coloured bottle. Prepare fresh reagent every fortnight.
5. Standard phosphate solution: Dissolve 439 mg potassium dihydrogen phosphate in water, add 10 mL of 10 N H_2SO_4 and make up to 1 L with water (1 mL = 0.1 mg P). Add 0.5 mL chloroform as preservative.
   Dilute 10 mL of the earlier mentioned stock solution to 500 mL with water and use as working standard (1 mL = 20 μg P)
6. 0.2 M Solution of magnesium acetate = 4.289 g in 100 mL water.

Procedure

1. Pipette out 2 mL of respective (acid or alkaline) buffer in a test tube.
2. Add 0.5 mL magnesium acetate solution.

3. Add 2 mL of β-glycerophosphate solution.
4. Make up the volume to 7 mL with the buffer.
5. Add 1 mL enzyme extract and incubate for 1 h at 37°C.
6. Stop the reaction by the addition of 2 mL of 10% TCA.
7. For the control add the enzyme after incubation.
8. Centrifuge at 10,000 × *g* for 10 min.
9. Pipette out different known aliquots of supernatant of test, and control into test tubes. Make up the volume in each tube to 5 mL by adding water.
10. Add 1 mL of ammonium molybdate solution to each tube and mix thoroughly.
11. Add 0.4 mL of ANSA reagent to each tube and mix well. After standing for 10 min dilute to 10 mL. Read the intensity of blue colour at 660 nm against a reagent blank.
12. Pipette out different aliquots of phosphate solution (5–50 μg P) into test tubes and treat as described above. A reagent blank is also prepared with water in place of phosphate solution. Plot absorbance vs. phosphorus concentration. Read the phosphorus content from the calibration curve and express enzyme activity as the amount of phosphorus released in unit time per milligram protein.

Considerations

- Any trace of phosphate impurity in the distilled water interferes with the determination. This is easily noticed from the reagent blank which should actually be colourless.
- After developing, the blue colour intensifies with the progress of time. It is preferable to read the tubes 10 min after adding ANSA reagent.

10.3.17 Amylases (α-1,4 Glucan 4-Glucanohydrolase EC 3.2.1.1) and (α-1,4 Glucan Maltohydrolase EC 3.2.1.2) (Kruger 1972)

α-Amylase causes endo-cleavage of substrates and hydrolyses α-1,4 linkages in a random manner. It has the ability to by-pass α-1,6 branch points. The viscosity reduction of the substrate is fast but the production of reducing is slow. β-Amylase hydrolyses alternate bonds from the non-reducing end of the substrate. The enzyme degrades amylase, amylopectin or glycogen in an exo- or stepwise fashion by hydrolyzing alternate glycosidic bonds. The end product is β-maltose. β-Amylase is incapable of bypassing branch points i.e. 1,6-glycosidic linkages in amylopectin and glycogen. This results in about 55% conversion of amylopectin to maltose. The other product is a large limit dextrin. The viscosity reduction of the substrate due to β-amylase action is slow but the production of reducing sugars is fast.

Principle

The reducing sugars produced by the action of α- and/or β-amylase react with dinitrosalicylic acid (DNS) and reduce it to a brown coloured product, nitroamino-salicylic acid.

Reagents

1. Sodium acetate buffer, 0.1 M pH 4.7.
2. Starch, 1% solution: Prepare a fresh solution by dissolving 1 g starch in 100 mL acetate buffer. Slightly warm, if necessary.
3. DNS reagent.
4. 40% Rochelle salt solution (Potassium Sodium Tartrate).
5. Maltose solution: Dissolve 50 mg maltose in 50 mL distilled water in a standard flask and store it in a refrigerator.
6. Extraction of amylases: Extract 1 g of sample material with 5–10 volumes of ice-cold 10 mM calcium chloride solution overnight at 4°C or for 3 h at room temperature. Centrifuge the extract at 54,000 × *g* at 4°C for 20 min. The supernatant is used as enzyme source.
7. Extraction of β-amylases: The free β-amylase is extracted from acetone defatted sample material in 66 mM phosphate buffer (pH 7.0) containing 0.5 M NaCl. The extract is centrifuged at 20,000 rpm for 15 min. The supernatant is used as a source of free β-amylase. The pellet is then extracted with phosphate buffer containing 0.5% 2-mercaptoethanol. The clear extract is used as source of bound β-amylase. All operations are carried out 4°C.

Procedure

1. Pipette out 1 mL of starch solution and 1 mL of properly diluted enzyme in test tube.
2. Incubate it at 27°C for 15 min.
3. Stop the reaction by the addition of 2 mL of DNS reagent.
4. Heat the solution in a boiling water bath for 5 min.
5. While the tubes are warm, add 1 mL potassium sodium tartrate solution.
6. Then cool it in running tap water.
7. Make up the volume to 10 mL by addition of 6 mL water.
8. Read the absorbance at 560 nm.
9. Terminate the reaction at zero time in the control tubes.
10. Prepare a standard graph with 0–100 μg maltose.

Calculation

A unit of α- or β-amylase is expressed as milligram of maltose produced during 5 min incubation with 1% starch.

Considerations

- The extraction procedure given is suitable for cereal grains. There are a variety of extraction procedures used depending upon the source material. For instance, the plant tissue is extracted in pre-cooled 20% aqueous glycerol and the filtrate is used as enzyme source of amylases.

10.3.18 Cellulases (1,4-(1,3:1,4): β-D-Glucan 4-glucanohydrolase EC 3.2.1.4) (Hinton and Pressey 1974)

Hydrolysis of crystalline cellulose is a complex process. A minimum of three different types of enzymes are believed to be involved.

1. Endo-β-1,4 glucanase (C_x-cellulase).
2. Exo-β-1,4 glucanase (C_1-celluiase).
3. β-glucosidase (cellobiase).

Initiation of hydrolysis of native cellulose is effected by C_1 enzyme. This enzyme is an exo-β-1,4 glucanase. Exo- glucanase splits alternate bonds from the non-reducing end of cellulose chain yielding cellobiose. The endo-glucanase is distinguished by the mechanism of their attack on carboxymethyl cellulose. It dose not act on native cellulose. β-glucosidases play an important function in the degradation of cellulose by hydrolyzing cellobiose which is an inhibitor of exo-glucanase. Only organisms producing C_1-cellulose (exo-glucanase) are capable of hydrolyzing native cellulose (filter paper, cotton etc).

10.3.18.1 Assay of C_x-cellulose (Endo β-1,4 Glucanase) (Viscometric Method)

Principle

Endo-β-1,4 glucanase acts on CMC and hydrolyses the β-1,4 glucosidic bonds in a random manner. As a result, the viscosity of CMC solution is reduced. This is measured in an Ostwald viscometer.

Reagents

- Citrate-phosphate buffer 0.1 M (pH 6.0)
- Carboxymethyl cellulose 0.5% solution: Dissolve 0.5 g sodium carboxymethyl cellulose in hot water. Adjust to pH 6.0
- Chloramphenicol-cyclohexamide solution: Dissolve 25 mg each of chloramphenicol and cyclohexamide in 20 mL water
- Ostwald viscometer

Procedure

1. Pipette out 3 mL carboxymethyl cellulose solution and 1 mL citrate-phosphate buffer into a test tube.
2. Add 1 mL of enzyme extract.
3. Add 0.1 mL chloramphenicol-cyclohexamide solution to prevent microbial contamination.
4. Incubate at 37°C for 16 h.
5. After incubation, heat the solution in boiling water for 3 min, cool and then centrifuge at 8,000 × *g* for 20 min.
6. Run a control which contains denatured enzyme (heat the enzyme extract for 3 min in boiling water).
7. Draw 5 mL portion of control and test supernatant solution and measure the viscosity in an Ostwald viscometer.

Calculation

The percent loss of viscosity is interpreted as proportional to the cellulose activity. Calculate the percent reduction in viscosity as below:

$$V = \frac{T_0 - T}{T - T_{H_2O}} \times 100$$

where, V – percent loss in viscosity, T_0 – flow time in seconds of zero time, T – flow time after incubation and T_{H_2O} – flow time of water.

10.3.18.2 β (1–4) Glucanase Assay (C_x) (Colorimetric)

Principle

The production of reducing sugar (glucose) due to cellulolytic activity is measured by DNS method.

Reagents

1. Sodium citrate buffer 0.1 M (pH 5.0).
2. Carboxymethyl cellulose 1%: Dissolve 1 g carboxymethyl cellulose in 100 mL Sodium Citrate Buffer 0.1 M (pH 5.0).
3. DNS reagent.
4. 40% Rochelle salt solution (Potassium sodium tartrate).

Procedure

1. Pipette out 0.45 mL of 1% CMC solution at a temperature of 55°C and 0.05 mL of enzyme extract.
2. Incubate the mixture at 55°C for 15 min.

3. Immediately after removing the enzyme substrate mixture from the bath add 0.5 mL DNS reagent.
4. Heat the mixture in a boiling water both for 5 min.
5. While the tubes are warm, add 0.1 mL potassium sodium tartrate solution.
6. Cool to room temperature.
7. Add water to make 5 mL volume.
8. Measure the absorbance at 540 nm.
9. Prepare a standard graph with glucose in the concentration range 50 μg to 100 μg/mL.

Calculation

Express the enzyme activity as the milligram glucose released per minute per milligram protein.

10.3.18.3 C_1 and C_x-Cellulase (Combined Assay)

Reagents

1. Citrate-phosphate buffer 0.1 M (pH 5.8).
2. Filter paper disc.
3. Cut the Whatman filter paper No. 1 with a paper punch (7 mm diameter) to ensure same surface area of substrate in reaction tube.

Procedure

1. Add 0.5 mL enzyme extract to 32 mg of dry Whatman No. 1 filter paper.
2. Incubate the mixture for 1 h at 50°C.
3. Follow steps 3–9 in the C_x-cellulose colorimetric assay.

10.3.19 Polygalacturonase Activity (Poly 1,4-α-D-Galacturonide Glycanohydrolase EC 3.2.1.15) (Pressey and Avants 1976)

In assessing fruit ripening process and shelf life of fruits, assay of polygalacturonase serves as a good indicator. PG activity is also measured in diseased plants, since a large number of pathogens have been reported to produce pectic enzymes to macerate the plant tissues. PG activity can be assayed by both reductometric and viscometric methods. Polygalacturonase (PG) is implicated in the softening of many fruits such as tomatoes, peaches, pears, avocados and mangoes. The enzyme is essentially a hydrolase and the mode of action is attributed to two enzymes exo-PG (PGII; EC 3.2.1.15) and endo-PG (PGI). Exo-PG acts by removing galacturonic acid units from the non-reducing end of polygalacturonic acid

(PGA) while endo-PG hydrolyses PGA randomly, giving rise to short chain oligogalacturonic acids which can ultimate be hydrolysed to a mixture of tri-, di-, and monogalacturonic acids, depending on the source of the enzyme considered. Fruits such as peaches, tomatoes, papayas and cucumbers were found to have both enzymes. Avocado and carrot, on the other hand, have only one PG enzyme each, PGI and PGII, respectively.

Principle

Polygalacturonases I and II are capable of releasing reducing groups from the substrate polygalacturonic acid or pectic acid. The quantum of reducing groups released is the measure of the activity of the enzyme. The reducing groups in the assay mixture are analyzed by the conventional arsenomolybdate method. An enzyme unit is defined as that amount which catalyzes the formation of 1 μmol of reducing group/h.

Reagents

1. Polyethylene glycol
2. 0.2% $Na_2S_2O_2$ (Sodium disulphite or Sodium metabisulphite)
3. 0.5 M NaCl
4. PM-10 Membrane
5. Sephadex G-100
6. 0.15 M NaCl
7. 0.2 M Tris–acetate buffers – pH 4.5 and 5.5
8. 0.01 M $CaCl_2$
9. 1% solution of PGA in Tris–acetate buffer (PGA I)
10. *Enzyme extract*: Slice the plant sample (100 g) and blend in 100 mL cold H_2O containing 12 g polyethylene glycol 4000 and 0.2 g Na_2S_2O. Centrifuge the slurry at 8,000 × *g* for 20 min at 2°C. Discard the supernatant. Disperse the sediment in 2 L cold water containing 0.2% $Na_2S_2O_2$ and centrifuge. Repeat the washing step and suspend the insoluble residue in 2 L of 0.5 M NaCl. Stir the mixture in the cold for 2 h and centrifuge. Concentrate the supernatant by ultrafiltration to 30 mL using PM-10 membrane. Use this concentrate as enzyme source for PG assay. If separation of PGI and PGII is desired proceed as given below.

    Prepare a column of Sephadex G 100 (5 × 90 cm) and equilibrate with 0.15 M NaCl. Apply 30 mL of the enzyme concentrate in 0.15 M NaCl to the top of the column. Elute the column with 0.15 M NaCl and collect 20 mL fractions. The peaks 17 and 27 from this column show PGI and PGII activities, respectively. Pool the fractions under each peak, concentrate by ultrafiltration to 10 mL and use as enzyme source

Procedure

1. The assay mixture For PGI contains.

    0.2 mL enzyme in 0.15 M NaCl
    0.2 mL 0.5 M Tris–acetate buffer pH 4.5

0.1 mL 0.01 M $CaCl_2$
0.5 mL 1% solution of PGA I

2. Prepare a blank for each sample by boiling the reaction mixtures before the addition of substrate.
3. Incubate for 1 h at 37°C.
4. Stop the reaction by heating at 100°C for 3 min.
5. Take 0.5 mL of the reaction mixture and analyze for reducing sugar by Nelson–Somogyi method.

For assaying PGII – Use the buffer pH 5.5 instead of pH 4.5 for the reaction mixture as well as for preparing the substrate.

Calculation

Calculate the micromole of sugar units released and find out the enzyme units per milligram of protein as per the definition given in the principle section individually for PGI and PGII.

Considerations

- An alternative procedure for extraction and assay for PG can also be used. Blend 500 g of cleaned and sliced plant sample with 500 mL of 50 mM potassium phosphate buffer with pH 7.5 and containing 6% (w/v) $(NH_4)_2SO_4$ 0.5% (v/v) Triton X-100 and 1.5% (w/v) PVP. Stir the homogenate for 30 min at 11,000 rpm at 2°C. Discard the pellet and precipitate PG from the supernatant with $(NH_4)_2SO_4$ to 80% saturation. Gummy material may float, if the sample is a fruit. Stir for 1 h and then filter through cotton wool. Dissolve the residue in 150 mL of 1% NaCl solution. Filter the slurry through a Whatman No. 1 filter paper, prelayered with PVP. Dialyse the filtrate against 1% NaCl overnight. Centrifuge at 15,000 rpm for 50 min at 2°C to clarify the dialysate. Concentrate to 1/3 volume by ultrafiltration using YM-10 membrane with 50 psi pressure of the compressed air.

  Determine PG activity with the reaction mixture containing 0.5 mL of enzyme source, 0.5 mL of 50 mM Tris–acetate buffer (pH 5.2) and 1 mL of 1% sodium polypectate. The substrate solution should also be at pH 5.2. For reagent blank, heat 0.5 mL of enzyme in a boiling water bath for 3 min before adding the buffer and the substrate. Incubate both at 37°C for 24 h. Terminate the reaction by heating the tubes in a boiling water bath for 3 min. Measure the reducing group in an aliquot of the mixture by the Nelson–Somogyi method. Define one unit of enzyme as the amount of enzyme which produces 1 μmol reducing groups/mL of enzyme/24 h. Specific activity is expressed as units of enzymes per milligram protein.
- Use galacturonic acid as standard for measuring the reducing groups through the Nelson–Somogyi method.

10.3.20 Estimation of Papain (Papainase EC 3.4.22.2) (Arnon 1970)

Papain, one of the plant proteases, is present in the papaya latex. Though the latex contains other proteases, papain is the most studied one because of its abundance. It is largely used in brewing and food industries. It is also used in cosmetics and pharmaceutical industries.

Principle

An artificial substrate Benzoyl, L-Arginine *p*-nitroanilide (BA*p*NA) is hydrolysed by papain and the colour intensity of the released *p*-nitroaniline is measured at 410 nm. With the molar extinction coefficient of *p*-nitroaniline, the activity of the enzyme is calculated. Alternatively, the colour intensity of the released *p*-nitroaniline in the sample may be compared with the standard *p*-nitroaniline.

Reagents

1. Tris–HCl buffer 50 mM (pH 7.5): Dissolve 605 mg Tris in 50 mL of distilled water. Adjust to pH 7.5 with 0.05 N hydrochloric acid and make up volume to 100 mL.
2. To the same buffer (100 mL) add 87.8 mg of cysteine hydrochloride (0.005 M) and 74.4 mg of EDTA (0.002 M) and dissolve completely.
3. Buffered substrate solution: Dissolve 43.5 mg BA*p*NA in 1 mL of dimethyl sulphoxide and make up the volume to 100 mL with Tris–HCl buffer containing 5 mM cysteine and 2 mM ethylenediamine tetra acetate.
4. Acetic acid 30%.
5. Enzyme source: Dissolve papain to a concentration of 0.1 mg/mL distilled water (for crude latex 100 mg/mL will give measurable level of papain).

Procedure

1. Pipette out 0.5 mL of papain solution into a test tube.
2. Make up the volume to 1.0 mL with Tris–HCl buffer.
3. Add 5 mL of substrate solution.
4. Incubate for 25 min at 25°C.
5. Terminate the enzyme reaction by adding 1 mL of 30% acetic acid.
6. Measure the absorbance of the released *p*-nitroaniline at 410 nm against a control (without the enzyme) in a spectrophotometer.

Calculation

$$\text{Molar extinction coefficient of } p\text{-nitroaniline } (E_{1\,\text{cm}}^{1\text{M}}) = 8,800/1\ \text{mol/cm}$$

$$1\ \text{mmol/L} = \frac{8,800}{1,000} = 8.8\text{A}$$

$$1\ \mu\text{mol/L} = \frac{8.8}{1,000} = 0.0088\text{A}$$

$$1\ \mu\text{mol/mL} = 8.8$$

If the absorbance is 8.8, amount of *p*-nitroaniline = 1 μmol/mL

If *X* is the absorbance of the sample, amount of *p*-nitroaniline is

$$\frac{1\ \mu\text{mol/mL}}{8.8} \times X$$

$$\frac{X\ \mu\text{mol/mL}}{8.8}$$

This is for 1 mL of the sample for 25 min.

For 7 mL (volume of the assay mixture) of the sample solution it is

$$\frac{X\ \mu\text{mol}}{8.8} \times \frac{7.0}{25(\text{min})}$$

The assay mixture contains 0.5 mL of the enzyme.

Therefore, Activity of the enzyme in 1 mL is

$$\frac{X\ \mu\text{mol} \times 7.0}{0.5 \times 25}$$

(micromole for *p*-nitroaniline released per minute)

From this calculate the activity per gram sample. For specific activity, determine the protein content of the enzyme solution by Lowry's method and express per milligram protein.

Considerations

- If *p*-nitroaniline is available, a standard graph may be drawn and the quantity of *p*-nitroaniline released by the enzyme can then be calculated.
- Papain can also be estimated using casein, albumin or benzoylarginine ethylester as substrate where the unit of enzyme activity will differ. But for comparison of papain activity among varieties, any one of the substrates may be used.

10.3.21 Phosphoenol Pyruvate Carboxylase (Orthophosphate: Oxaloacetate Carboxylase Phosphorylative EC 4.1.1.31) (Arnon 1970)

This enzyme is involved in the fixation of carbon dioxide in C_4 plants like sorghum, maize, sugarcane etc. It has a high affinity to CO_2 molecule. The reaction catalyzed by this enzyme is given below.

$$\text{Phosphoenol pyruvate} + CO_2 \xrightarrow{\text{PEP case}} \text{Oxaloacetic acid}$$

Principle

The oxaloacetic acid formed by the action PEP case is coupled with MDH and followed by NADH oxidation. Since the sources of PEP case contain enough MDH there is no need to add MDH exogenously.

Reagents

1. Tris–HCl buffer 0.1 M (pH 7.8)
2. Magnesium chloride (10 μM) 203.5 mg/50 mL water
3. Sodium bicarbonate (10 mM) 84 mg/100 mL water
4. Phosphoenolpyruvate (5 mM) 23.28 mg/100 mL water
5. NADH (0.4 μM) 5.32 mg/10 mL water
6. Enzyme extract – same as used for MDH

Procedure

1. Mix various recipes as shown below:

S. No	Reagents	Test (mL)	Blank (mL)
1.	Tris–HCl buffer pH 7.8	0.8	1.3
2.	Magnesium chloride	0.5	0.5
3.	Sodium bicarbonate	0.5	0.5
4.	Phosphoenol pyruvate	0.5	0.5
5.	Enzyme extract	0.2	0.2
6.	NADH	0.5	0

2. Proceed steps 2–4 as in MDH.

Calculation

Calculate as for MDH.

10.3.22 *Ribulose Biphosphate Carboxylase (3-Phospho D-Glycerate Carboxylyase EC 4.1.1.39) (Kung et al. 1980)*

Ribulase bisphosphate carboxylase (RuBP-carboxylase) is the key enzyme which catalyzes the fixation of CO_2 in plants. It catalyzes the combination of RuBP and CO_2 resulting in the formation of 3-phosphoglyceric acid. Portion of 3-phosphoglyceric acid is recycled form of RuBP. The plants which produce C_3 acids (3PGA) are termed as C_3 plants which include wheat, spinach, bean etc. The fact that RuBP-carboxylase constitutes about 50% of soluble leaf protein shows how important this enzyme is to the plant. RuBP-carboxylase also functions as an oxygenase at elevated levels of O_2 which seems to be the prime reaction for photorespiration.

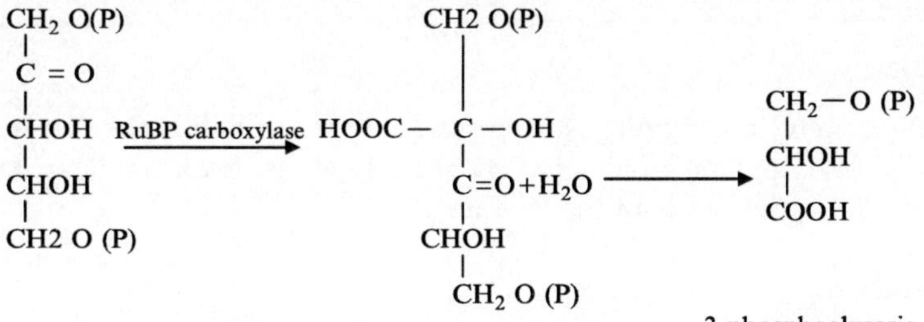

Ribulosse 1,5 - biphosphate (RuBP)

Unstable intermediate

3 phosphoglyceric acid (3 PGA)

Principle

RuBP-carboxylase is assayed radiometrically. The enzyme is made to utilize labelled CO_2 as the substrate and the radioactivity in the products is counted as a measure of enzyme activity.

Reagents

1. HEPES buffer 25 mM, pH 7.8:
   Dissolve 650 mg sodium salt of HEPES in water, adjust to pH 7.8 with NaOH and make up to 100 mL
2. RuBP 1 mM:
   Dissolve 4 mg tetra sodium salt of RuBP in 10 mL water
3. $MgCl_2$ 1 M:
   Dissolve 952 mg in 100 mL of water
4. Dithiothreitol (DTT), 50 mM:
   Dissolve 771 mg in 100 mg in 100 mL water
5. $NaH^{14}CO_3$ – sp. activity 1 μCi/μmol
6. 10% Acetic acid (v/v)

Procedure

1. Prepare a reaction mixture with the following:

   0.10 mL HEPES buffer, 0.25 mM, pH 7.8
   0.05 mL RuBP; 1 mM
   0.05 mL $MgCl_2$, 100 mM
   0.05 mL DTT, 50 mM
   0.20 mL enzyme extract

2. Illuminate the reaction mixture at 25°C for 2 min.
3. Initiate the reaction by adding 1 μCi/μmol of $NaH^{14}CO_3$.
4. Incubate the assays for 10 min.
5. Stop the reaction by adding 0.1 mL of 10% acetic acid.
6. Transfer known aliquots on to Whatman No. 3 filter paper discs, dry and count for radioactivity.
7. Run a blank similarly without RuBP and instead add 0.05 mL of buffer.

Calculation

The RuBP-carboxylase activity is expressed in terms of millimole CO_2 or micromole CO_2 per kilogram protein per seconds. It is also expressed per kilogram chlorophyll or square metre of the leaf area.

Considerations

- The enzyme assay can also be conveniently carried out directly in liquid scintillation vials sealed with serum stoppers. The initiation of the reaction with $NaH^{14}CO_3$ may be done by injecting it through the serum stopper. Likewise, acetic acid may be injected to stop the reaction. The vials are unstoppered in the fume hood and the contents taken to dryness preferably in a forced draft oven at 90°C. One millilitre water is added followed by 12 mL Liquid Scintillation Cocktail (LSC) solution. The vials are capped and the acid stable (3PGA) radioactivity is determined by Liquid Scintillation Counting.
- The specific radioactivity of the $NaH^{14}CO_3$ stock is determined by adding a 30 μL aliquot (15 μmol) to an LSC vial containing 1 mL water previously made basic with three drops of ethanolamine, followed by 12 mL of LSC solution.
- Various buffers and reaction mixtures are also successfully used for the assay of RuBP case.

10.3.23 Phenylalanine Ammonia Lyase (L-Phenylalanine Ammonia Lyase EC 4.3.1.5) (Brueske 1980)

Phenylalanine ammonia lyase (PAL) is responsible for the conversion of L-Phenylalanine to *trans*-cinnamic acid. Cinnamic acid serves as a precursor for the biosynthesis of coumarins, isoflavanoids and lignin. These compounds play an important role in pest and disease resistance mechanism. Changes in

PAL activity accompanying fungal, viral and bacterial infections of plants have been reported.

Principle

PAL activity is determined spectrophotometrically by following the formation *trans*-cinnamic acid which exhibits an increase in absorbance at 290 nm (crude enzyme) 270 nm (purified enzyme).

Reagents

1. Borate buffer 0.2 M (pH 8.7).
2. L-Phenylalanine 0.1 M:
   Dissolve 165 mg L-Phenylalanine in 10 mL water and adjust to pH 8.7 with 0.1 N KOH.
3. 1 M Trichloroacetic acid:
   Dissolve 16.3 g in 100 mL water.
4. Enzyme extract:
   Homogenize 500 mg of the plant material in 5 mL of cold 25 mM borate-HCl buffer pH 8.8 containing 5 mM mercaptoethanol (0.4 mL/L). Centrifuge the homogenate at 12,000 × *g* for 20 min. Use the supernatant as enzyme source.

Procedure

1. Pipette out 0.5 mL borate, 0.2 mL enzyme solution and 1.3 mL water in a test tube.
2. Initiate the reaction by the addition of 1 mL L-Phenylalanine solution.
3. Incubate for 30–60 min at 32°C.
4. Stop the reaction by the addition of 0.5 mL of 1 M trichloroacetic acid.
5. Run a control in which add phenylalanine after trichloroacetic acid.
6. Measure the absorbance at 290 nm.
7. Prepare a standard graph for *trans*-cinnamic acid.

Calculation

Express the reaction rate as micromole *trans*-cinnamic acid formed per milligram protein per minute.

10.3.24 Glutamine Synthetase (L-Glutamate: Ammonia Ligase EC 6.3.1.2) (Pateman 1969)

Glutamine synthetase (GS) appears throughout the plant and animal kingdoms. It catalyzes the following reaction. This enzyme has high affinity for ammonia. High concentration of this enzyme in the nodules is probably related to a role in assimilating ammonia formed as a result of nitrogen fixation.

$$\text{L-Glutamate} + NH_3 + \text{ATP} \xrightarrow{Mn^{2+}} \text{L-Glutamine} + \text{ADP} + \text{Pi}$$

Principle

The activity of the enzyme is measured by estimating the production of inorganic phosphate. GS also catalyzes the ∂-glutamyl transfer reaction.

$$\text{Glutamine} + \text{hydroxylamine} \xrightarrow[\text{[Arsenate]}]{\text{ADP, Mn}^{2+}} \text{Glutamyl hydroxamate}$$

Hence, it can be assayed by measuring the production of ∂-glutamyl hydroxamate. The latter method is described below. The ∂-glutamyl hydroxamate is made to react with ferric chloride to produce brown colour in acidic medium. Activity in the transferase assay generally gives rates several times greater than those obtained in the synthetase assay (measuring Pi) and hence, this method is often used for measurement of GS in relatively crude preparations. When the activity is measured in the presence of Mn^{2+}, it represents total glutamine synthetase activity (adenylated and unadenylated forms). The biologically active unadenylated form may be measured by inhibiting the adenylated form by the addition of 60 mM Mg^{2+}.

Reagents

1. Prepare the following reagents in 20 mM Tris–HCl buffer (pH 8.0)
2. L-Glutamine 0.2 mM (700 mg/12 mL)
3. Sodium arsenate 20 mM (500 mg/10 mL) (Disodium Hydrogen Arsenate)
4. MnCl2 3 mM (83 mg/10 mL)
5. Hydroxylamine 50 mM (278 mg/10 mL)
6. Adenosine diphosphate 1 mM (40 mg/10 mL)
7. Ferric chloride reagent: Dissolve 10 g trichloroacetic acid and 8 g ferric chloride in 250 mL of 0.5 N hydrochloric acid
8. Enzyme extract: Extract 1 g plant material in 5 mL of 50 mM imidazole-acetate buffer (pH 7.8) containing 0.5 mM EDTA, 1 mM DTT, 2 mM $MnCl_2$ and 20% glycerol at 4°C. Centrifuge at 10,000 × *g* for 30 min. If purification is required, precipitate the enzyme with $(NH_4)_2SO_4$ at 60% saturation. Resuspend the precipitate in extraction buffer. Desalt over Sephadex G 25.

Procedure

1. Pipette out the reagents (mL) as mentioned in the order below:

Glutamine	2.0
Sodium arsenate	0.5
$MnCl_2$	0.3
Hydroxylamine	0.5
ADP	0.5
Enzyme extract	0.2

2. For blank, add 2 mL 20 mM Tris–HCl buffer instead of glutamine.

3. Incubate the reaction mixture for 30 min at 37°C.
4. Stop the reaction by adding 1 mL of ferric chloride reagent.
5. Measure the brown colour developed at 540 nm.
6. Prepare a range of standards containing 100–500 μg ∂-glutamyl hydroxamate in 4 mL buffer solution and develop the colour by adding 1 mL of ferric chloride reagent.

Calculation

Find out the amount of ∂-glutamyl hydroxamate formed in the reaction using the standard graph. Express the enzyme activity as nanomole ∂-glutamyl hydroxamate formed per minute per milligram protein.

10.3.25 Asparagine Synthetase (L-Aspartate; Ammonia Ligase EC 6.3.1.4) (Pateman 1969)

The Asparagine synthetase (AS) reaction involves the amidation of aspartate by either glutamine or ammonia. Asparagines synthetase is found in germinating cotyledons and legume plants.

1. $\text{Aspartate} + \text{glutamine} + \text{ATP} \xrightarrow{Mg^{2+}} \text{Asparagines} + \text{glutamate} + \text{AMP} + \text{PPi}$
2. $\text{Aspartate} + NH_3 + \text{ATP} \xrightarrow{Mg^{2+}} \text{Asparagines} + H_2O + \text{AMP} + \text{PPi}$

Principle

The enzyme is measured by substituting hydroxylamine for ammonia in the earlier given reaction.

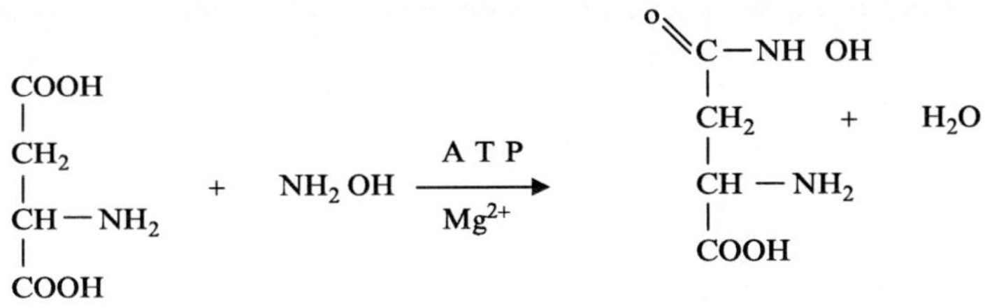

The amount of aspartyl hydroxamic acid formed is determined colorimetrically with ferric chloride reagent.

Reagents

1. Tris-NH_2OH-$MnCl_2$ solution:
   Weigh 2.42 g Tris base (0.25 M), 11.1 g hydroxylamine hydrochloride (2 M) and 47 mg $MnCl_2 \cdot 4H_2O$ and dissolve in about 40 mL water. Adjust to pH 6.4 by

the addition of 8 N KOH. Make up the final volume to 80 mL. Prepare this reagent fresh.
2. Potassium of Sodium ATP, 0.1 M:
   Dissolve 551 mg ATP disodium salt in 10 mL water and neutralize it. Prepare fresh solution.
3. L-Aspartic acid, 0.1 M, pH 6.4:
   Dissolve 133 mg of L-Aspartic acid in 10 mL water and adjust to pH 6.4.
4. Ferric chloride reagent:
   Mix 125 mL of 20% TCA (20 g TCA in 100 mL), 50 g $FeCl_3 \cdot 6H_2O$ and 28 mL of Conc. HCl and make up to a final volume of 500 mL with water. This is a stable reagent.
5. Enzyme extract
   Homogenize 1 g chilled plant materials in 10 mL 100 mM Tris–HCl, pH 8.5 buffer containing 15% (by volume) glycerol, 56 mM mercaptoethanol and 4 mM KCN. Pass the homogenate through four layers of cheese cloth and centrifuge the filtrate at 15,000 × *g* for 30 min.

Procedure

1. Pipette out 0.4 mL of the Tris–NH_2OH–$MnCl_2$ solution, 0.1 mL of ATP, 0.2 mL of enzyme. 0.2 mL of L-Aspartic acid and 0.1 mL distilled water in the order given.
2. Incubate at 37°C for 10 min.
3. Stop the reaction by the addition of 3 mL ferric chloride reagent.
4. Centrifuge, and remove the supernatant solution for measurement.
5. Measure the absorbance of the supernatant solution at 540 nm.
6. Use 1 mL water and 3 mL of ferric chloride reagent for zero adjustment.
7. Run a control in which aspartate is omitted from the reaction mixture.
8. Prepare a standard curve by taking 0–2.5 μmol of β-aspartyl hydroxamate.

Calculation

The enzyme activity is expressed as micromole of β-aspartyl hydroxamate formed per milligram protein per minute. The concentration of β-aspartyl hydroxamate can also be obtained by multiplying the assay absorbance value by a factor of 6.1, if the synthetic one is not available for standardization.

10.3.26 Lactate Dehydrogenase (L-Lactate: NAD Oxidoreductase, EC 1.1.1.27)

The activity of LDH of animal origin can be measured spectrophotometrically with pyruvate and NADH. Measurements of the activity by the colour reaction of the 2,4-dinitrophenylhydrazone formed from the unreacted pyruvate are difficult because the NADH also forms hydrazone which absorbs in the same region.

The lactate dehydrogenase in serum, which may originate from liver, heart, skeletal muscle, erythrocytes, etc., are not only different from one another, but themselves consist of several enzymatically active fractions which can be separated from each other. Such enzymes which differ in their protein structure and therefore in the optimum conditions for their action, but have the same specificity (in this case towards L-lactate) are termed "isoenzymes". A method is described which gives approximately optimum conditions for the measurement of LDH from heart and liver.

The most important characteristics for the assay of activity of the individual lactate dehydrogenases are their different substrate and pH optima. Measurements which are not carried out under optimum conditions naturally result in values for the activity which are too low. The optimum conditions for the measurements of the enzyme in serum after myocardial infarction, in liver damage, blood diseases and tumours have been established.

Principle

UV-Assay with Pyruvate and NADH

$$\text{Puruvate} + \text{NADH} + \text{H}^+ \underset{}{\overset{\text{LDH}}{\rightleftarrows}} \text{Lactate} + \text{NAD}^+$$

The LDH activity is determined by rate of oxidation of NADH.

Equipment

1. Spectrophotometer
2. Water bath (25°C)
3. Stopwatch

Reagents

1. Potassium dihydrogen phosphate, KH_2PO_4
2. Dipotassium hydrogen phosphate, K_2HPO_4
3. Sodium pyruvate
4. Reduced nicotinamide-adenine dinucleotide, NADH disodium salt, NADH-Na_2
5. Sodium bicarbonate

Preparation of Sodium

1. Phosphate/pyruvate (50 mM phosphate, pH 7.5; 0.63 mM pyruvate):
   Dissolve 700 mg. K_2HPO_4. 90 mg. KH_2PO_4 and 6.2 mg. Sodium pyruvate in distilled water and make up to 90 mL.
2. Reduced nicotinamide-adenine dinucleotide (ca. 11.3 mM β-NADH):
   Dissolve 14 mg. NADH-Na_2 and 15 mg. $NaHCO_3$ in 1.5 mL distilled water. Prepare all solution with fresh, doubly distilled water.

Procedure

Assay

Bring the phosphate/pyruvate solution and the serum to 25°C before the assay.

Pipette into cuvettes		Concentration in assay mixture
Phosphate/pyruvate solution (I)	3.00 mL	48 mM Buffer
NADH solution (II)	0.05 mL	0.6 mM Pyruvate
Serum	0.10 mL	0.18 mM NADH

Mix immediately, read extinction and start stopwatch. Repeat readings at exactly 1, 2 and 3 min and calculate the mean extinction change.

The ΔE/minute values at wavelength: 340 nm

Calculations

Wavelength: 340 nm
Volume activity = $5.064 \times \Delta E$/minute [U/I]

10.3.27 Glutamate–Oxaloacetate Transaminase

The GOT activity can be measured calorimetrically

Principle

After a fixed time, the 2,4-dinitrophenylhydrazone of the reaction product, oxaloacetate, formed according to equation is determined colorimetrically in alkaline solution. Some of the oxaloacetate spontaneously decarboxylates to pyruvate. The assay mixture therefore contains oxaloacetate, pyruvate and 2-oxoglutarate, all of which form 2,4-dinitrophenylhydrazones with absorption maxima at different wavelengths. During the reaction the concentration of 2-oxoglutarate decreases, while that of oxaloacetate and pyruvate increases. To keep low the contribution of the 2-oxoglutarate hydrazone to the colour, the measurements are made at wavelengths (about 500–550 nm) higher than the absorption maximum which allows the greatest differentiation between the extinctions of the three hydrazones.

Equipment

1. Photometer suitable for measurements at 500 and 550 nm
2. Water bath (37°C)
3. Stopwatch

10.3.27.1 Reagents

1. Potassium dihydrogen phosphate, KH_2PO_4
2. Dipotassium hydrogen phosphate, K_2HPO_4

3. L-Aspartic acid Sodium salt
4. 2-Oxoglutaric acid Free acid
5. Sodium pyruvate
6. 2,4-Dinitrophenylhydrazine
7. Hydrochloric acid, 1 N
8. Sodium hydroxide

Preparation of Solutions (for 50 determinations)

To prevent the growth of micro-organisms in the solutions, sterilize the containers. Prepare all solutions with fresh, doubly distilled water.

1. Buffer/substrate (0.1 M phosphate buffer, pH 7.4; 0.1 M aspartate ; 2 mM 2-oxoglutarate): Dissolve 1.50 g K_2HPO_4, 0.20 g. KH_2PO_4, 30 mg. 2-oxoglutaric acid, 1.57 g. L-aspartate, mono-sodium salt (or 1.32 g L-aspartic acid) in 70 mL distilled water, check pH with glass electrode, if necessary, adjust to pH 7.4 with NaOH (solution III) and dilute to 100 mL with distilled water.
2. Chromogen (1 mM 2,4-dinitrophenylhdrazine): Dissolve 20 mg. 2,4-dinitrophenylhydrazine in 1 N HCl and make up to 100 mL.
3. Sodium hydroxide (0.4 N): Dissolve 16 g. NaOH in distilled water and make up to 1,000 mL.
4. Pyruvate (2 mM): Dissolve 22.0 mg. Sodium pyruvate in distilled water and make up to 100 mL.

Stability of Solutions

NaOH (III) and chromogen solution (II) are stable indefinitely if well stoppered. Store buffer-substrate solution (I) and pyruvate solution (IV) at ca. 4°C. After addition of a few drops of chloroform to prevent bacterial growth, these solutions are stable for more than 2 months.

Procedure

Assay System

Wavelength: 500–550 nm; light path: 1 cm; 37°C; incubation volume: 1.20 mL; final volume: 12.2 mL; read against blank containing distilled water instead of sample.

Pipette into a test tube			Concentration in assay mixture
Buffer/substrate solution	(I)	1.00 mL	80 mM Phosphate 80 mM L-Aspartate 1.6 mM 2-Oxogulatarate
Serum		0.20 mL	
Mix and incubate for exactly 60 min at 37°C in water bath			
Chromogen solution	(II)	1.00 mL	0.45 mM
Mix and allow to stand for 20 min at room temperature			
NaOH	(III)	10.00 mL	0.33 mM
Mix and after 5 min read extinction against blank			

With results over 70 U/l serum dilute 1:10 with physiological saline and repeat measurements.

Calculation

Standard Curve and Table of Values:

Pipette successively into test tubes:

Test tube No.	Sodium pyruvate solution (IV) (mL)	Buffer/substrate solution (I) (mL)
1	0.00	1.00
2	0.05	0.95
3	0.10	0.90
4	0.15	0.85
5	0.20	0.80
6	0.25	0.75

Pipette into each tube 0.2 mL distilled water and 1.0 mL chromogen solution (II), mix and after 20 min add 10 mL NaOH (III). Mix and after 5 min read the extinctions of tubes 2–6 against tube 1. Plot the extinctions (ordinate) against mU/mL (abscissa).

By comparison with the UV method the following relationship is obtained

Tube No. 2 = 10.5 U/l serum
Tube No. 3 = 21 U/l serum
Tube No. 4 = 32 U/l serum
Tube No. 5 = 48 U/l serum
Tube No. 6 = 70 U/l serum

By direct comparison of the two methods the following table was constructed for measurements at 546 nm:

Extinction	U/l	Extinction	U/l
0.020	4	0.160	35
0.040	8	0.180	41
0.060	12	0.200	50
0.080	6	0.220	59
0.100	20	0.240	70
0.120	24	0.260	83
0.140	30		

Evaluation

When the measurements are made at 546 nm, obtain the U/l from the earlier mentioned table.

When a standard curve has been prepared with pyruvate solution (IV), convert the extinctions obtained for serum into U/l.

If control sera of different activities have been used, draw a standard curve (abscissa: U/I; ordinate: extinctions) and convert the extinctions obtained for serum to U/I with the aid of this curve.

10.3.28 Glutamate–Pyruvate Transaminase

Glutamate–Pyruvate Transaminase (GPT) activity can be measured by assay with 2,4-dinitrophenylhydrazine.

Principle

After a fixed time the pyruvate formed from L-alanine and 2-oxoglutarate is determined colorimetrically by treating the 2,4-dinitrophenylhydrazone with alkali. The residual 2-oxoglutarate also forms a hydrazone, but its absorption maximum in alkaline solution is different from that of the pyruvate hydrazone. To keep the contribution of the 2-oxoglutarate hydrazone colour to the extinction low, the measurements are made between 500 and 550 nm instead of at the absorption maximum of the pyruvate hydrazone.

Equipment

1. Spectrophotometer for measurements between 500 and 550 nm
2. Water bath (37°C)
3. Stopwatch

Reagent

1. Potassium dihydrogen phosphate, KH_2PO_4
2. Dipotassium hydrogen phosphate, K_2HPO_4
3. DL-Alanine
4. 2-Oxoglutaric acid
5. Sodium pyruvate
6. 2,4-Dinitrophenylhydrazine
7. Hydrochloric acid, 1 N
8. Sodium hydroxide

Preparation of Solutions (for ca. 50 determination)

Prepare all solutions with fresh, doubly distilled water.

1. Buffer/substrate solution (0.1 M phosphate buffer, pH 7.4; 0.2 M DL-alanine; 2 mm 2-oxoglutarate):
   Dissolve 1.50 g. K_2HPO_4, 0.20 g. KH_2PO_4, 0.030 g. 2-oxoglutaric acid and 1.78 g. DL-alanine in distilled water and make up to 100 mL. Check the pH with a pH meter, if necessary, adjust to pH 7.4 with NaOH (solution III).

2. Chromogen solution (1 mm 2,4-dinitrophenylydrazine):
   Dissolve 20 mg. 2,4-dinitrophenylhydrazone in 1 N HCl and make up to 100 mL.
3. Sodium hydroxide (0.4 N): Dissolve 16 g. NaOH in distilled water and make up to 1,000 mL.
4. Pyruvate solution (2 mM): Dissolve 22.0 mg. Sodium pyruvate in distilled water and make up to 100 mL.

Procedure

Assay system

Wavelength: 500–550 nm; light path: 1 cm; incubation temperature: 37°C; incubation volume: 1.20 mL; final volume: 12.2 mL; room temperature; read against a blank containing distilled water instead of serum.

Pipette into a test tube			Concentration in assay mixture
Buffer/substrate solution	(I)	1.00 mL	80 mM Phosphate 80 mM L-Aspartate 1.6 mM 2-Oxogulatarate
Serum		0.20 mL	
Mix and incubate for exactly 60 min at 37°C in water bath			
Chromogen solution	(II)	1.00 mL	0.45 mM
Mix and allow to stand for 20 min at room temperature			
NaOH	(III)	10.00 mL	0.33 mM
Mix and after 5 min read extinction against blank			

With results over 70 U/I serum dilute 1:10 with physiological saline and repeat measurements.

Calculations

Standard Curve and Table of Values

Pipette into test tubes:

Test tube No.	Pyruvate solution (IV) (mL)	Buffer/substrate solution (I) (mL)
1	0.0	1.0
2	0.1	0.9
3	0.2	0.8
4	0.3	0.7
5	0.4	0.6
6	0.5	0.5

Pipette into each tube 0.2 mL distilled water and 1.0 mL chromogen solution (II), mix and after 20 min at room temperature add 10.0 mL NaOH solution (III). Mix and after 5 min read the extinctions of test tubes 2–6 against tube 1. Plot extinctions (ordinate) against GPT units (abscissa). The following relationship was found by direct comparison with the UV method (lactate dehydrogenase as indicator enzyme).

Test tube No. 2 = 13 U/l serum
Test tube No. 3 = 28 U/l serum
Test tube No. 4 = 48 U/l serum
Test tube No. 5 = 78 U/l serum
Test tube No. 6 = 102 U/l serum

By direct comparison of the two methods the following table was constructed for measurements at 546 nm:

Extinction	U/l	Extinction	U/l
0.025	2.5	0.225	35
0.050	5.5	0.250	41
0.075	9.0	0.275	47
0.100	12	0.300	54
0.125	17	0.325	61
0.150	21	0.350	70
0.175	25	0.375	80
0.200	30		

If the measurements are made at 546 nm read off the U/l from the table. If the standard curve has been prepared with pyruvate solution (IV) convert the extinction obtained for serum to U/l with the aid of this standard curve. If control sera of different activity have been used, plot the standard curve (abscissa: U/l, ordinate) extinction and convert the extinctions obtained for the serum sample to U/l with the aid of this standard curve.

10.3.29 Phosphatases

Phosphatases catalyze the hydrolytic cleavage of phosphoric acid esters. They are designated either "acid" or "alkaline" phosphatases according to their pH optima.

Alkaline phosphatases (Orthophosphoric-monoester phosphohydrolase, alkaline optimum, EC 3.1.3.1) occur in practically all animal and human tissues. Bile and osteoblasts have particularly high activity. Several isoenzymes can be demonstrated in serum and tissue extracts. Raised values in human serum are nearly always due to diseases accompanied by increased osteoblast activity or involvement of the liver or bile ducts.

Acid phosphatase (Orthophosphoric-monoester phosphohydrolase, acid optimum, EC 3.1.3.2) are also found in nearly all human and animals cells. High activity is found in erythrocytes and particularly in prostatic tissue. By means of electrophoresis three active fractions have been separated in serum and up to 17 fractions in various tissue extracts.

10.3.29.1 Acid and Alkaline Phosphatase in Serum (Two-Point Method)

Principle

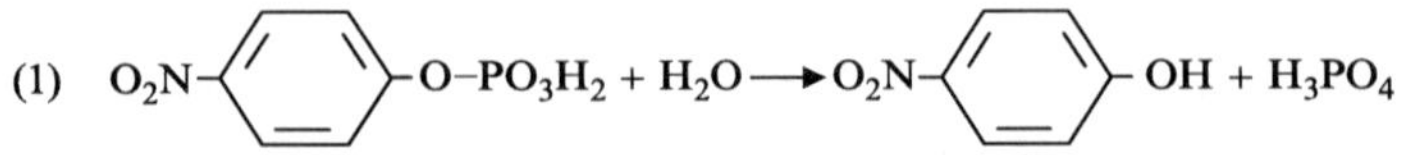

The amount of 4-nitrophenol liberated/unit time, as determined in an alkaline solution at 400–420 nm, is a measure of the phosphatase activity.

The reaction is stopped by the addition of NaOH. In the determination of acid phosphatase, the alkali results in the formation of a yellow colour due to the 4-nitrophenol liberation. Acid phosphatase of prostatic origin can be distinguished by the enzyme in the presence and absence of tartrate.

Equipment

1. Spectrophotometer (400–420 nm)
2. Water bath 25°C
3. Stopwatch

Reagents

1. Citric acid, $C_6H_8O_7 \cdot H_2O$
2. Sodium citrate, $C_6H_5Na_3O_7 \cdot 5H_2O$
3. Sodium tartrate, $C_4H_4Na_2O_6 \cdot 2H_2O$, A.R
4. Diethanolamine, $C_4H_{11}NO_2$
5. 4-Nitrophenylphosphate
   Disodium salt, $C_6H_4NNa_2O_6P \cdot 6H_2O$, A.R
6. Sodium hydroxide, 0.05 and 0.1 N
7. Hydrochloric acid, 0.1 N

Preparation of Solutions

1. Acid buffer/substrate solution (50 mM citrate buffer, pH 4.8; 5.5 mM 4-nitrophenylphosphate):
   Dissolve 0.41 g citric acid + 1.125 g sodium citrate + 0.203 g 4-nitrophenylphosphate in doubly distilled water to 100 mL. Check pH with a glass electrode.
2. Alkaline buffer/substrate solution (0.1 M diethanolamine, pH 9.8; 1.25 mM 4-nitrophenylhosphate):
   Dissolve 1.052 g. diethanolamine + 8.0 mL 0.1 N HCl + 46.4 mg. 4-nitrophenylphosphate in 85 mL doubly distilled water, adjust to pH 9.8 by addition of 0.1 N HCl (glass electrode) and dilute to 100 mL with distilled water.
3. Tartrate (0.4 M):
   Dissolve 9.2 g. sodium tartrate in distilled water and make up to 100 mL.

Procedure

Collection, Treatment and Stability of Sample

Collection of sample: Allow blood to flow from the vein through a cannula (No. 1 or larger) into a centrifuge tube. Centrifuge for 10 min at ca. 3,000 × *g* to obtain serum. Use only non-haemolysed, fresh serum, especially for the assay of acid phosphatases.

Stability of enzyme in sample: The stability of acid phosphatase is dependent on the pH of the serum. Although the enzyme is stable for 2 weeks in serum at pH 5–6 at room temperature, the activity decreases to 50% after 7 days in non-acidified serum. Addition of 5 mg sodium hydrogen sulphate ($NaHSO_4{\cdot}H_2O$) to 1 mL serum stabilizes it sufficiently for dispatch by post.

The alkaline phosphatases in serum are stable for at least 1 week at room temperature providing that bacterial contamination is excluded.

Assay System

Wavelength: H 405 nm (400–420 nm); light path: 1 cm; 25°C. Incubation volume: 1.25; final volume: 3.2 mL. Read against blank (without tartrate) to which the sample is add after the NaOH. Bring the buffer/substrate solution to 25°C before the start of the assay.

10.3.29.2 Acid Phosphatases

Pipette into 12 mL test tubes		Sample A	Sample B	Concentration in assay mixture
Buffer/substrate solution, pH 4.8	(I)	1.0 mL	1.0 mL	41.6 mM Citric 4.58 mM 4-Nitrophenylphosphate
Tartrate solution	(III)	–	0.05 mL	16 mm Tartrate
Serum		0.2 mL	0.2 mL	
Mix and incubate for exactly 30 min in a water bath				
0.1 N NaOH		2.0 mL	2.0 mL	0.062 N NaOH
Measure extinctions. The increase in extinction ΔE over the blank is used for the calculations. The yellow colour is stable for several hours.				

Calculations

The extinction coefficient of 4-nitrophenol in alkaline solution is 18.5 cm^2. Therefore under the above conditions:

$$\text{Volume activity} = \frac{\Delta E \times 3.2 \times 1{,}000}{30 \times 18.5 \times 0.2} = \Delta E \times 28.8(\text{U/l})$$

The activity of prostatic phosphatase is obtained by the difference between the activities of sample A and B (the total acid phosphatase activity is measured in A). If prostatic phosphatase only is to be determined, prepare reaction mixtures A and B and read A against B.

The volume of the reaction mixture in B in 50 μL is larger than in A, but this difference can be neglected in the calculations. If the measurements cannot be made at 405 nm, it is necessary to prepare a standard curve.

10.3.29.3 Alkaline Phosphatases

For the estimation of alkaline phosphatase activity, read against blank to which the sample is added after the NaOH. Incubation volume: 2.05 mL; final volume: 12.05 mL

Pipette into test tubes			Concentration in assay mixture
Buffer/substrate solution, pH 9.8	(II)	2.0 mL	97.5 mM diethanolamine 1.21 mM 4-nitrophenylphosphate
Serum		0.05 mL	
Mix and incubate for exactly 30 min			
0.05 N NaOH		10.0 mL	0.041 N NaOH
Read extinction. The increase in extinction ΔE above the bank is used for the calculations. The yellow colour is stable for several hours			

With extinctions above 0.800, dilute 0.1 mL serum with 0.9 mL physiological NaCl solution and assay again. Make the calculations with the help of standard curve.

Calculation

$$\text{Volume activity} = \frac{\Delta E \times 12.05 \times 1,000}{30 \times 18.5 \times 0.05} = \Delta E \times 434\ [\text{U/l}]$$

Normal Values

Total acid phosphatase in serum: up to 5.5 U/l (25°C). Alkaline phosphatase in serum: 18–63 U/l (25°C) (normal adults 18–85 years). The upper normal limit of alkaline phosphatase in serum of children is 150 U/l (25°C).

Chapter 11
Isoenzyme Analysis

The term isozyme is restricted to those forms of an enzyme with similar enzymatic activity occurring within a single species, as a result of the presence of more than one structural gene.

As per the definition, the following categories are regarded as isozymes.

1. Genetically – independent proteins arising from the presence of multiple gene loci.
2. Enzyme variants from occurrence of allelic genes at a particular gene locus – these isozymes are called allelozymes.
3. Heteropolymers (Non-covalent hybrid molecules of two or more different polypeptide chains).

The enzymatic activities of the members of a family of enzymes are similar, but by no means identical in all cases. Differences are observed in pH optima or in sensitivity to inhibitors or to denaturing agents, in their aminoacid sequence, substrate affinity, V_{max} and/or regulatory properties. The different properties between isozymes determined by multiple gene loci presumably account for their emergence during evolution and their functional significance in present species. Therefore, from their discovery isozymes have been linked with cell differentiation and development, suggesting a progressive adaptation to changing environment. Yet in many cases, no clear reasons for the existence of isozymes have been found. Since isozymes are tissue-, stage- and species-specific they have been described as efficient genetic markers.

Isozymes are generally made up of a number of subunits, and it is the varying combination of the subunits which gives rise to isozymes. An isozyme may be constituted by the association of similar subunits or dissimilar subunits. They are an essential feature of the biochemical organization of living things. A number of major biological problems such as evolution of population, the transformation, the regulation of gene expression and metabolic regulation in differentiated tissues are understood in the light of isozymes. They may arise due to heterogeneity in proteins, changes in conformation, genetic mechanisms, etc.

R. Katoch, *Analytical Techniques in Biochemistry and Molecular Biology*,
DOI 10.1007/978-1-4419-9785-2_11, © Springer Science+Business Media, LLC 2011

11.1 General Methodology for Isozyme Analysis

For isozymic analysis, the sample extract is electrophoresed in starch or polyacrylamide buffered (non-denaturing) slab gels at a low temperature (4–8°C). Each lane should be loaded with equal amount of proteins after normalizing the protein content in extract in as small volume as possible (25–50 μl). After electrophoresis the gel is incubated in a solution containing all the necessary components for enzyme reaction. The colored reacton products stain the gel where the enzymes are located.

1. Conduct electrophoresis in polyacrylamide gels (with no SDS) at low temperature as described under Polyacrylamide Disc gel electrophoresis.
2. Immediately after electrophoresis, incubate the gel in the substrate solution (s). The zones where the enzymes are located in the gel are visualized due to the appearance of colored reaction products. After sufficient incubation period, stop the reaction by adding appropriate stop solution and photograph the zymogram. Otherwise, the relative position of each visualized band in the gel may be drawn schematically for easy reference.
3. Enzyme extraction and stain for various isoenzymes are described subsequently.

11.2 Isozyme Analysis of Esterase (Smith et al. 1970)

Enzyme extract: Homogenize the sample material in fivefold volume of 10 mM sodium phosphate buffer (pH 9.5), 1mM EDTA Na^{2++} and 1 mM 2-mercaptoethanol. Centrifuge the homogenate at 10,000 × *g* for 10 min and use the supernatant as enzyme source. All operations are at 0–4°C. Incubate the gel in a solution given below at 37°C for 20–30 min, in dark.

Sodium dihydrogen phosphate	2.8 g
Disodium hydrogen phosphate	1.1 g
Fast blue RR salt	0.2 g
Alpha-naphthyl acetate	0.03 g
Water for volume	200 ml

Stop the enzyme reaction by adding a mixture of methanol: water: acetic acid: ethyl alcohol in the ratio 10:10:2:1.

11.3 Peroxidase (Reddy and Gasber 1971)

Enzyme extract: As for the assay of enzyme Peroxidase.

Conduct the electrophoresis as described for the other experiments.

After gel-electrophoresis, incubate the gel in the following solutions.

Benzidine	2.08 g
Acetic acid	18 ml
Hydrogen peroxide (3%)	100 ml
Water	80 ml

Bright blue colored bands appear in gel. When the bands are stained sufficiently, arrest the reaction by immersing the gel into a large volume of 0.67% sodium hydroxide or 7% acetic acid solution for 10 min.

11.4 Polyphenol Oxidase (PPO) (Jayaraman et al. 1987)

Enzyme extract: The PPO is extracted by homogenizing the material in 0.01 M potassium phosphate buffer (pH 7.0) containing 1% non-ionic detergent (Tween 80) at 0°C for 15 min. Centrifuge the homogenate at 20,000 × *g* for 15 min at 0°C in a refrigerated centrifuge. Use the supernatant as enzyme source.

Equilibrate the gel for 30 min in 0.1% p-phenylenediamine in 0.1 M potassium phosphate buffer (pH 7.0) followed by 10 mm catechol in the same buffer.

11.5 Acid Phosphatase (De and Roy 1984)

Enzyme extract: See for the preparation of extract for the enzyme Phosphatases.

After electrophoresis wash the gel 3–4 times in 0.1 M acetate buffer (pH 5) by changing the buffer every 15 min (to lower the pH of the gel to 5). Incubate the gel at 37°C for 2–3 h in the following solution.

1 – Napthyl phosphate	0.05 g
Fast blue RR	0.05 g
Sodium chloride	1.0 g
10% magnesium chloride	0.5 ml
0.1 M Acetate buffer (ph 5.0) to	50 ml

Reddish-brown bands visible in the gel. Fix the gel using 50% ethanol.

11.6 Glutamine Synthetase (Atkins et al. 1984)

Enzyme extract: As used for the assay of enzyme Glutamine synthetase.

Mix the following solutions:

		ml
Tris Sigma 7–9 (Sigma T. 1328)	0.1 M	36
Imidazole-HCl	1 M	10
Hydroxylamine HCl	0.8 M	10
Magnesium	3 M	0.4
Monosodium glutamate	0.8 M	21

Adjust for pH 7.7 with NaOH.

Incubate the gel at 32°C for 10 min.

Add 14.5 ml of 0.2 M ATP solution to the above mixture and continue the incubation for 50 min. Terminate the reaction by adding 80 ml of mixture consisting of ferric chloride hexahydrate (55 g/L), trichloroacetic acid (20 g/L) and Conc. HCI (21 ml/L).

The isozymes appear as purple bands against yellow background and are not permanent unless stored at 4°C.

11.7 Phosphoenol Pyruvate Carboxylase (Francis and Gnanam 1979)

Enzyme extract: Homogenize the plant tissue in ice-cold 100 mM Tris-HCI (pH 7.8), 10 mM magnesium chloride, 10 mM sodium bicarbonate, 2% polyvinyl pyrrolidone, 1 mM EDTA and 15 mM 2-mercaptoethanol. Centrifuge at 30,000 × *g* for 45 min and use the supernatant.

Conduct the electrophoretic run as described. After electrophoresis incubate the gel in a solution of 100 mM Tris-HCI (pH 8.0), 10 mm magnesium chloride, 200 mM calcium chloride, 10 mM sodium bicarbonate and 5 mM phosphoenol pyruvate for 30 min at 40°C. During the incubation white bands of precipitate develop indicating the position of enzymes on the gel. The phosphate released by the enzyme action is precipitated as white calcium phosphate in the gel and is stable for many hours.

11.8 Glutamate Dehydrogenase (Nash and Davies 1975)

Enzyme extract: Same as prepared for the assay of enzyme Glutamate dehydrogenase.

Conduct the electrophoretic run as described. After electrophoresis incubate the gels in a solution containing 8 mM monosodium glutamate, 0.2 mM methylthiazol tetrazolium (MTT) or nitrobluetetrazolium (NBT), 0.1 mM phenazine methosulphate, 0.2 mM NAD, 1.2 mM sodium cyanide in 200 mM Tris-HCI buffer pH 7.5. The staining solution for glutamate dehydrogenase (NADP-dependant) contains 0.2 M NADP instead of NAD.

11.9 Indolacetic Acid Oxidase

Enzyme extract: As prepared for the assay of the enzyme.

Incubation of the gel: Incubate the gel in staining mixture of 1 mg potassium indoleacetate, 0.08 mg 2,3,6-trichlorophenol and 2 mg Fast blue BB per ml in 60 mM phosphate buffer (pH 6.0) at 30°C for sufficient time (preferably overnight).

11.10 Malate Dehydrogenase (Honold et al. 1966)

Enzyme extract: Prepare the extract as described for the assay of the enzyme.

Staining of the gel after electrophoresis: First pre-incubate the gels for 15–20 min in 200 mM Tris-HCl (pH 7.5) buffer. Then transfer the gel to a solution containing 16 mM L-malate, 0.2 mM NAD, 0.25 mM methylthiazol tetrazolium, 0.8 mM phenazine methosulfate, 4 mM $MgCl_2$ and 1.2 mM sodium cyanide in 200 mM Tris-HCl (pH 7.5) buffer. To stain MDH (NADP-dependant), use NADP instead of NAD.

Chapter 12
Chromatographic Separations

12.1 Separation and Identification of Amino Acids by Descending Paper Chromatography

12.1.1 Principle

Amino acids in a given mixture are separated on the basis of differences in their solubilities and hence differential partitioning coefficients in a binary solvent system. The amino acids with higher solubilities in stationary phase move slowly as compared to those with higher solubilities in the mobile phase. The separated amino acids are detected by spraying the air dried chromatogram with ninhydrin reagent. All amino acids give purple or bluish purple colour on reaction with ninhydrin except proline and hydroxyproline, which give a yellow coloured product. The reactions leading to the formation of purple complexes are given below (Fig. 12.1).

12.1.2 Reagents

1. Whatman No. 1 filter paper sheet.
2. Micropipette.
3. Oven (105°C)
4. Drier
5. Sprayer
6. Chromatographic chamber saturated with water vapours.
7. Developing solvent: Butanol, acetic acid and water in the ratio of 4:1:5 in a separting funnel and mix it thoroughly. Allow the phases to separate out completely. Use the lower aqueous phase for saturating the chamber. The upper organic phase is used as mobile phase.

R. Katoch, *Analytical Techniques in Biochemistry and Molecular Biology*,
DOI 10.1007/978-1-4419-9785-2_12, © Springer Science+Business Media, LLC 2011

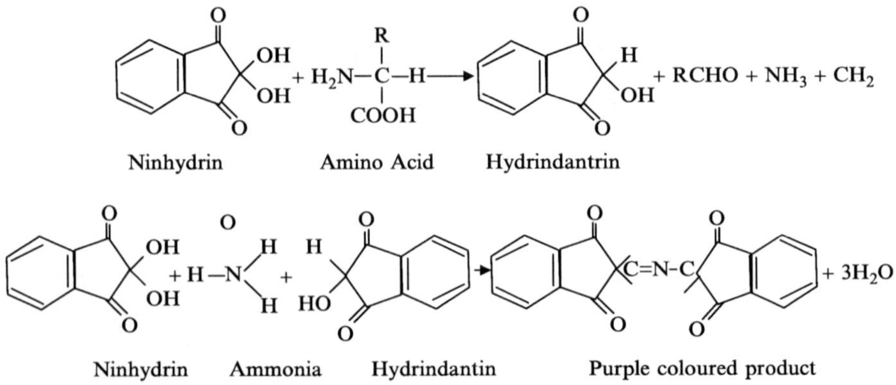

Fig. 12.1 Reaction for ninhydrin test

8. Ninhydrin spray reagent: Prepare fresh by dissolving 0.2 g ninhydrin in 100 ml acetone.
9. Standard amino acids: Prepare solutions of authentic samples of amino acids such as methionine, tryptophan, alanine, glycine, threonine, etc. (1 mg/ml of 10% *iso*-propanol).
10. A sample containing mixture of unknown amino acids.

12.1.3 Procedure

1. Take Whatman No. 1 filter paper and lay it on a rough filter paper. Throughout the experiment care should be taken not to handle the filter paper with naked hands and for this purpose either gloves should be used or it should be handled with the help of folded piece of rough filter paper.
2. Fold the Whatman No. 1 filter paper about 2.0–2 cm from one edge. Reverse the paper and again fold it 2 cm further down from the first fold.
3. Draw a line across the filter paper with a lead pencil at a distance of about 2 cm from the second fold. Put circular marks along this line at a distance of 2.5 cm from each other.
4. With the help of a micropipette or micro-syringe apply 20 μl of solution of each standard amino acid on a separate mark. Also apply spot of the sample or mixture to be analyzed, preferably on the mark at centre of this base line. The size of the spot should be as small as possible so that the developed spots are compact and do not overlap.
5. Hang the filter paper in a chromatographic chamber which has previously been saturated with aqueous phase of the solvent system. This is done by keeping Petri plates containing the aqueous phase at the bottom of the chamber. The paper is hung from the trough/tray and a glass rod is kept to hold it in place. Care should be taken to ensure that the base line should not get submerged

when the mobile phase is added to the trough otherwise the spotted material would get dissolved in the solvent.

6. Close the chamber firmly so that it is airtight. Allow sufficient time for cellulose fibres of the paper to get fully hydrated.
7. Pour the mobile phase through the holes provided on the lid of the chamber into the trough. Replace the rubber bungs in the hole and allow the mobile phase to run down the paper till the solvent front reaches about 5 cm from the opposite edge.
8. Remove the paper and mark the solvent front with lead pencil and let it dry at room temperature.
9. Spray the filter paper (chromatogram) with ninhydrin reagent and after drying it at room temperature, transfer it to an oven at 105°C for 5–10 min.
10. Blue or purple coloured spots would appear on the paper. Mark the boundary of each spot with lead pencil.
11. Measure the distance between the centre of the spots and also the distance of the solvent front from the base line.
12. Calculate the R_f value of standard amino acids as well as those in the given mixture or sample.

$$R_f = \frac{\text{Distance traveled by unknown amino acid}}{\text{Distance traveled by the solvent system}}.$$

13. Identify the amino acids in the mixture or sample by comparing their R_f values with those of the reference standards.

12.1.4 *Observations and Calculations*

Distance travelled by the solvent front from base line = x cm
Distance travelled by glycine from base line = a cm
Distance travelled by alanine from base line = b cm
Distance travelled by threonine from base line = c cm
Distance travelled by methionine from base line = d cm
Distance travelled by spot no.1 in sample from base line = a cm

The sample contains glycine since R_f value of spot no. 1 is identical to that of glycine standard.

12.1.5 *Note*

- The chromatography should be carried out in temperature controlled room because any fluctuation in the temperature would cause the uneven flow of the solvent and may alter the R_f value.

- The paper should not be touched with naked hands because sweat on hands contains significant amount of amino acids.
- The spots of the applied sample should be as compact as possible. Larger the spot, poorer will be the resolution.
- At the time of fixing paper in chromatography chamber, it should be ensured that the base line on which the sample has been applied does not dip into the solvent otherwise the sample might get washed away in the solvent.
- Allow sufficient time for filter paper to absorb sufficient water (which will act as a stationary phase) before pouring the solvent into the tank. Inadequate conditioning or equilibration will result in improper or poor quality resolution.
- Solvent front should advance in a straight line and should not be zig-zag or sloping but should be parallel to the base line.
- Dry the paper thoroughly before spraying with the detection reagent. Wet paper may interfere with the appearance of evenly shaped compact spots.

12.2 Separation and Identification of Amino Acids by Ascending Paper Chromatography

12.2.1 Reagents and Materials

Same as given in above experiment except that cylindrical chromatography chambers are needed for this experiment.

12.2.2 Procedure

1. Take Whatman No. 1 filter paper sheet of appropriate size so that it can be rolled into a cylinder and can be accommodated in the cylindrical chromatographic jar.
2. Draw a base line 2 cm from one of the breadthwise edge of the paper. Put small circular marks along the base line in such a way that the distance from the edge of the paper and the first spot and the distance between the adjacent spots is not less than 2.5 cm.
3. Apply 20 μl aliquots of the standard amino acids and of the sample on different spots. Diameter of the spotted material should be as small as possible and, if required, the applied solution may be dried prior to loading additional volume.
4. Roll the paper into a cylinder; fasten its edges with a paper clip. Pour sufficient volume of the mobile phase into the chromatographic jar which has been earlier saturated with water vapours by lining the tank with filter paper saturated with aqueous phase of the solvent system.

5. Gently place the rolled filter paper upright in the jar ensuring that it does not touch the sides of the chamber and at the same time taking care that the base line where the spots have been applied does not dip into the solvent.
6. Close the tank with n airtight lid or a glass plate to which sufficient amount of silicon grease has been applied.
7. Leave the set up undisturbed and allow the solvent to move up fill it reaches about 5 cm from the upper edge.
8. Remove the chromatogram from the chamber and air dry it.
9. Spray the paper with ninhydrin reagent and let it dry again at room temperature prior to transferring it to an oven at 105°C for 5–10 min. Locate the position of amino acids from the bluish or purple coloured spots on the chromatogram.
10. Calculate the R_f values of the standard amino acids and those is the sample or mixture as described in above experiment.
11. Identify the amino acids in the mixture or sample by comparing R_f values with those of applied standard amino acids.

12.3 Separation and Identification of the Amino Acids by Two-Dimensional Paper Chromatography

12.3.1 *Principle*

Amino acids having very close R_f values in a particular solvent system may appear as a single or overlapping spots in a single dimensional chromatography and may be mistaken as one component. They can be separated into individual components by developing the chromatogram again in a direction perpendicular to the first run in a second solvent system in which they have different R_f values. The main limitation of this method is that only one spot either of the sample or of a standard amino acid can be applied on each filter paper sheet necessitating running of a large number of chromatograms for the standard amino acids.

12.3.2 *Reagents*

1. As in previous experiment with an additional chromatographic chamber for the second solvent system.
2. Solvent system No.2: Phenol: water (80:20, *w*/*v*) is used as second solvent system. Add 125 ml of water to 500 g to phenol. Add a few drops of ammonia (0.88%) to this mixture just before use (CAUTION-Phenol is corrosive and can cause burns on the skin).
3. Standard amino acids: Prepare 1% solution of standard amino acids such as aspartate, glycine, serine, arginine in 10% *iso*-propanol (*u*/*v*).

12.3.3 Procedure

1. Lay the chromatographic paper sheet flat on the rough filter paper using gloves.
2. Draw a base line 5 cm from one of the edges of the paper.
3. Draw another line perpendicular to the first line again 5 cm away from the adjacent edge.
4. Apply 60 μl of the sample solution or given mixture containing unknown amino acids at the point of intersection of these two lines. The sample should be applied in small volumes at a time with the help of a micropipette with intermittent drying to ensure that the zone of applied solution is as small as possible.
5. Repeat the same procedure for a mixture of three standard amino acids using a separate chromatographic sheet for each mixture. The composition of mixture of standard amino acids should be such that each amino acid is present in at least two different mixtures so that its identity can be established.
6. Hang the paper in the chromatographic tank whose interior has previously been saturated with aqueous phase of solvent system No.1 (Butanol: Acetic acid: Water mixture is 4:1:5).
7. After allowing an equilibration period of half an hour, pour the solvent No.1 into the trough of the chamber and let it run till t\it is about 10 cm from the opposite edge of the paper.
8. Take the paper out, air dry it, turn it at 90°C angle and now develop the paper in the second chromatographic chamber using the solvent system No.2 (Phenol: Water).
9. Remove it when the solvent has travelled upto about 10 cm from the opposite end.
10. Dry it at room temperature and spray it with ninhydrin reagent. After air drying it, keep the chromatogram in an oven at 105°C for 10 min. Mark the blue and/or purple coloured zones which appear on the paper.
11. Calculate the R_f value of the standard amino acids and those in given mixture as given in previous experiment in both the solvents. From these values identify the amino acids in the given mixture.

12.4 Identification of Lipids by TLC

12.4.1 Principle

Lipids generally exists as lipoprotein complexes and these need to be isolated. Lipids, being soluble in non-polar organic solvents and proteins being soluble in polar aqueous solvents, the efficient lipid extraction can be achieved only when an aqueous solvent like ethanol or methanol is included in the non-polar organic solvent like chloroform and diethyl ether. This would help in breaking the lipoprotein complexes. Extracted lipid components can be separated on TLC base on their differential mobility along the porous stationary phase such as silica gel and these can be located by spraying the plates with either 2',7'-dichlorofluorescein or 50% sulfuric acid.

12.4.2 Reagents and Materials

1. TLC tank
2. Oven set at 110°C
3. Glass plates (20 × 20 cm) for TLC
4. Ultraviolet lamp
5. Solvent system: Petroleum ether (b.p. 60–70°C) or hexane: diethyl ether: glacial acetic acid (80:20:1, *u/v*)
6. Lipid standards: Various lipids such as cholesterol acetate, vitamin A palmitate, triacyl glycerol (e.g., trioleate, tripalmitate, tristearate)
7. Sulfuric acid (50%, *u/v*)
8. 2',7'-dichlorofluorescein: Prepare 0.2% solution of 2',7'-dichlorofluorescein in 95% *(u/v)* ethanol

12.4.3 Procedure

1. *Extraction of lipids from sample*: Grind 1 g of the tissue in the extraction solvent [either ethyl ether: ethanol (3:1) or chloroform: methanol (2:1)] in pestle and mortar. Transfer the homogenate to a separating funnel. Shake the contents vigorously and allow it to stand till the two phases have completely separated. Drain out the lower organic layer which contains the lipids. Evaporate the solvent under organic layer which contains the lipids. Evaporate the solvent under vacuum and keep the concentrated lipid extract protected from light under N_2 almosphere.
2. Prepare the TLC plates using Silica Gel G as the adsorbent as described in previous experiment.
3. Activate the TLC plates at 110°C for 30 min, cool them in a desiccator and spot the lipid samples, standards as well as unknown sample.
4. Develop the plates in the solvent system consisting of petroleum ether or hexane: ethyl ether: glacial acetic acid (80:20:1) till the solvent has travelled upto 1 cm from the opposite side of the plate.
5. Remove the plate and allow it to air dry.
6. Locate the lipid spots by either of the following methods:

   (i) Spray the plate with 2',7'-dichlorofluorescein and examine it under UV light. Lipids show up as green fluorescent regions against the dark background.

(ii) Spray the plate carefully with 50% H_2SO_4 and heat it in an oven at 110°C for 10 min. Areas containing lipids gel charred and appear as black spots.

7. Calculate the R_f value of the lipid components in the sample and identify them by comparing their R_f values with lipid standards.

12.5 Separation of Pigments by Adsorption Column Chromatography

12.5.1 Principle

Different pigments get adsorbed to alumina to different extents. They can be selectively desorbed by using mobile phase of increasing polarity in a stepwise manner.

12.5.2 Materials and Reagents

1. Leaves/flowers
2. Pestle and mortar
3. Glass column
4. Whatman No.1 filter paper
5. Alumina
6. Benzene: methanol (2:1)
7. Sodium sulphate (anhydrous)
8. Acetone

12.5.3 Procedure

(a) *Preparation of extract*:

1. Homogenize 5 g leaves or flowers in a pestle and mortar, using sand as an abrasive in 20 ml of benzene: methanol (2:1) adding a small amount of this extractant at a time.
2. Filter the extract through Whatman No.1 filter paper and transfer the filtrate to a separating funnel.
3. Add 10 ml of water to the filtrate and after shaking the contents and allowing the phase to separate out, drain out the lower aqueous methanol layer. Repeat this step. Avoid very vigorous shaking.
4. Collect the benzene layer in a beaker and add small amount of solid anhydrous Na_2SO_4 to remove the traces of moisture.
5. Decant the clear benzene layer to another beaker and concentrate the extract by evaporating the solvent on a boiling water bath.

(b) *Column preparation*:

1. Mount a burette or a glass column vertically on a burette stand with the help of clamps.
2. Place lightly a plug of glass wool at the base of the burette and close the stopcock or outlet at the bottom of the column.
3. Take 5 g of alumina (adsorbent) previously dried at 120°C for 8 h and prepare its slurry in benzene. Pour the slurry carefully into the column or burette, by gentle tapping of the column so that no air bubbles get trapped in the adsorbent.
4. Allow the adsorbent to settle by opening the outlet. After the adsorbent has completely settled, add 20 ml of benzene and let it pass out of the column. Care should be taken not to let the adsorbent to get dried. When about 1 cm layer of the solvent remains at the top of the chromatographic bed, close the outlet.

Sample application:

1. Allow the solvent at the surface of the column to drain out slowly and transfer the leaf or flower extract with the help of pipette without disturbing the surface of the column adsorbent. Let it enter into the column and then add a few drops of benzene to wash the traces of the extract sticking to the wall of the column.
2. Add 20 ml more benzene to wash out the column of any unadsorbed material.

(c) *Column development*:

1. For desorption of the adsorbed substance change the polarity of the solvent in a stepwise manner. After 20 ml of benzene has passed through the column add 10 ml of 5% acetone (*u/v*) in benzene and let it percolate through the column and collect 1–2 ml fractions of the effluent from the outlet. Continue increasing the concentration of acetone in benzene at every succeeding step. Finally pass pure acetone through the column.

   - Note the change in the colour of the collected fractions. In case of the leaf extract, the initial fractions are colourless followed by yellow coloured and then by the green coloured ones. The colourless fractions do not contain any pigments but it is quite possible that these fractions may contain some UV absorbing materials.

12.6 Separation and Identification of Sugars by Adsorption TLC

12.6.1 Principle

Sugars get separated on the basis of differential adsorption onto silica gel. The sugars which have higher affinity for stationary phase are adsorbed more strongly

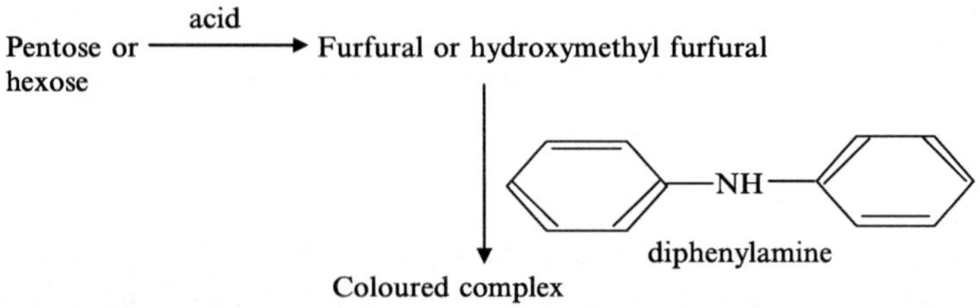

Fig. 12.2 Detection of sugars on TLC plate

and hence they migrate slowly when mobile phase moves over them. On the other hand, those having lower affinity for stationary phase are weakly adsorbed and are more easily carried by the mobile phase. The separated sugars are then located as coloured zones by spraying TLC plates with aniline diphenylamine reagent (Fig. 12.2).

12.6.2 Reagents and Materials

1. TLC chromatographic tank
2. Glass plates (20 × 20 cm)
3. Spreader
4. Micropipette
5. Oven maintained at 105°C
6. Hair drier
7. Sprayer or atomizer
8. Solvent system: Prepare a mixture of ethyl acetate: *iso*-propanol: water: pyridine (26:14:7:2,*u*/*v*)
9. Standard sugar solutions: Prepare 1% solution of standard sugars such as glucose, ribose, lactose, etc. in 10% *iso*-propanol (w/v). For mixture in which the sugars have to be identified, mix the sugar solutions in equal proportions.
10. Aniline diphenylamine reagent: Mix 5 volumes of 1% aniline and 5 volumes of 1% dihenylamine in acetone with 1 volume of 85% phosphoric acid

12.6.3 Procedure

1. Place thoroughly cleaned and dried glass plates (20 × 20 cm) on a flat plastic tray side by side with no gap between the two adjacent plates.
2. Prepare slurry of the Silica Gel G (stationary phase) free of clumps in water or in an appropriate buffer.
3. Spread a uniform layer of 250 μm thickness with the help of a spreader or an applicator by moving it from one end of the tray to its other end.

4. Activate the plates by keeping them at 105°C for 30 min. Allow the plates to cool in a desiccators before use.
5. Gently put marks in a straight line with the help of a pin at a distance of about 2 cm from one edge of the plate. The adjacent marks should be about 1.5–2.0 cm apart from each other. Extreme care should be taken that silica does not get scratched off while putting these marks.
6. Carefully apply the solution of individual standard sugars and the mixture or alcoholic extract of the sample on the separate marked sots.
7. Gently put marks or draw a line 1 cm from the opposite edge
8. Place the plate in chromatographic tank which has already been equilibrated with the solvent taking care that base line on which samples have been applied does not dip into the solvent.
9. Close the chromatographic tank with air tight lid and allow the solvent to ascend along the plate by capillary action till the solvent reaches the marked line on the upper side of the plate. This may take about 90 min.
10. Remove the plate from chromatographic tank and let it dry at room temperature.
11. For determining the location of sugars on the TLC plates, spray it with freshly prepared aniline-diphenylamine reagent ensuring that silica gel in not removed or blown off while spraying.
12. Place the plates in hot air oven at 100°C for 10 min. Appearance of bluish spots on the white background indicates presence of sugars at that region of the plate.
13. Measure the distance from the base line to the centre of the coloured spotand calculate the R_f value of each sugar as described above in other experiments.
14. Identify the sugars in the given mixture or sample by comparing their R_f values with those of sugar standards.

12.6.4 Note

- Thickness of the layer should be uniform throughout the length of the plate.
- Thoroughly cleaned glass plates free of any greasy spots should be used.
- The slurry of the chromatographic media should be free of any clumps.
- The TLC plates should be activated at recommended temperature and duration. Poor resolution of components occurs on over or under activated plates.
- The layer of chromatographic media should not get scraped off at the time of putting marks of application of samples.
- Size of the applied spot should be as small as possible. If large volume of the sample has to be spotted, then it should be done in small aliquots with an intermittent drying. Overloading of the sample should be avoided.
- The chromatographic tank should be airtight and chromatography should be performed under temperature controlled conditions.

12.7 Separation of Amino Acids by Ion Exchange Column Chromatography

12.7.1 Principle

Ion exchange chromatography can be used for separation of substance which possess a net electrical charge. Anion exchangers reversibly bind negatively changed compounds through electrostatic forces whereas positively changed molecules interact with cation exchangers. Different compounds are held by ion exchangers with varying strengths depending upon charge. Amino acids are amphoteric substances and have a net charge of zero at isoelectric pH. At pH below the isoelectric point, the amino acids are positively charged. At very low pH of 1.0 almost all amino acids (including acidic amino acids) exist as cation and hence these can be separated on a cation exchanger.

12.7.2 Materials and Reagents

1. Chromatographic column (2.5 × 25 cm)
2. Spectrophotometer/Colorimeter
3. Dowex-50 resin in 0.05 M citrate buffer pH 3.0
4. Citrate buffer 0.05 M, pH 3.0
5. Citrate buffer 0.05 M, pH 6.0
6. Citrate buffer 0.05 M, pH 9.0
7. Amino acids (aspartic acid, alanine, lysine and histidine; 2 mg of each amino acid/ml of 0.1 M HCI, pH 1.0)
8. 4 N HCI
9. Ninhydrin reagent

12.7.3 Procedure

(a) *Preparation of ion exchanger*: Dowex-50 is a cation exchanger. Before packing it into the column, it must be fully saturated with H^+ first and this H^+ form can be converted to Na+ form so that the resin can function as cation exchanger.

1. Suspend 10 g of Dowex-50 into sufficient volume of 4 N HCI for 15 min to ensure that the resin is saturated with H^+ ions.
2. Filter the suspension and wash it repeatedly with distilled or deionized water till the pH of the filtrate is neutral.
3. Transfer the resin into 2 N NaOH for 15 min to get Na^+ form. Wash it till the pH of the filtrate is neutral.

(b) *Equilibration of the resin*:

1. Suspend the resin in citrate buffer (pH 3.0) and allow it to stand for 1 h.
2. Mount the column upright and pour the suspension with help of a glass rod while tapping the column gently.
3. Allow the suspension to settle down, open the outlet and pass two to three bed volumes of the citrate buffer (0.05 M, pH 3.0) through the column. This will fully equilibrate the resin to pH 3.0. When only a thin layer of the buffer remains at the top of the resin, stop the flow by closing column outlet.

(c) *Sample loading*:

1. Open the outlet and let the buffer at the top drain into the column surface. Close the stopcock.
2. Load the column with 1 ml of mixture of amino acids in 0.1 M HCI having a pH value of 1.0.
3. Add a small amount of buffer to wash the traces of the sample from inside walls of the column and when the level just reaches the surface, close the stopcock.

(d) *Development of column*:

All the amino acids would be in cationic form at pH 1.0 and so would be bound to the cation exchanger. Gradient elution using increasing pH and ionic strength facilitates sequential elution of the bound amino acids.

12.8 Concentration of Dilute Protein Solutions with Sephadex G-25

12.8.1 Principle

Separation is based on the fact that proteins are macromolecules whereas, the salts are low molecular weight substances. When the sample is passed through a column packed with Sephadex G-10 or G-25, proteins remain totally excluded from the gel and move with the void volume while salts enter into the gel particles and take longer time to get eluted.

12.8.2 Materials and Reagents

1. 0.1 M Tris HCI buffer (pH 7.0)
2. Glass column
3. Sephadex G-25
4. Sodium phosphate
5. Bovine serum albumin

12.8.3 Procedure

1. Suspend 5 g of Sephadex G-25 in 0.1 M Tris-HCI buffer (pH 7.0) and swell it by keeping it for 4–5 h at room temperature with intermittent stirring.
2. Decant the excess of buffer alongwith any suspended fine particles to obtain slurry of appropriate thickness.
3. Fix the column upright on a burette stand with the help of clamps.
4. Keep the outlet of the column closed, place a plug of glass wool at the base of the column and pour a small volume of the buffer or water into the column.
5. Pour the slurry gradually into it along the inner surface with gentle tapping to expel any air bubbles.
6. Allow the chromatographic media to settle down evenly and then open the outlet to drain excess liquid from the column.
7. Place a filter paper disc or a nylon gauze on the surface of the packed bed to prevent disturbance of the upper layer while loading the sample or feeding the eluent into the column.
8. Prepare a mixture of 10 mg of bovine serum albumin and 40 mg of sodium phosphate in 2 ml Tris-HCI buffer (pH 7.0)
9. Apply it to chromatography column by any one of the two procedures:
   - The mobile phase at the top of the packing is drained out till the bed surface gets exposed. Close the outlet and gently apply the sample uniformly over the bed surface with pipette and the loaded sample is then allowed to just enter into the column by opening the outlet. A small amount of mobile phase (or buffer) is added to wash the traces of the sample into the column.
   - In the second method, sucrose or glycerol, upto the concentration of 1% is added in the sample to increase its density. This sample is applied just above the surface of bed directly through the layer of the buffer in the column bed. Since the sample has higher density, it automatically settles on surface of the gel. Then open the outlet to facilitate entry of the sample into the column. When using this procedure, it is advisable to ensure that addition of glycerol or sucrose does not interfere with the separation and subsequent analysis of the separated compounds.
10. Add sufficient amount of buffer on top of the column and connect it to the buffer reservoir.
11. Collect fraction (2 ml each) either manually or using an automatic fraction collector. Determine the protein content either by monitoring absorbance at 280 nm or by Lowry's method and phosphate ions in each of the fractions.
12. Plot a graph of concentration of protein and phosphate vs. fraction number or elution volume (Fig. 12.3).

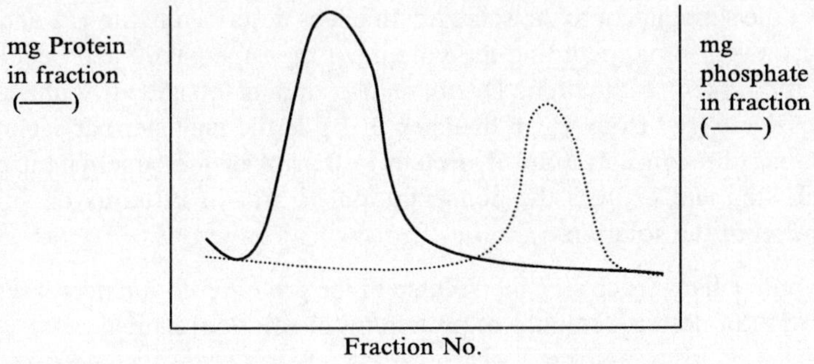

Fig. 12.3 Elution profile of protein and phosphate on Sephadex G-25 column

12.9 Concentration of Dilute Protein by Column Chromatography using Sephadex G-25

12.9.1 Principle

When the dry beads of Sephadex G-25 are added to a dilute protein solution, they start swelling and in the process absorb water. The proteins being macromolecules are excluded from the swollen gel and hence remain in the solution. Due to absorption of water, volume of the solution decreases without affecting amount of high molecular weight solutes such as proteins in it resulting in concentration of the solution.

12.9.2 Reagents

1. Bovine serum albumin
2. NaCI (1 mM)
3. Sephadex G-25

12.9.3 Procedure

1. Dissolve 20 mg bovine serum albumin in 100 ml of 1 mM NaCl. Retain 1 ml of this solution for protein estimation by Lowry's/Bradford method.
2. Add 5 g of dry Sephadex G-25 (coarse) to the remaining protein solution.
3. Let it stand at room temperature for 30 min to swell and then centrifuge it at 3,000 × *g* for 10 min. Carefully decant the supernatant into the measuring cylinder and note its volume. Again retain 1 ml for protein estimation.

4. Subject the supernatant to the same treatment as described in Steps 2 and 3 twice or thrice every time recording the volume of the supernatant and keeping 1 ml aside for protein estimation. Decrease the amount of added Sephadex G-25 progressively (say from 5 g in first step to 1 g in the final step) at each step.
5. Determine the concentration of protein in 0.5 ml of the supernatant obtained at each step and express the concentration of protein in terms of amount of protein/ml of the solution.

It will be noted that at each step the volume of the protein solution decreases but the concentration of protein (amount of protein/ml of solution) increases.

12.10 Determination of Molecular Weight of Protein by Gel Filtration

12.10.1 Principle

During gel filtration, solutes are separated primarily on the basis of their molecular size. Due to molecular sieving effect, the large molecules are eluted from the column first followed by compounds of smaller molecular mass. A plot between K_d or elution volume vs. $\log_{10}$ molecular weight gives a straight line. Molecular weight of a given protein can be established from its elution volume through gel filtration column which has previously been calibrated with standard marker proteins of known molecular weight.

12.10.2 Materials and Reagents

1. Glass column (2.5 × 70)
2. Sephadex G-100
3. HEPES–NaOH buffer (20 mM; pH 8.0)
4. $MgCl_2$
5. DTT
6. Glycerol
7. Standard protein markers:

   (i) β-amylase (200 Kd)
   (ii) Alcohol dehydrogenase (150 Kd)
   (iii) Bovine serum albumin (66 Kd)
   (iv) Carbonic anhydrase (29 Kd)
   (v) Cytochrome c (12.4 Kd)
   (vi) Blue dextran

12.10.3 Procedure

1. Suspend 15 g of Sephadex G-100 mM HEPES-NaOH (pH 8.0) buffer containing 5 mM $MgCI_2$ and 5 mM DTT for 5 h in boiling water bath.
2. Allow it to cool and pack it into the glass column taking all suggested precautions for avoiding entrapment of any air bubbles in the gel bed.
3. Equilibrate the column by passing buffer equivalent to 2–3 volume of the bed volume.
4. Find out the void volume (*Vo*) of the column by determining elution volume of blue dextran solution (2 mg/ml) through the column. Again pass 2 bed volumes of the starting buffer.
5. Apply the mixture of the standard maker proteins of known molecular weight and elute the column with buffer.
6. Collect fractions of 2 ml each and determine the protein content in each of these fractions either by Lowry's method or by recording absorbance at 280 nm.
7. Determine the elution volume of the standard proteins and prepare a graph of graph of $\log_{10}$ molecular weight vs. V_e or $\left\{K_d = \frac{V_e - V_0}{V_i}\right\}$. Again pass two volume of the starting buffer.
8. Layer the sample containing protein whose molecular weight has to be determined. Elute it and collect the fractions of 2 ml each. Test each fraction for the presence for the presence of protein and determine its V_e.
9. Determine the molecular weight of the given protein from the calibration curve prepared in Step 7 as shown below (Fig. 12.4).

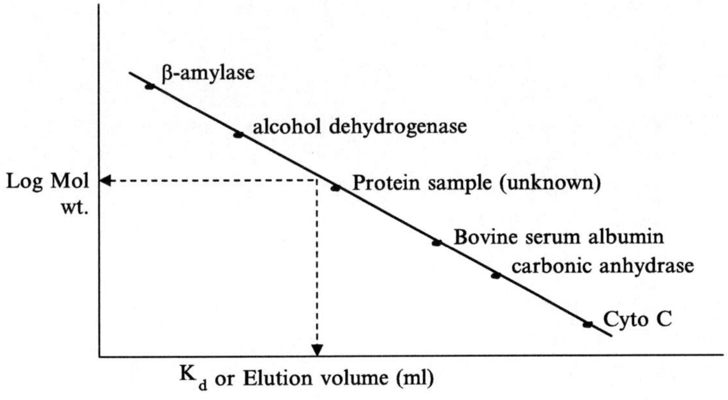

Fig. 12.4 Relationship between elution volume or K_d vs. molecular weight of proteins

Chapter 13
Methods for Nutritional Quality Evaluation of Food Materials

A good and balanced-diet consists of all the nutrients required for the metabolism, *viz.*, carbohydrates, protein, oils and fats, minerals and vitamins in right proportion as per the needs of different age groups. Cereals such as rice, wheat, maize and barley and millets provide protein and energy, while pulses and oilseeds are the rich sources of protein and energy, respectively. The diets based on cereals and pulses are generally deficient in one or the other nutrient. Protein-calorie malnutrition is wide spread among majority of population in developing countries. In order to effectively control protein-calorie malnutrition, it is essential to know the nutritional requirements as well as nutritive quality of various food grains. The presence of various anti-nutritional factors like proteolytic enzyme inhibitors, polyphenols, phytic acid and lathyrogens in cereals and pulses can result in impaired growth or produce acute and lethal effects and also reduce the availability of otherwise good proteins in the diet. The presence of excessive amounts of fibre and tannins (polyphenols) in cereals might be responsible for their reduced digestibility and lower biological value. Likewise, the quality of oil is also an important area concerning human health. The oilseed like mustard may have abnormal content of erucic acid or glucosinolates. Therefore, nutritional evaluation of food stuffs becomes important for ensuring good and healthy diet.

13.1 Determination of Iron (Jackson 1973)

(a) Total Iron

The total iron is determined by reduction to ferrous with hydroxylamine hydrochloride. The ferrous form reacts with ortho-phenanthroline, a chelating compound of intense red colour is formed which can be measured at 490 nm.

R. Katoch, *Analytical Techniques in Biochemistry and Molecular Biology*,
DOI 10.1007/978-1-4419-9785-2_13, © Springer Science+Business Media, LLC 2011

Equipment and Glassware

1. Spectrophotometer
2. Hot plate

Reagents

1. 10% Hydroxylamine hydrochloride solution: Dissolve 10 g of hydroxylamine hydrochloride in 100 ml distilled water.
2. 1.5% Ortho-phenanthroline solution: Dissolve 1.5 g of ortho-phenanthroline in 100 ml of 95% ethanol.
3. Tri acid mixture (1:1:4, HNO_3:H_2SO_4:$HClO_4$ v/v).
4. Standard iron solution: Dissolve 0.7023 g of ferrous ammonium sulphate [$Fe(NH_4)_2(SO_4)_2 \cdot 6H_2O$] in water then add 20 ml of 0.6 N warm HCl, if necessary. Make up the volume to 1 l. This solution contains 100 ppm of iron. Dilute 10 ml of the solution to 100 ml with water. This solution contains 10 ppm of Fe. Take aliquots (0.5, 1, 2, 3, 4, 6 ml) for preparation of iron calibration curve giving 5–60 μg of iron.

Procedure

1. Weigh 0.5 g dried fine powdered sample in a conical flask. Add 10 ml triacid mixture.
2. Keep it overnight, digest next day on hot plate at 180–200°C till the dense white fumes of H_2SO_4 and $HClO_4$ are completely evolved.
3. Dilute the content of the flask with water, filter through Whatman No. 42 filter paper in 100 ml volumetric flask. Wash the filter paper with water and make up the volume.
4. Take 5 ml of the aliquot in 25 ml volumetric flask, add 2 ml of 10% hydroxylamine hydrochloride and 1 ml of 1.5% ortho-phenanthroline solution.
5. Mix well and make up the volume with distilled water and read it at 490 nm. The colour reaches maximum strength immediately and is stable for months.
6. Run simultaneously blank with all the reagents but without the sample solution.
7. The concentration of the unknown solution is read from the standard curve drawn.

(b) Available Iron

Available iron is estimated after digesting the sample with pronase and trypsin, and developing the colour using ortho-phenanthroline reagent. The intensity of the colour is measured at 490 nm.

Equipment and Glassware

Same as given for iron estimation.

Chemicals

1. Sodium barbiturate buffer (0.02 M, pH 7.0)
2. Trypsin
3. Pronase

4. Pepsin
5. Trichloroacetic acid (2.5%)

Procedure

1. Weigh finely ground dried sample in a digestion tube.
2. Add 5 ml water, shake and boil for 5 min.
3. Add 1 ml of 0.02 M barbiturate buffer (pH 7.0) containing 0.15 M NaCl. To it add 5 mg pronase and 1 mg of trypsin.
4. Shake for 18 h and add 1 ml of 2.5% TCA. Centrifuge.
5. Add 5 ml of triacid mixture (prepared as given for total iron estimation) both in residue and supernatant along with 100 mg dried original sample.
6. Digest next day on a hot plate at 180–200°C till the dense white fumes of H_2SO_4 and $HClO_4$ are evolved completely.
7. Dilute the content with distilled water, filter through Whatman No. 42 filter paper. Make up the volume to 100 ml.
8. Take 5 ml aliquot from each of residue, supernatant and original sample in 25 ml volumetric flask.
9. Add 2 ml of 10% hydroxylamine hydrochloride and 1 ml of 1.5% solution of ortho-phenanthroline reagent.
10. Make up the volume to 25 ml and read at 490 nm.

$$\text{Iron availability (Residue)} = \frac{\text{Total iron in sample} - \text{Total iron in ppt}}{\text{Total iron in sample}} \times 100$$

$$\text{Iron availability (supernatant)} = \frac{\text{Total iron in supernatant}}{\text{Total iron in sample}} \times 100.$$

Mean of both values are used as final value.

13.2 Estimation of Protein Content by Micro-Kjeldahl Method (AOAC 1970)

The most commonly used methods for protein estimation are:

(a) Kjeldahl method
(b) Biuret method
(c) Dye-binding method

However, the auto-analyzer could also be used for the estimation of proteins.

(a) Kjeldahl method

The micro-Kjeldahl method is most reliable method for nitrogen estimation. It is used in a variety of forms. The percentage of protein can be determined by first determining the nitrogen content and multiplying the per cent nitrogen by the factor 6.25 (maize, millet, sorghum, pulses), 5.95 (rice) and 5.7 (wheat).

Principle

The nitrogen in the protein or any organic material is determined by digesting the samples in concentrated sulphuric acid at high temperature with catalyst to convent it into ammonium sulphate. This salt on steam distillation with excess of NaOH liberates ammonia which is collected in boric acid solution. The ammonium borate formed is then titrated with standard hydrochloric acid.

Glassware

(1) Micro-Kjeldahl digestion unit
(2) Micro-Kjeldahl distillation unit
(3) Digestion tubes/flasks
(4) Measuring cylinders
(5) Volumetric flasks

Reagents

1. Sulphuric acid, Sp. gr. 1.84, nitrogen free.
2. Catalyst mixture-Grind together 99.0 g of K_2SO_4, 4.1 g of HgO and 0.8 g $CuSO_4$.
3. Sodium hydroxide and sodium thiosulphate solution – Dissolve 50 g NaOH and 5 g $Na_2S_2O_3{\cdot}5H_2O$ in 50 ml distilled water and dilute to 100 ml.
4. Boric acid solution – Dissolve 4 g of boric acid in warm distilled water and dilute to 100 ml.
5. Standard Hydrochloric acid solution – 0.02 N.
6. Mixed indicator – Mix One-part of 0.2% methyl red in ethanol with 5 parts of 0.2% bromocresol green in ethanol.

Procedure

1. Weigh 50 mg of sample and place in a digestion tube/flask. Add 1 g of catalyst mixture and 5 ml of concentrated sulphuric acid.
2. Digest the sample using digestion unit till the solution becomes colourless (approximately 40–60 min at 370°C).
3. Cool and add minimum quantity of water to dissolve the solid. Transfer the digest to a steam distillation apparatus.
4. Take a 100-ml conical flask containing 10 ml of boric acid. Add 2–4 drops of mixed indicator dye, and place the flask beneath the condenser with delivery tip immersed in the solution.
5. Transfer digest to distillation apparatus and add 5–10 ml sodium hydroxide-sodium thiosulphate solution to the digest and steam distill until the first appearance of violet colour, as the end point.
6. Run a blank containing the same quantities of all the reagents but without sample for every set of nitrogen determination. Standard check sample of known protein content may also be included in the run.

Calculation

Calculate percentage of nitrogen (N) as follows:

$$\% \text{ N} = \frac{\left[\begin{array}{l}\text{ml HCI used in\_ml HCI used} \\ \text{determination in blank}\end{array}\right] \times \text{Normality of HCI } \times 14.00 \times 100}{\text{mg sample}}$$

%protein = % N × Factor for a given sample.

13.3 Estimation of Tryptophan (Spies and Chambers 1994)

Principle

Tryptophan is one of the limiting amino acids in food grains. Tryptophan and *p*-dimethyl amino benzaldehyde react in acid medium to form a condensation product. This condensation product is oxidized by sodium nitrite to yield a blue colour, which can be measured colorimetrically at 660 nm.

Glassware

1. Conical flasks (50–100 ml)
2. Volumetric flasks (50 and 1,000 ml)

Reagents

1. 19 N H_2SO_4 solution.
2. 21.4 N H_2SO_4 solution.
3. 0.045% $NaNO_2$ solution: Dissolve 45 mg $NaNO_2$ in 1,000 ml distilled water.
4. *p*-dimethyl amino benzaldehyde.
5. Standard tryptophan solution (100 μg per ml):

   Dissolve 10 mg of DL-tryptophan in 100 ml of distilled water (Add 2–4 drops of 10% NaOH for solubilization).

Procedure

1. Weigh 15–20 mg of fine powdered defatted flour in each of three different 50 ml conical flasks. In two flasks add 30 mg of *p*-dimethyl amino benzaldehyde. Third flask serves as blank.
2. To all the above flasks, add 10 ml of 19 N H_2SO_4 solution. Keep the flasks in dark at 30°C for 20 h. Remove the flask from dark and add 0.1 ml of 0.045% $NaNO_2$ solution to each flask. Mix properly and gently, keep it for half an hour at room temperature and measure the absorbance of blue colour developed at 660 nm.

Standard Curve

Standard curve of tryptophan is prepared by using different concentrations (10–100 μg/ml) of standard tryptophan solution. Adjust the volume to 0.6 ml by adding distilled water, add 9.4 ml H_2SO_4 (21.4 N) very slowly, mix the content gently and further proceed as per the steps described above.

Calculation

$$\text{Tryptophan } (\mu g/g) = \frac{\mu g \text{ tryptophan from standard curve } \times 1,000}{\text{Weight of the sample (mg)}}.$$

Precautions

1. Sulphuric acid should be added slowly by the sides of the flasks.
2. After the addition of sulphuric acid into the sample, keep it in dark.

13.4 Estimation of Methionine (Mc Carthy and Paoille 1959)

Methionine is one of the sulphur containing amino acids and is limited in legumes. Grain protein is first hydrolysed under mild acidic condition and the liberated methionine gives yellow colour compound with nitroprusside under alkaline condition which becomes red on acidification. Glycine is added to the reaction mixture in order to inhibit colour formation with other amino acids.

Equipment and Glassware

1. Autoclave
2. Spectrophotometer
3. Volumetric flask of 100 ml capacity
4. Conical flask, 250 ml

Reagents

1. 2 N HCl solution
2. 10 N NaOH solution
3. 10% NaOH solution (w/v)
4. 10% Sodium nitroprusside solution (w/v)
5. 3% glycine solution (w/v)
6. Concentrated orthophosphoric acid
7. Activated charcoal
8. Standard solution of methionine (1 mg per ml): Dissolve 100 mg of DL-methionine in 5 ml of 20% HCl and dilute with water to 100 ml.

Procedure

1. Weight 2 g of defatted sample into a 250-ml conical flask. To this add 25 ml of 2 N HCl and mix well. Autoclave at 15 lb pressure for 1 h.
2. Add a pinch of activated charcoal to the autoclaved sample. Heat to boiling and filter when hot. Wash the charcoal with hot water. Also Collect the washings.
3. Neutralize the filtrate with 10 N NaOH to bring the pH to 6.5. Make up the volume to 100 ml with water.
4. Transfer 50 ml of the hydrolysate in 250 ml conical flask and add 6 ml of 10% NaOH followed by 0.3 ml of 10% sodium nitroprusside solution. Keep aside the contents for 10 min with occasional shaking. Add 2 ml of 3% glycine solution after 10 min. Allow it to stand for 10 min.
5. Add 4 ml of concentrated orthophosphoric acid and shake the contents vigorously.
6. Read the intensity of colour after 10 min at 520 nm against a blank prepared by omitting sodium nitroprusside.

Standard Curve

Prepare standard curve by taking different concentrations of methionine (0, 2, 4, 6, 8 and 10 mg) in duplicate and water to make up the total volume to 50 ml. Develop the colour in the same way as mentioned above. Finally, draw the standard curve.

Calculations

$$\%\,\text{methionine} = \frac{\text{mg methionine from standard curve}}{\text{Weight of sample}} \times 100.$$

13.5 Estimation of Starch (Clegg 1956)

Principle

Starch is extracted with perchloric acid after removal of sugars by extracting the sample flour with hot 70% alcohol. In hot acidic medium starch is hydrolysed to glucose and dehydrated to hydroxymethyl furfural. This compound when reacts with anthrone forms green colour.

Equipment and Glassware

1. Spectrophotometer
2. Measuring flasks of various capacities (10, 25, 50, 100 ml)
3. Measuring cylinders (50, 100 ml)
4. Pipettes

Reagents

1. Anthrone reagent: Dissolve 200 mg anthrone in 100 ml of ice-cold 95% sulphuric acid.
2. 70% ethanol.
3. 52% perchloric acid.
4. Standard glucose solution: Dissolve 200 mg glucose in 100 ml distilled water. Standard (100 μg/ml) – 10 ml of standard stock solution of glucose is diluted to 100 ml with distilled water.

Procedure

1. Extract oven dried sample flour (100 mg) with hot 70% ethanol. Centrifuge and retain the residue. Wash the residue with 70% hot ethanol. Dry the residue.
2. To the residue add 5.0 ml of water and 6.5 ml of 52% perchloric acid.
3. Shake the content for 5 min, centrifuge and collect the supernatant.
4. Repeat the extraction using 5 ml fresh perchloric acid for 10 min and centrifuge to collect the supernatant.
5. Re-extract the pellet using 5 ml fresh perchloric acid and shake for 30 min.
6. Pool the supernatants and make up the volume to 100 ml with water.
7. Take suitable aliquot for glucose estimation. Add 5 ml of anthrone reagent. Heat on boiling water bath for 10 min.
8. Cool rapidly and read the intensity of colour formed at 620 nm.
9. Find out glucose content using the standard curve. Multiply the value by a factor 0.9 as 0.9 g of starch yields 1 g of glucose on hydrolysis.

13.6 Estimation of Amylose (Williams et al. 1970)

Starch is composed of two components namely amylose and amylopectin. Amylose is a linear and non-branched polymer of glucose, joined together by α 1–4 glucosidic linkages. Amylase when reacts with iodine, which is absorbed within the helical coils, produces a blue coloured complex. The blue colour is measured colorimetrically.

Equipment and Glassware

1. Centrifuge
2. Spectrophotometer
3. Centrifuge tubes
4. Volumetric flask, 100 and 50 ml

Reagents

1. Stock iodine solution: Dissolve 20 g of KI and 2 of resublimed I_2 in minimum amount of water and make up the volume to 100 ml.
2. Iodine reagent: Dilute 10 ml of stock solution 1–100 ml at the time of use.

3. KOH (0.5 N).
4. HCl (0.1 N).

Procedure

1. Take 20 mg of dry, finely ground material (minimum 60 mesh). Add 10 ml of 0.5 N KOH and stir thoroughly with a magnetic stirrer for 10 min.
2. Make up the volume to 100 ml.
3. Take 10 ml of aliquot into 50 ml volumetric flask and add 5 ml of 0.1 N HCl and 0.5 ml of 1_2 reagent 2. Make up the volume.
4. Read the colour at 625 nm.
5. Calculate the amount of amylase from standard curve prepared by using standard solution containing different amounts of amylase (0.2–2 mg) and amylopectin (0.8–8 mg) keeping the ratio of amylase and amylopectin constant (1:4 w/w).
6. Amylopectin content can be calculated by subtracting the amylase from total starch content.

Calculation

Absorbance corresponding to 10 ml of the test solution = y mg
Amylose/g dry wt = $y \times 10 \times 50$ mg.

13.7 Estimation of Gluten in Wheat (Misra and Gupta 1995)

When water is mixed with wheat flour and the contents are kneaded, a cohesive mass of dough is formed. This mass on washing removes starch, bran and yields a visco-elastic gum like material known as gluten. Gluten is mainly composed of gliadin and glutenin proteins.

Equipment and Glassware

1. Beaker
2. Glass rod
3. Sieve (100 mesh)
4. Oven
5. Analytical balance

Reagents

1. KI solution: Dissolve 200 mg iodine and 2 g KI in water and make up to 100 ml.

Procedure

1. Take 10 g of wheat sample in a beaker.
2. Add to it 7 ml distilled water and make a dough with the help of glass rod for around 2 min.

3. After proper mixing, make a small ball of the dough and immerse it in a beaker containing water.
4. The ball is left for 30 min.
5. Place the beaker under tap water; put a sieve (100 mesh) on it. Now the dough is washed gently under running water till a chewing gum type cohesive mass separates out. Bran will settle over the sieve and starch will pass down to the beaker.
6. To ensure complete removal of bran, wash the extract gluten with excess of water by stretching inside the fingers.
7. To check whether whole of the starch is washed out or not, add KI to last extract. Absence of any violet colour indicates completes removal of starch.
8. Squeeze out the adhered water from the extracted gluten and weigh. This will be the weight of wet gluten. Dry at 105°C for 6 h and weigh to get the yield of dry gluten.

13.8 Estimation of Lysine (Tsai et al. 1972)

Lysine is an essential amino acid generally deficient in cereals. The method described is for non-pigmented cereals (maize, sorghum etc.). The sample is hydrolysed with papain or other proteolytic enzyme releasing free amino acids and low molecular weight peptides. The α-amino group of the free amino acids are blocked with copper, while the ε-amino group of lysine remains free. The ε-dinitropyridyl derivative of lysine is then formed on reaction with 2-chloro-3, 5-dinitropyridine. Excess of 2-chloro-3 5-dinitropyridine is removed with ethyl acetate and the colour of ε-dinitropyridyl derivative is read at 390 nm.

Equipment and Glassware

1. Spectrophotometer
2. Centrifuge
3. Centrifuge tubes, 15 ml
4. Screw-cap glass vials, 30 ml
5. Incubator

Reagents

1. Copper phosphate reagent: The reagent is a suspension of copper phosphate in borate buffer.

    (i) Borate buffer (0.5 M, pH 9.0)
    (ii) Copper phosphate suspension:

    Solution A: 2.8 g $CuCl_2{\cdot}2H_2O$ dissolved in 100 ml distilled water
    Solution B: 13.8 g $Na_3PO_4{\cdot}12H_2O$ dissolved in 200 ml distilled water

    Pour solution A into B with swirling and centrifuge (300 × *g*) for 5 min. to collect the precipitate. Discard the supernatant. Resuspend the pellet 3 times in 15 ml of borate buffer and centrifuge after each suspension. After the third washing, resuspend the pellet in 80 ml of borate buffer. The reagent can be used for 2 weeks.

2. Pyridine reagent: Just before use, prepare a 3% solution of 2-chloro 3,5 dinitroypyridine in methanol.
3. Papain solution: Papain is dissolved in 0.1 N sodium acetate buffer (pH 7.0) to a final concentration of 4 mg/ml. Prepare the enzyme solution daily. Filter the solution, if required.
4. Sodium carbonate buffer: 0.05 M, pH 9.0.
5. Amino acid mixture: (in milligrams): alanine, 3; arginine, 50; aspartic acid, 60; cystine, 20; glutamic acid, 300; glycine, 40; histidine, 30; isoleucine, 30; leucine, 50; threonine, 30; tyrosine, 30 and valine, 40. Dissolve 100 mg of this mixture in 10 ml of carbonate buffer (0.05 M, pH 9.0).
6. HCl (1.2 N) solution.
7. Ethyl acetate.

Standard Curve

Prepare a standard curve in a range of 0–200 μg lysine/ml. Dissolve 62.5 mg of pure lysine mono-hydrochloride in 20 ml of carbonate buffer (0.05 M, pH 9.0). Prepare the solution by dilution with carbonate buffer to get the following concentrations 0, 250, 500, 750 and 1,000 of lysine (μg/ml). From each of these solution take 1 ml and add 4 ml of papain (5 mg/ml) solution. Now the respective concentration of lysine will be 0, 50, 100, 150 and 200 μg/ml. Pipette out 1 ml of each solution into centrifuge tube, add 0.5 ml of copper phosphate suspension. Follow the same procedure as described below for the sample.

Procedure

1. Weigh 100 mg defatted finely group sample into a glass vial. Add 5 ml of papain solution. Cap the vials and mix well so that the sample is totally wet. Incubate at 65°C in the oven overnight. Include a blank (5 ml of papain solution without any sample) with every group of samples.
2. Remove the sample from the oven, shake well, cool to room temperature centrifuge and decant the clear digest.
3. From the supernatant fraction transfer 1 ml into a 15-ml centrifuge tube and add 0.5 ml of carbonate buffer and 0.5 ml of copper phosphate suspension.
4. Shake the mixture for 5 min in a vortex mixer and centrifuge to precipitate the excess copper phosphate.
5. Pipette out 1 ml of the supernatant into a 30-m test tube and add 0.1 ml of the pyridine reagent and mix well. Cap the tubes with rubber/velvet corks and shake occasionally for 2 h at room temperature.
6. Add 5 ml of 1.2 N HCl and mix well by vortexing.
7. Add 5 ml of ethyl acetate, invert capped tubes at least 10 times, then remove the top layer (ethyl acetate) using a syringe adapted with a polyethylene tube. This step should be repeated 3 times.

8. Determine the absorbance of the aqueous phase in a colorimeter/spectrophotometer at 390 nm. Relate absorbance to the corresponding lysine value on the standard curve.

Calculation

Plot absorbance reading against μg standard. Read off μg of sample from curve.

$$\%\,\text{Lysine in sample} = \frac{\mu\text{g lysine from standard curve}}{\mu\text{g sample}} \times 100$$

or

$$\text{Lysine content (g per 16 g N)} = \frac{\mu\text{g lysine from standard curve} \times 0.16}{\%\,\text{in the sample}}.$$

13.9 Protein Fractionation in Cereals (Landry and Moureaux 1970)

Seed proteins have been classified into five major groups based on their solubility (Mendel and Osbome). These include: water-soluble albumins, salt soluble globulins, alcohol soluble prolamins, alkali soluble gluteline and residue (insoluble) proteins. Protein fractionation procedure as given by Landry and Moureaux (1970) separates 5 fractions namely: albumins + globulins, prolamin, prolamin-like, glutelins-like and true glutelin which are obtained by following the scheme described herein. This method is good for maize, sorghum and barley. However, it has been found that the extractability of prolamin is greater for barley when 50% isopropanol with 2% mercaptoethanol is used instead of 70% isopropanol containing 0.6% mercaptoethanol.

Although solubility is one of the classic criteria for separating seed proteins, it is not an ideal classification as the solubility is critically dependent on a number of conditions like the exact nature of the solvent, the ratio of solvent to sample, the hydration state of sample, the temperature of the solvent, fineness of the sample, etc (Table 13.1).

Equipment and Glassware

1. Shaker
2. Centrifuge
3. Kjeldahl digestion and distillation set
4. pH meter
5. Conical flask
6. Centrifuge tubes

Table 13.1 Landry-Moureaux fractionation sequence

Fraction	Solvent	Extraction temperature	Time of agitation (min)	Protein fraction
1	NaCl, 0.5 M	4°C	60	Albumins and globulins
			30	
			30	
	Water		15	
			15	
II	Isopropanol, 70%	20°C	30	Prolamin
			30	
			30	
III	Isopropanol, 70% + 2-ME, 0.6% (v/v)	20°C	30	Prolamin-like
IV	Borate buffer, 0.05 M	20°C	60	Glutelin-like
	pH 10 + 2-ME, 0.6%(v/v)	20°C	30	
			15	
V	Borate buffer pH 10 + 2-ME, 0.6% (v/v) + SDS, 0.5% (v/v)		60	Glutelin
			30	
			15	

Reagents

As specified in the scheme.

Procedure

1. Take 2 g defatted finely ground sample (100 mesh) in a stoppered conical flask or centrifuge tube.
2. Add 20 ml 0.5 M NaCl solution and shake for 60 min at 4°C.
3. Centrifuge at 6,000 × *g* for 5 min and collect the supernatant.
4. Re-extract the residue by following the same extraction procedure twice except that the shaking should be done for 30 min each time.
5. Then extract the residue twice by using distilled water in the same manner.
6. Collect all the supernatants and make up volume to 100 ml with saline. This represents fraction I which contains albumins and globulins.
7. Extract prolamin, prolamin-like, glutelin-like and glutelin from the residue obtained in step 6 at 20°C following the same procedure but using solvents and shaking time specified in scheme given earlier. The residue obtained at the last extraction should also be analysed and represented as residue protein.
8. To obtain albumins and globulins from fraction I, add equal volume of 10% TCA and allow it to remain cold for 30 min then centrifuge at 6,000 × *g* for 15 min at 4°C. The precipitate obtained represents albumins + globulins, while the supernatant will contain non-protein nitrogen (free amino acids and peptides).
9. Estimate nitrogen in all the fractions after digestions and distillation following micro-Kjeldahl method.

Precautions

1. It is advisable to check for the completeness of extraction of different protein fractions. It can be done easily by monitoring absorbance of the last extraction at 280 nm. If considerable absorbance is observed, continue the extraction further with the same solvent.
2. If there is any delay in the estimation of N, then few drops of toluene should be added to each fraction.
3. Reagents for fractions I and II are stable, while for fractions III, IV and V are unstable, and these should be prepared not more than 1 week before use.

13.10 Estimation of Crude Fibre (AOAC 1965)

Principle

Crude fibre consists mainly of cellulose and lignin (97%) plus some minerals. It can be estimated by treatment of the sample first with acid and subsequently with alkali. Oxidative hydrolytic degradation of the native cellulose and lignin occur. The residue obtained after final filtration is weighed incinerated, cooled and weighed again. The loss in weight gives the crude fibre content.

Equipment and Glassware

1. Extracting apparatus
2. Hot plate
3. Oven
4. Muffle furnace
5. Tall spoutless beakers, 600 ml
6. Dessicator
7. Filtering device
8. Filtering cloth (muslin cloth)

Reagents

1. Sulfuric acid solution (0.255 ± 0.005 N) containing 1.25 g H_2SO_4 per 100 ml.
2. Sodium hydroxide solution. (0.313 ± 0.005 N) containing 1.25 g NaOH per 100 ml, free or nearly so, form Na_2CO_3. Strength of acid and alkali solutions must be accurately checked.
3. Methyl alcohol, 95%.
4. Petroleum ether or diethyl ether.

Procedure

1. Extract 2 g of ground dry matter with ethyl ether or petroleum ether to remove fat. If material contains less than 1% fat, the extraction may be omitted.

2. Add 200 m boiling H_2SO_4 solution, immediately connect digestion flask to condenser and heat. (Contents of flask must come to boiling within 1 min and boiling must continue briskly for 30 min. Keep rotating the flask).
3. After 30 min remove flask, filter immediately through muslin cloth and wash with boiling water until washings are no longer acidic. Test with $BaCl_2$ solution.
4. Boil with 200 ml of NaOH solution. Connect flask with reflux condenser and boil for 30 min. Rotate flask frequently until sample is thoroughly wetted.
5. After 30 min remove flask and filter. If filtering cloth is used, thoroughly wash residue with boiling water and transfer to crucible.
6. For material difficult to filter, after 30 min boiling, filter through funnel using, vacuum and wash with hot 10% K_2SO_4 solution (K_2SO_4 solution may be added during filtering whenever filtration becomes difficult).
7. Return residue to digestion flask, thoroughly washing all residues from cloth with hot K_2SO_4 solution. Filter into crucible.
8. After thorough washing with boiling water, wash with 15 ml alcohol. Dry crucible and contents at 130°C to constant weight.
9. Cool the crucible in a dessicator and weigh.
10. Ignite contents of crucible in muffle furnace until carbonaceous matter is consumed (approximately 20 min).
11. Cool in dessicator and weigh.
12. Report loss in weights crude fibre.

Calculation

$$\text{Crude fibre }\% = \frac{\text{Loss in weight}}{\text{Weight of sample}} \times 100.$$

13.11 Estimation of Dietary Fibre (Vansoest 1963)

Principle

Dietary fibre is composed mainly of cellulose, hemicellulose and pectin. It is broken down into formic, acetic and galacturonic acid in large intestine. These hydrolytic products are poorly absorbed by the intestinal wall and rapidly broken down *via* intense and metabolic activity and transferred to intestinal microflora. Hemicellulose is determined by subtracting the value of acid detergent fibre (ADF) from neutral detergent fibre (NDF). NDF is made up of cellulose, silica and lignin.

Equipment and Glassware

1. Neutral detergent solution

   A. Dissolve 18.61 g disodium ethylene diaminotetracetate (EDTA) dehydrate, 6.81 g $Na_2B_4O_7{\cdot}10H_2O$ (Sodium borate decahydrate) in 500 ml of water. Heat till it dissolves.
   B. Add 30 g sodium lauryl sulphate and 10 ml of 2-ethoxyethanol.

C. Dissolve 4.56 g of disodium hydrogen phosphate anhydrous in 100 ml of distilled water, heat it if needed.
   Mix solution A, B and C. Check the pH (6.9–7.1). Make the volume to 1 l.

2. Acid detergent solution: Dissolve 20 g cetyl trimethyl ammonium bromide (CTAB) in 1 l of 1 N H_2SO_4.
3. Acetone

Procedure

1. Weigh 500 mg of defatted finely ground sample (100 mesh) and transfer it to 1 l long without spout.
2. Add 100 ml neutral detergent solution for the determination of neutral detergent fibre or acid detergent solution for the determination of acid detergent fibre.
3. Reflux for 1 h at 60°C.
4. Filter through previously weighed sintered crucible attached to vacuum. Washing of the residue with acetone is done after complete filtration.
5. Dry at 110°C for 24 h or till constant weight is obtained.

Calculation

$$\text{ADF or NDF \%} = \frac{\text{Loss in weight after drying}}{\text{Weight of Sample}} \times 100$$

$$\text{Hemicellulose} = \text{NDF} - \text{ADF}.$$

13.12 Estimation of Oil Content in Oilseeds (AOAC 1965)

Oil is an important ingredient of food. Estimation of oil in oilseed is generally carried out using the following two methods –

(a) Soxhlet method
(b) Cold percolation method

(a) Soxhlet Method

Principle

The principle is based on the extraction of oil using non-polar solvents *viz.*, ether, hexane, petroleum ether (40–60°C), chloroform:methanol (2:1). It involves repeated extraction of oil. The solvent is then distilled off completely. The oil is dried, weighed and the % of oil is calculated.

Equipment and Glassware

1. Soxhlet extractor assembly.
2. Absorbent cotton.

3. Whatman filter paper/Thimble.
4. Volumetric flasks, 100 ml.
5. Distillation unit, hot plate/heating mentle.

Reagent

Hexane or diethyl ether/petroleum ether (40–60°C).

Procedure

1. Fold a piece of filter paper and make into a sample packet in such a way to hold the seed meal (2.5 g) depending on oil content.
2. Place the sample packet or the thimble (alternatively the sample can be put in a thimble also) into extract of soxhlet apparatus, after placing some cotton at the bottom. A piece of cotton is placed at the top to evenly distribute the solvent as it drops on the sample during extraction.
3. Add organic solvent, two and a half times the capacity of the extractor and extract oil for a period of 6 h, or for a longer period till the solvent in the extractor becomes colourless.
4. Put off the heaters and allow cooling. Flash evaporate the solvent or distill. Keep it in oven at 70°C for 10 min. Cool at room temperature.
5. Weigh the flask after removing moisture. Repeat heating until constant weight is recorded.

Calculation

$$\%\,\text{oil in the sample} = \frac{\text{Weight of oil (g)} \times 100}{\text{Weight of sample (g)}}.$$

(b) Cold Percolation Method (Kartha and Sethi 1957)

Principle

The extraction of oil from the sample involves the use of non-polar solvents *viz.* hexane, petroleum ether (40–60°C) which percolates through a column containing a mixture of sample powder and anhydrous sodium sulphate. The oil gets eluted by the solvent and is collected in a conical flask.

Equipment and Glassware

1. Glass percolator.
2. Volumetric flask.
3. Distillation unit.
4. Cotton wool.
5. Hot plate/heating mentle/water bath.

Reagents

1. Anhydrous sodium sulphate
2. CCl_4 or hexane or petroleum ether (40–60°C) or solvent ether.

Procedure

1. Grind 1–2 g of seed material to a fine powder with the help of 20–30 g of anhydrous sodium sulphate and mix well.
2. Plug the percolator by putting cotton wool at the bottom of it. Put approximately 2 g of sodium sulphate (anhydrous) over the plug.
3. Pack the powdered material slowly inside the percolator carefully. Ensure that whole of the material is packed properly, pack little cotton over the material and cover with a layer of anhydrous sodium sulphate.
4. Add 10 ml solvent in the mortar and pestle to ensure complete removal of the material and transfer it into the percolator.
5. Keep a flask below the percolator and fill the percolator with the solvent.
6. Let the solvent tickle down under gravity. The solvent will flow down after extracting oil from the material.
7. Add more solvent, to fill the percolator and allow it to pass through the column. Repeat this thrice.
8. Flash evaporate or distill off the solvent using heating mentle/water bath. When 5 ml of solvent is left in the flask (round bottom in case of flash evaporator/ distillation), remove it and transfer it into a weighed conical flask and keep in the oven till the solvent is almost completely evaporated. Weigh to constant weight after keeping in dessicator for a few minutes.

Calculation

$$\text{Weight of flask + Glass bead} = X\,\text{g}$$

$$\text{Weight of flask + Glass bead + oil} = Y\,\text{g}$$

$$\text{Weight of oil} = Y - X\,\text{g}$$

$$\%\ \text{oil in the sample} = \frac{(Y - X) \times 100}{\text{weight of the material (g)}}.$$

13.13 Estimation of Fatty Acids by Gas–Liquid Chromatography (GLC) Morrison and Smith (1964); Sharma et al. (1981)

Analysis of complex fatty acid mixtures can be carried out by GLC. In this procedure the fatty acids are first converted into a volatile form, usually their methyl esters. The esters of fatty acids are identified by comparing with a set of standard fatty acid esters and quantified by the method of triangulation.

Reagents

1. Sodium methoxide (0.5 N): Dissolve sodium methoxide in dried methanol BF_3 – methanol reagent.
2. Hexane/Petroleum ether 40–60°C (Spectroscopic grade).
3. Sodium sulphate (anhydrous)
   All solvents should be anhydrous.

GLC Conditions

- Column – 10% DEGS (diethylene glycol succinate) in chromosorb P or W (60–80 mesh).
- Detector – Flame ionization detector (FID).
- Carrier gas – Nitrogen/Argon with a flow rate of 40–50 ml/min. hydrogen (0.5 kg/cm^2) and air (0.8 kg/cm^2) are also used in this detector as a fuel.
- Column temperature – 170–200°C.
- Detector temperature – 230°C.
- Injector port temperature – 230°C.
- Recorder speed – 1 cm/min.

Temperature programming: Temperature programming is resorted to when the mixture to be analysed contains components of widely varying chain lengths. Elution temperature rather than retention time forms this basis of identification. There is an approximate linear relationship chain. However, there are practical problems with temperature programmed GLC. Programming alters the flow rate of carrier gas through the columns and the detector. Polar columns baseline drift becomes critical and must be compensated by dual column operation or by an electronic baseline corrector. Temperature programming shortens the life of polyester columns.

Generally, the injector and detector temperature are kept 50°C above column temperature.

Procedure

1. Grind 0.1 g of oilseed with 5 ml of 0.5 N sodium methoxide in instalment of 1 ml each in a pestle-mortar. Transfer the contents to an air-tight screw capped 15 ml vial. Wash the pestle-mortar twice with 1 ml of 0.5 N sodium methoxide and transfer into vial. Take 20 mg oil (1–2 drops of oil) with 5 ml of 0.5 N sodium methoxide in screw capped vial.
2. Keep the vial in a boiling water bath for 10–15 min.
3. Cool to room temperature, add 1–2 drops BF_3 – methanol reagent and again heat for 5 min.
4. Cool and add 1–2 ml of hexane, shake and wait till hexane layer separates out. Take the appropriate amount (2–3 μl) of ethyl ester formed from hexane layer and inject GLC.
5. Measure the retention time and identify fatty acids by comparing with the retention time of standard. The area of each peak should be calculated by

measuring peak height and width at half height (Triangulation method). Calculate the percentage fatty acids.

$$\text{Peak area} = \text{Peak height} \times \text{width at half height.}$$

13.14 Determination of Lipase and Lipoxygenase Activity (Sardar and Joseph 1992; Shekhar and Reddy 1982)

The storage and other environmental parameters influence the nutritional status of food grains especially; the oilseeds and/or oil bearing materials. The two enzymes, namely; "Lipase" and "Lipoxygenase" are of special significance as both together contribute to the deterioration of fats and development of off flavours in soybean and brown rice during storage. The enzymes liase (Triacylglycerol acylhydrolase EC 3.1.1.3) hydrolyses triacylglycerols to release free fatty acids and glycerol which makes the food stuffs more susceptible to oxidative rancidity generally caused by lipoxygenase. Lipoxygenase (Linoleate:oxygen oxidoreductase, EC 1.13.11.12) is one of the primary catalysts of oxidation. It is implicated with heme proteins and specially catalyzes the oxidation of methylene interrupted unsaturated fatty acids such as linoleic, linolenic and arachidonic acid to their respective peroxides. In view of the involvement of lipase and lipoxygenase in deterioration of oil/fat, the methods of the determination of activity of these two enzymes are described below:

A. Lipase Activity

Principle

The quantity of fatty acids released in unit time is measured by titrating it against a standard alkali solution. Lipase activity can be expressed either in terms of 0.05 N NaOH used to neutralize free fatty acids librated from 1 g material or as micromoles of fatty acids released per minute.

Equipment and Glassware

1. Water bath with thermostat
2. Sonicator
3. Centrifuge

Reagents

1. Olive oil
2. Polyvinyl alcohol
3. Tris-Cl buffer (50 mM, pH 8.0)
4. Tris-Cl buffer (50 mM, pH 8.3)
5. Sodium hydroxide (0.05 N)

6. Acetone:Ethanol (1:1 v/v)
7. Phenolphthalein indicator

Preparation of Substrate

Take 100 ml of olive oil, add 2.5 g of polyvinyl alcohol, sonicate it, till an emulsion is formed.

Procedure

1. Grind 2 g of defatted sample thoroughly using acid washed sand, with 20 ml of 50 mM Tris-Cl buffer (pH 8.0) centrifuge at 20,000 × *g* for 30 min.
2. Collect the supernatant, which is used as crude intracellular lipase extract.
3. Prepare the reaction mixture as follows:

   (i) 2.5 ml of emulsified substrate
   (ii) 2 ml of 50 mM Tris-Cl buffer (pH 8.3)
   (iii) 1 ml of enzyme extract incubate at 37°C for 30 min

4. Stop the reaction after 30 min by adding 10 ml of acetone:ethanol (1:1) mixture.
5. Titrate the librated fatty acid with 0.05 N NaOH, using phenolphthalein indicator.
6. Express the result in terms of ml of 0.05 N NaOH used to neutralize free fatty acids librated from 1 g material.

Calculate in terms of units as one unit of lipase activity is defined as the amount of enzyme that liberate 1 macro mole of fatty acids/min/ml at 37°C.

1 Unit = μ moles of fatty acids released/min.

B. Lipoxygenase Activity

Principle

The enzyme extract (crude or purified) is allowed to react with the substrate (Linoleic acid). The rate of maximal A_{234} nm per minute between 1 and 3 min interval is used for calculating the specific activity.

Equipments and Glassware

1. Spectrophotometer
2. Refrigerated centrifuge
3. Centrifuge tubes
4. Muslin cloth
5. Volumetric flask (50 ml)

Assay of Lipoxygenase in Rice Sample

Procedure

A. Preparation of Enzyme Extract

1. Suspend 1 g of defatted rice flour in 4 ml of 0.1 M sodium phosphate buffer (pH 6.8) at 3–5°C for 45 min.

2. Squeeze the suspension through four layers of muslin cloth.
3. Centrifuge at 10,000 × *g* for 30 min to remove all the debris. Collect the supernatant.

B. Partial Purification and Assay

1. Adjust the pH of the supernatant to 5.2 with 1 M acetic acid. Centrifuge the turbid solutin at 10,000 × *g* for 30 min.
2. Adjust the pH of the supernatant to 6.8 with 1 M NaOH.
3. Add ammonium sulphate to the solution to 50% saturation.
4. Adjust the pH immediately to 6.8 with 2 N ammonia solution.
5. Centrifuge, collect the precipitate, dissolve in water and dialyze against water.
6. Prepare the substrate by ultrasonically dispersing 50 μl to Tween 20 and 5 ml double distilled water containing 35 μl linoleic acid (~35 mg).
7. Keep the solution at pH 9.0 by adding 0.2 M NaOH until all the linoleic acid gets dissolve and the pH remains stable.
8. After adjusting the pH to 6.5–7.0 by adding 0.2 M HCl, 0.1 M phosphate buffer (pH 6.5–7.0) is added to a total volume of 100 ml.
9. The substrate solution is flushed with and kept under a nitrogen atmosphere.
10. Take 2.0 ml substrate solution; add 100 μl of diluted enzyme extract and 0.9 ml of 0.1 M phosphate buffer (pH 6.5–7.0).
11. The increase in absorbance at 234 nm is measured for 1 min at 25°C (maximal ΔA_{234} per min).

The activity can be expressed as μmol hydroperoxide formed/min/mg protein using a molar extinction coefficient of 25,000/M/cm. The specific activity is expressed in terms of unit/mg protein.

13.15 Estimation of Carotenoid Pigments (Ranganna 1986)

The carotenoids are an abundant group of naturally occurring pigments, present in all green tissues; they are the constituents of the chloroplast and are responsible for most of the yellow to red colours of flowers and fruits. Carotenoid hydrocarbons are called carotenes, whereas derivatives containing oxygen functions are the xanthophylls. They may be acyclic (lycopene) or contain 5–6-membered rings at one or both ends of the molecule (β –carotene, lutein). Mango, papaya, leafy vegetables, pumpkin, wheat and durum wheat are rich sources of carotenes. Tomato, watermelon and apricot contain mainly lycopene, while pigment of red peppers is xanthophylls.

β-carotene is the precursor of vitamin A and is nutritionally important in human diet. Carbon structure of vitamin A is the same as that of half of – carotene, oxidation of the latter at the midpoint produces 2 molecules of vitamin A, but oxidation is not very efficient since 1 unit of carotene is equivalent to 0.6 unit of vitamin A. β-carotene is unstable in light, particularly light of short wavelength and ultraviolet. It is also unstable in acidic or alkaline medium and rapidly destroyed when heated in the presence of oxygen. Oxidation seems to be the main cause of destruction. The carotenes must be protected from oxidation and light. They are also sensitive to auto-oxidation. Yellow colour of the pigment is mainly because of conjugated double bonds, which are destroyed by light. The methods for the estimation of carotenoid pigments in fruits, vegetables and wheat are described below.

A. Estimation of Carotenoids in Fruits and Vegetables

(a) Total Carotenoids

Principle

Analysis of total carotenoids is based on the extraction of crude pigment mixture in a lipid solvent and measurement of its optical density at 460 nm. The pigment content is expressed as β-carotene. The sample is extracted in acetone which dissolves both the fat and water-soluble pigments. The acetone extract is then taken in petroleum ether layer. The fat soluble carotenoids pass from acetone to the petroleum ether leaving all the rest of the pigments in the acetone.

Equipments and Glassware

1. Separatory funnel
2. Pestle and mortar
3. Analytical balance
4. Spectrophotometer
5. Conical flasks

Reagents

1. Acetone
2. Petroleum ether
3. MgO (magnesium oxide)/alumina neutral (AI_2O_3)
4. Sodium sulphate (anhydrous)

Procedure

1. Weigh 1–2 g of fresh sample and grind it with acetone using acid and alkali washed sand in a mortar with pestle.
2. Pour the extract in a 250-ml conical flask.
3. Continue the extraction of the residue with acetone till residue is colourless (3 times).
4. Pool all the extracts in conical flask and transfer it into a separatory funnel.
5. Add 10–15 ml of petroleum ether in separatory funnel and shake thoroughly.
6. The yellow pigment is then transferred into the petroleum ether by diluting the acetone with water containing 5% sodium sulphate.
7. Keep on adding petroleum ether until all colour gets transferred into the petroleum ether layer.
8. Make up the volume with petroleum ether and measure the intensity of the colour at 460 nm. The results are expressed in terms of β-carotene as mg/100 g of the material.

(b) β-carotene

The β-carotene is usually estimated by Chromatographic separation.

Procedure

A. Weigh samples according to their expected β-carotene content (Spinach 1 g, carrot 2 g, peas 10 g and cauliflower 20 g).
B. Grind weighed sample with pestle and mortar with acetone till all the yellow colour is extracted and the material becomes white.
C. Transfer the extract to a separatory funnel; add petroleum ether to the acetone extract till all the colour taken up by the petroleum ether layer becomes colourless.
D. Vacuum distill the extract, till the volume is reduced to about 5 ml.
E. Load this carotene extract on to a column of alumina (10 × 1 cm) which has previously been activated by keeping in oven at 50–60°C for 48 h. the column also contains 3% anhydrous sodium sulphate.
F. Elute β-carotene with petroleum ether containing 3% acetone.
G. Note the volume of the eluate and measure the absorbance at 460 nm.

Alternatively, the column can be filled with supercell powder and magnesium oxide (MgO) in the ration 3:1 up to 10 cm in length and absorbent is pressed 2–3 times with plunger to ensure a tight column. Then place 1 cm of anhydrous Na_2SO_4 layer over top of the column. Wet the column with 25–50 ml of petroleum ether by leaving the last ml of petroleum ether above Na_2SO_4 layer. A 5–10-ml sample is then loaded.

Calculation

$$\beta\text{ - carotene (mg/100 g)} = \frac{3.1206 \times A460 \times \text{Vol. made up or dilution} \times 100}{\text{Weight of sample} \times 1{,}000}.$$

B. Estimation of Carotenoids in Wheat (AOAC 1962)

(a) β-Carotene

Yellow colour in durum wheats imparts attractive appearance to the pasta product and therefore, majority of the pasta consumers prefer the yellow pigment. Xanthophylls and specially β-carotene contributes to the colour production in the semolina. High lipoxygenase activity has been described to be responsible for an appreciable loss to β-carotene. Linoleic acid acts as a source of substrate for the enzyme to form hydroperoxides which ultimately oxidize the pigment. β-carotene acts as a preservative. Durum wheat endosperm contains twice the concentration of β-carotene than that of *T.aestivum.* Since β-carotene is highly susceptible to oxidation, precaution has to be taken for its proper determination.

Method 1

Equipment and Glassware

1. Spectrophotometer
2. Erlenmeyer flask
3. Volumetric flasks, 50, 100 and 250 ml
4. Whatman No. 1 filter paper

Reagents

1. Water-saturated n-butanol (WSB): Prepare a solution n-butanol and water in a proportion of 6:2 (v/v) and shake vigorously. Use the clear upper layer after separation of the phases
2. Diethyl ether
3. Synthetic β-carotene, crystalline

Procedure

1. Preparation of extract: Weigh 10 g of sample and disperse it in 50 ml of water-saturated n-butanol to give a homogenous suspension. Shake gently and allow it to stand overnight (16 h) at room temperature under the dark. Shake and filter completely through the filter paper (Whatman No. 1) into a 100-ml volumetric flask.
2. Standard solution of β-carotene: In a 100-ml volumetric flask, weigh 25 mg of β-carotene. Dissolve it in diethyl ether and make up to the mark with diethyl ether, 20 ml of this homogenous solution (= 5 mg β-carotene) is pipetted into a 250-ml volumetric flask. Make up to the mark with water-saturated n-butanol. Take 25 ml of this solution and place in a 100-ml volumetric flask and make up with water-saturated n-butanol. This standard solution has the following concentration. 1 ml = 0.005 mg = 5 μg β-carotene
3. Preparation of calibration curve: Prepare suitable dilutions of the standard solution with water-saturated n-butanol in calibrated 10 ml volumetric flasks (e.g. from 0.5 to 3 ml of standard solution in 10 ml). Measure the absorbance, of each dilution and establish a calibration curve (β-carotene in 10 ml of solution as a function of absorbance).
4. Determination of β-carotene content: Measure the absorbance of the clear filtrate at 440 nm. Unfiltered water-saturated n-butanol may be used as blank. Evaluation of the contents is based on a β-carotene calibration curve (related to the β-carotene content in a 10 mL solution).

Calculation

The yellow pigment, Yp, expressed as milligrams of β-carotene in 100 g dry matter, is

$$\mathrm{Yp} = \frac{a \times 5}{100 - H},$$

where,
a = β-carotene content of a 10-ml extract equivalent to 2 g of the test sample, in mg.
H = moisture content of the test sample, expressed as percentage by mass.

Method 2

This method can be used for screening a large number of samples for their β-carotene content.

Glassware and Reagents

Same as described previously.

Procedure

Preparation of extract

1. Weigh 8 g semolina or flour into 150 ml glass-stoppered Erlenmeyer flask and add 40 ml water-saturated n-butanol. Shake contents for 1 min and let it stand for 16 h. Shake and filter through Whatman No. 1 filter paper into 50 ml volumetric flask.
2. Measure transmission of extract in colorimeter at 440 nm employing standard containing reagent water-saturated n-butanol. To obtain the values for carotene content (ppm) the transmission reading of unknown sample is put in the equation.

$$\beta - \text{carotene (ppm)} = 0.174 + 16.57\,L,$$

where, L = Transmission (apparent density).

As described earlier yellow pigment is essentially a preferable feature of durum wheats. Range of β-carotene is generally 4–8 ppm but durums with less than 5 ppm of β-carotene are not acceptable in the international market. Some countries offer little preference to yellow pigment.

13.16 Estimation of Lycopene

Lycopene is responsible for the red colour of tomato, fleshy part of water melon, fruits and vegetables. It is a carotene with the formula $C_{40}H_{56}$. Colour of tomato due to lycopene has a great role in consumer acceptability.

Principle

The carotenoids in the sample are extracted in acetone and then taken up in petroleum ether. Lycopene has absorption maxima at 473 and 503 nm. One mole of lycopene when dissolved in 1 l light petroleum (40–60°C) and measured in a spectrophotometer at 503 nm in 1 cm light path gives an absorbance of 17.2×10^4. Therefore, a concentration 3.1206 μg lycopene/mL gives unit absorbance.

Materials

1. Acetone
2. Petroleum ether 40–60 (AR)
3. Anhydrous sodium sulphate
4. 5% Sodium sulphate

Procedure

1. Take sample of 3–4 tomato fruits and pulp it well to a smooth consistency in a waring blender.
2. Weigh 5–10 g of this pulp.
3. Extract the pulp repeatedly with acetone using pestle and mortar or a waring blender until the residue is colourless.
4. Pool the acetone extracts and transfer to a separating funnel containing about 20 mL petroleum ether and mix gently.
5. Add about 20 mL of 5% sodium sulphate solution and shake the separating funnel gently. Volume of petroleum ether might be reduced during these processes because of its evaporation. Therefore, add 20 ml more of petroleum ether to the separating funnel for clear separation of two layers. Most of the colour will be noticed in the upper petroleum ether layer.
6. Separate the two phases and re-extract the lower aqueous phase with additional 20 ml petroleum ether until the aqueous phase is colourless.
7. Pool the petroleum ether extracts and wash once with a little distilled water.
8. Pour the washed petroleum ether extract containing carotenoids into a brown bottle containing about 10 g anhydrous sodium sulphate. Keep it aside for 30 min or longer.
9. Decant the petroleum ether extract into a 100-ml volumetric flask through a funnel containing cotton wool. Wash sodium sulphate slurry with petroleum ether until it is colourless and transfer the washings to the volumetric flask.
10. Make up the volume and measure the absorbance in a spectrophotometer at 503 nm using petroleum ether as blank.

Calculation

Absorbance (1 unit) = 3.1206 μg lycopene/ml.

$$\text{mg lycopenein 100 g sample} = \frac{31.206 \times \text{Absorbance}}{\text{Wt. of sample (g)}}.$$

13.17 Estimation of Chlorophylls

The chlorophylls are the essential pigments for photosynthesis and are present in chloroplasts as green pigments in all photosynthetic plant tissues. They are bound loosely to proteins but are readily extracted in organic solvents such as acetone or ether.

Chemically, each chlorophyll molecule contains a porphyrin (tetrapyrole) head with a chelated magnesium atom at the centre and a long-chain hydrocarbon (phytol) side chain attached through a carboxylic acid group. There are at least five types of chlorophylls in plants. Chlorophylls a and b occur in higher plants, ferns and mosses. Chlorophylls c, d and e are only found in algae and in certain bacteria.

Principle

Chlorophyll is extracted in 80% acetone and are read at 663 and 645 nm using spectrophotometer. Using the absorption coefficients, the amount of chlorophyll is calculated.

Materials

80% Acetone (prechilled).

Procedure

1. Weigh 1 g of finely cut leaf or fruit tissue of sample into a clean mortar.
2. Grind the tissue to a fine pulp with the addition of 20 ml of 80% acetone.
3. Centrifuge at 5,000 rpm for 5 min and transfer the supernatant to a 100-ml volumetric flask.
4. Grind the residue with 20 ml of 80% acetone, centrifuge and transfer the supernatant to the same volumetric flask.
5. Repeat this procedure until the residue is colourless. Wash the mortar and pestle thoroughly with 80% acetone and collect the clear washings in the volumetric flask.
6. Make up the volume to 100 ml with 80% acetone.
7. Read the absorbance of the solution at 645, 663 and 652 nm against the solvent (80% acetone) blank.

   - The amount to tissue taken for extraction may be varied. Accordingly amount of 80% acetone used may be altered so that the final extract has volume based on 10 mg plant material extracted in 1 ml of acetone.

Calculation

Calculate the amount of chlorophyll present in the extract mg chlorophyll per g tissue using the following equations:

- $\text{mg chlorphyll a/g tissue} = 12.7(A_{663}) - 2.69(A_{645}) \times \frac{v}{1,000 \times w}$
- $\text{mg chlorophyll b/g tissue} = 22.9(A_{645}) - 4.68(A_{663}) \times \frac{v}{1,000 \times w}$
- $\text{mg total chlorophyll/g tissue} = 20.2(A_{645}) - 8.02(A_{663}) \times \frac{v}{1,000 \times w}$,

where A = absorbance at specific wavelengths,
V = final volume of chlorophyll extract in 80% acetone
and W = fresh weight of tissue extracted.

13.18 Estimation of Curcumin

Turmeric rhizome contains a number of pigments. It is used as a natural dye in food industries and in cosmetic and pharmaceutical products as an antimicrobial principle. The powder contains a large number of aromatic compounds; curcumin is the major compound responsible for the characteristic colour in Indian recipes. In pure form it is an orange yellow crystalline powder, soluble in alcohol and glacial acetic acid. Curcumin content is used as measure of turmeric quality.

Principle

Curcumin is quantitatively extracted by refluxing the material in alcohol and is estimated spectrometrically at 425 nm.

Materials

1. Absolute alcohol
2. Stoppered flask and air condenser

Procedure

1. Dissolve 0.2–0.5 g of weighed, moisture-free turmeric powder in 250 ml of absolute ethanol.
2. Reflux the contents in the flask fitted with an air condenser over a heating mantle for 3–5 h, compensate alcohol loss if any due to evaporation by adding alcohol freshly into the flask.
3. Cool and decant the extract into a volumetric flask and make up the volume.
4. Dilute a suitable aliquot (1–2 ml) to 10 ml with absolute alcohol. Measure the intensity of yellow colour at 425 nm using spectrophotometer.

Calculation

Work out the curcumin content using the formula

$$\text{Curcumin content g/100 g} = \frac{0.0025 \times A_{425} \times \text{volume made up} \times \text{dilution factor} \times 100}{0.42 \times \text{weight of the sample (g)} \times 1{,}000}.$$

since 0.42 absorbance at 425 nm = 0.0025 g curcumin.

13.19 Estimation of Thiamine

Thiamine is one of the vitamins of B complex group whose deficiency is associated with disease known as beriberi. This vitamin usually occurs in outer layers of grains including cereals. Therefore, unpolished rice and foods made of whole wheat are

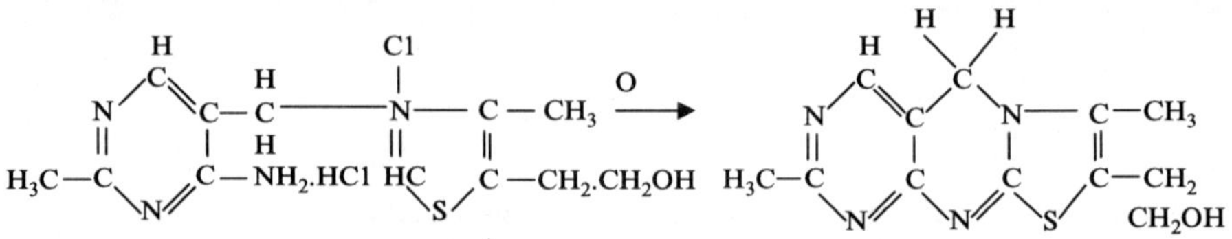

Fig. 13.1 Formation of thiochrome during estimation of thiamine

good sources of this vitamin. Thiamine is water-soluble, and overcooking may leach out or destroy thiamine originally present in the food sources. Among a number of chemical and micro-biological methods available for the estimation of thiamine, fluorimetric method is easier and popular (Fig. 13.1).

Principle

Alkaline potassium ferricyanide oxidizes thiamine to thiochrome which is a fluorescent compound. The thiochrome is extracted in isobutyl alcohol and measured in a fluorimeter.

Materials

1. 15% NaOH
2. 1% Potassium ferricyanide
3. Isobutyl alcohol
4. Anhydrous sodium sulphate
5. 0.1 N H_2SO_4
6. Standard thiamine hydrochloride stock solution: Dissolve 50 mg thiamine hydrochloride in 500 ml of about 0.1 N sulphuric acid containing 25% alcohol (100 μg/ml). Store this solution in a brown bottle in a refrigerator.
7. Working Standard Solution: Just before the experiment, dilute 5 ml of the stock solution to 100 ml with 0.1 N sulphuric acid and again dilute 5 ml of the second solution to 100 ml with 0.1 N H_2SO_4 (0.25 μg/ml) and use.

Procedure

Extraction of Thiamine

1. Weigh accurately 5 g finely ground sample into a 250-ml conical flask in duplicate.
2. Slowly add 10 mL 0.1 N H_2SO_4 without shaking; stopper the flask and let it stand overnight.

3. On next day shake the contents vigorously and filter through Whatman No.1 filter paper, discarding first 10–15 ml of filtrate.

Oxidation

1. Pipette out 10 ml of the extract in duplicate into 100 mL separating funnels.
2. Pipette out 10 ml of working standard (in 4–5 replicates).
3. Add 3 ml of 15% NaOH into each separating funnel immediately followed by four drops (0.2 ml) of ferricyanide solution.
4. Shake gently for exactly 30 s.
5. Add 15 ml of isobutanol rapidly from a quick delivery burette or a measuring cylinder.
6. Stopper immediately, shake vigorously for 60 s and allow the layers to separate.
7. Drain off the bottom layer carefully and discard it.
8. Add 1–2 g of sodium sulphate directly into the separating funnel, stopper and swirl gently to clarify the extract. If the extract is not clear, add a little more Na_2SO_4 and clarify.
9. Collect the clear extract from the top using a Pasteur pipette into a clean dry test tube.
10. Prepare a set of sample blanks by pipetting out 10 mL of the extract and follow the above procedure excluding addition of ferricyanide.
11. Prepare a blank for the standard (in duplicate) separately.
12. Select suitable primary (366 nm) and secondary filters as per the make of the fluorimeter.
13. Set fluorimeter by initially adjusting the standard blank to 0 reading and standard to 100. Then read the sample blank and sample. Since the light intensity sometimes changes progressively,5–6 readings should be recorded.

    Since the standard has to be read a number of times during the measurement it is convenient to oxidize the standard in 4 or 5 replicates and to combine all the oxidized extracts in one conical flask. Portions of this may be read at a time and discarded.

Calculation

$$\mu\text{g thiaminecontent in } 100\,\text{g sample} = \frac{0.25 \times 10}{a - a^1} \times \frac{(x - x^1) \times 100}{10} \times \frac{10}{5},$$

where

a = reading of standard = 100
a^1 = reading of standard bland = 0
x = reading of standard sample and
x^1 = reading of standard sample blank

13.20 Estimation of Riboflavin

Riboflavin (vitamin B_2) is present in milk, green leafy vegetables, egg etc. It is water-soluble and photosensitive vitamin.

Principle

Riboflavin fluoresces at wavelength 440–500 nm. The intensity of fluorescence is proportional to the concentration of riboflavin in dilute solution. The riboflavin is measured in terms of the difference in fluorescence before and after chemical reduction.

Reagents

1. 0.1 N H_2SO_4
2. Sodium acetate 2.5 M
3. Potassium Permanganate 4% (Prepare fresh every time)
4. Hydrogen peroxide 3%
5. Sodium hydrosulphite (solid)
6. *Riboflavin Standard*

Dry standard riboflavin over P_2O_5 in a desiccator for 24 h. Dissolve 50 mg in 1,500 ml water and 2.4 ml glacial acetic acid in a 2 l flask. Warm, cool and make up to volume. Store using toluene in amber bottles in a refrigerator (25 μg riboflavin).

Dilute 20 ml of the above riboflavin standard solution to 50 ml with water. For working solution, dilute 10 ml of the second riboflavin standard solution with water to 100 ml. Prepare this dilution fresh. Protect from light. (1 ml = 1 μg riboflavin).

Procedure

1. Weigh accurately 2–5 g of thoroughly mixed and ground sample into a 250-ml conical flask.
2. Add 75 ml 0.1 N H_2SO_4 and mix.
3. Autoclave at 15 lbs for 30 min or immerse the flask in boiling water for 30 min. Shake the flask every 5 min. Let it cool to room temperature.
4. Add 5 ml 2.5 M sodium acetate solution, mix and let stand for at least 1 h. The solution will have approximately pH near 4.5.
5. Transfer to a 100-ml volumetric flask and make up the volume with water. While transferring, wash the conical flask completely to transfer the material without any loss.
6. Filter through medium–fast paper such as Whatman No. 2 or No. 4 discarding the first 10–15 ml.

7. In test tubes of 1 in. diameter marked A and B with stirring bars carry out oxidation as follows:

Tube B	Low/blank tube A	Sample tube B	High/blank tube A	Sample
Sample solution (ml)	10	10	10	10
Standard solution (ml)	1	–	1	–
Water (ml)	1	2	–	1.0
$KMNO_4$(4%)[a]	0.5	0.5	1.0	1.0
Time gap (min)	2	2	4	4
H_2O_2(3%)[b]	0.5	0.5	1.0	1.0

[a]Stir samples after addition of permanganate
[b]Shake after adding peroxide until foaming is negligible

8. Set the fluorimeter so that glass standard or sodium fluorescein solution gives suitable galvanometer deflection, as directed in the instruction manual of the instrument. Determine fluoroscence of solutions A and B. Do not expose the solution for more than 10 s.
9. To solution B in the cuvette, add 20 mg sodium hydrosulphite, stir and note the blank fluorescence (C). Do not use reading (C) taken after colloidal sulphur begins to form.

Calculation

$$\text{Riboflavin } \mu\text{g}/100\,\text{g} = \frac{B - C}{A - B} \times \frac{R}{S} \times \frac{V}{V_1} \times 100,$$

where

A = reading of the sample plus riboflavin standard
B = reading of sample plus water
C = reading after adding sodium hydrosulphite
R = standard riboflavin added = μg/V_1 of sample solution

In this dilution $R = 1$; $V = 100$ and $V_1 = 10$.

13.21 Estimation of Niacin

Niacin is also known as nicotinamide or nicotinic acid depending upon the presence or absence of the amide group. It is also a water-soluble vitamin and present in all living organisms. It is physiologically very important since two coenzyme, NAD^+, and $NADP^+$ are derived for this vitamin.

Principle

Niacin reacts with cyanogen bromide to give a pyridinium compound which undergoes rearrangement forming derivatives. These derivatives couple with aromatic amines to give yellow coloured pigment. Under proper conditions, the intensity of the yellow colour produced is proportional to the amount of niacin present.

Materials

1. *Extraction*: Grind the sample (5 g) in 4 N sulphuric acid (30 ml) and steam it for 30 min. Cool and make up to 50 ml with distilled water. Filter through Whatman No.1 filter paper. Add 60% basic lead acetate to the filtrate (5–25 ml of filtrate). Adjust to pH 9.5 with a pH metre or thymol blue indicator, using 10 N sodium hydroxide and centrifuge. To the supernatant add 2 ml of Conc. H_2SO_4 and allow to stand for 1 h and again centrifuge and collect the supernatant. Add 5 ml 40% $ZnSO_4$ and adjust to pH 8.4 with 10 N NaOH. Centrifuge and collect the supernatant and adjust to pH 7. If precipitate appears, centrifuge and use the supernatant as the source material.
2. *Cyanogen bromide*: Keep the bromine bottle in an ice bath for a few minutes and then using a vacupet, pipette out 25 ml of bromine into a beaker or conical flask containing 500 ml distilled water. Take a 10% solution of sodium cyanide in a burette and deliver slowly to the bromine water with constant shaking until the solution becomes colourless.
3. Redistilled 4% aniline in absolute alcohol
4. *Standard*: Dissolve 10 mg niacin and make up to 100 ml with distilled water in a volumetric flask. Alternatively 100 mg is dissolved in 100 ml and then 10 ml of this is again diluted to 100 ml. One ml of the standard contains 100 μg niacin.

Procedure

1. Into a series of test tubes pipette out 0.1–0.5 ml of the standard niacin solution.
2. Take two or three different volumes of sample solution in other test tubes (0.1–0.5 ml).
3. Make up the volume to 6 ml with distilled water in all the tubes.
4. Add 3 ml of cyanogen bromide using a burette and shake the contents in the tube well. After 10 min add 1 ml of 4% aniline to each tube.
5. Read the yellow colour developed after 5 min at 420 nm against a reagent blank.

Calculation

Draw a standard graph with the readings of the standard niacin. Find out the niacin content in the sample volume taken and calculate for g weight of the sample.

13.22 Estimation of Ascorbic Acid

(1) Volumetric Method

Ascorbic acid/vitamin C is present in high amounts in fruits, fresh vegetables and other sources. It is a water-soluble and heat-labile vitamin.

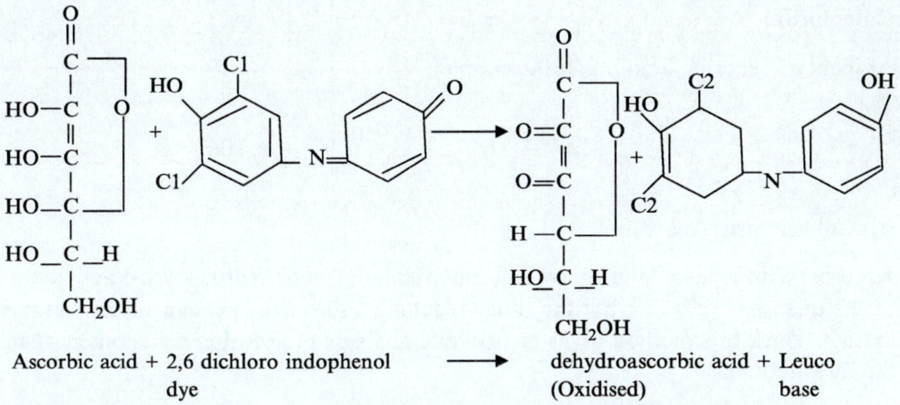

Fig. 13.2 Oxidation of ascorbic acid to dehydroascorbic acid during estimation

Principle

Ascorbic acid reduces the 2, 6-dichlorophenol indophenol dye to a colourless leuco-base. The ascorbic acid gets oxidized to dehydroascorbic acid. Though the dye is a blue coloured compound, the end point is the appearance of pink colour. The dye is pink coloured in acid medium. Oxalic acid is used as the titrating medium (Fig. 13.2).

Materials

1. Oxalic Acid 4%.
2. *Dye Solution*: Weigh 42 mg sodium bicarbonate into a small volume of distilled water. Dissolve 52 mg 2, 6-dichloro phenol indophenol to it and make up to 200 ml with distilled water.
3. *Stock Standard Solution*: Dissolve 100 mg ascorbic acid in 100 ml of 4% oxalic acid solution in a standard flask (1 mg/ml).
4. *Working Standard*: Dilute 10 ml of the stock solution to 100 ml with 4% oxalic acid. The concentration of working standard is 100 μg/ml.

Procedure

1. Pipette out 5 ml of the working standard solution into a 100-ml conical flask.
2. Add 10 ml of 4% oxalic acid and titrate against the dye (V_1 ml). End point is the appearance of pink colour which persists for a few minutes. The amount of the dye consumed is equivalent to the amount of ascorbic acid.
3. Extract the sample (0.5–5 g depending on the sample) in 4% oxalic acid and make up to a known volume (100 ml) and centrifuge.
4. Pipette out 5 ml of this supernatant, add 10 ml of 4% oxalic acid and titrate against the dye (V_2 ml).

Calculation

Amount of ascorbic acid mg/100g sample

$$= \frac{0.5\,\text{mg}}{V_1\,\text{ml}} \times \frac{V_2}{5\,\text{ml}} \times \frac{100\,\text{ml}}{\text{Wt. of the sample}} \times 100.$$

(ii) Colorimetric Analysis

Ascorbic acid is also determined colorimetrically. The dehydroascorbic acid alone reacts quantitatively and not the other reducing substances present in the sample extract. Thus, this method gives an accurate analysis of ascorbic acid content than the dye method.

Principle

Ascorbic acid is first dehydrogenated by bromination. The dehydroascorbic acid is then reacted with 2, 4 dinitrophenyl hydrazine to form osazone and dissolved in sulphuric acid to give an orange-red colour solution which is measured at 540 nm.

Materials

1. 4% Oxalic acid solution
2. 0.5 N Sulphuric acid
3. 2% 2, 4 Dinitrophenyl hydrazine (DNPH) reagent. Dissolve by heating 2 g DNPH in 100 ml 0.5 N H_2SO_4, filter before use.
4. 10% Thiourea solution
5. 80% Sulphuric acid
6. Bromine water: Dissolve 1–2 drops of liquor bromine in approximately 100 ml cool water.
7. *Ascorbic Acid Stock Solution:* Dissolve 100 mg ascorbic acid in 100 ml of 4% oxalic acid solution in a standard flask (1 mg/ml)
8. *Working Standard:* Dilute 10 ml of the stock solution to 100 ml with 4% oxalic acid. The concentration of working standard is 100 μg/ml.

Extraction

Grind 0.5–5 g of sample material either mechanically or using a pestle and mortar in 25–50 ml 4% oxalic acid solution. Centrifuge/filter and collect the liquid.

Transfer an aliquot (10 ml) to a conical flask and add bromine water dropwise with constant mixing. The enolic hydrogen atoms in ascorbic acid are removed by bromine. When the extract turns orange yellow due to excess bromine, expel it by blowing in air. Make up to a known volume (25 or 50 ml) with 4% oxalic acid solution. Similarly, convert 10 ml of stock ascorbic acid solution into dehydro form by bromination.

Procedure

1. Pipette out 1–100 μg standard dehydroascorbic solution into a series of tubes.
2. Also pipette out different aliquots (0.1–2 ml) of brominated sample extract.
3. Make up the volume in each tube to 3 ml by adding distilled water.
4. Add 1 ml of DNPH reagent followed by 1–2 drops of thiourea to each tube.
5. Set a blank as earlier but with water in place of ascorbic acid solution.
6. Mix the contents of the tubes thoroughly and incubate at 37°C for 3 h.
7. After incubation dissolve the orange-red osazone crystals formed by adding 7 ml of 80% sulphuric acid.
8. Measure absorbance at 540 nm.
9. Plot a graph ascorbic acid concentration vs. absorbance and calculate the ascorbic acid content in the sample.

Liquor bromine can cause burns: Prechill the ampoule containing bromine prior to cut open it.

13.23 Biological Evaluation of Protein Quality by Rat Feeding Experiments (Eggum 1973)

The nutritional quality of protein is best determined by in vivo experiments with rats which give more practical value than the chemical analysis of protein based on their amino acid composition which gives information about the likely deficiencies. However, this is not adequate to evaluate precisely the biological availability of the proteins for healthy growth of animals. Correct idea about the biological value can be obtained by measurements involving nitrogen balance. The biological value of a dietary protein is defined as the fraction of absorbed nitrogen retained in the body for maintenance and growth. The calculation of biological value (BV) therefore, requires an estimation of the amount of nitrogen absorbed into the body and the amount of absorbed nitrogen which is retained. This method is based upon direct determination of the amount of nitrogen in the faeces and urine and indirect determination of the fractions of the faecal nitrogen and of the urinary nitrogen that were of a dietary origin. The biological value of the protein is taken as the percentage of the absorbed nitrogen (Nitrogen intake – faecal nitrogen of dietary origin) that is not eliminated in the urine. This can be represented as:

$$\mathrm{BV} = \frac{\begin{array}{c}\text{N intake} - (\text{Faecal N} - \text{Metabolic N}) - \\ (\text{Urinary N} - \text{Endogenous N})\end{array}}{\text{N intake} - (\text{Faecal N} - \text{Metabolic N})} \times 100.$$

In the numerator faecal losses subtracted from the total intake are related to the part actually digested and the urinary losses reduced by its endogenous fraction before being subtracted. The numerator, therefore, represents the total nitrogen utilized including both that was used in maintenance and that incorporated into the growing tissues. Since the metabolic nitrogen is also subtracted from the total faecal output in the denominator, the biological value computed in the percentage of the digested nitrogen that is actually utilized. This provides a measure of the efficiency of absorbed protein for the combined function of growth and maintenance. The metabolic and endogenous nitrogen are estimated from the data obtained when rats are given a diet containing whole egg protein at a low concentration. It is assumed that the egg protein is completely digested and utilized by the growing rat so that faecal and urinary nitrogen excreted represents unavoidable metabolic and endogenous losses. The metabolic nitrogen in the faeces has been found to be related to the intake of dry feed and the endogenous urinary nitrogen to either the body weight or by a logarithmic function of the body weight.

Principle

The nitrogen content of the test diet is determined.

The feed consumption over a 5 days period is measured, leading to a calculation of the total nitrogen intake. The faecal and urinary nitrogen in these 5 days are also determined.

Based on nitrogen balance studies, Net Protein utilization (NPU), True Digestibility (TD), utilizable nitrogen (UN) and Retention coefficient (RC) can also be calculated.

$$\mathrm{TD} = \frac{\mathrm{N\,intake} - (\mathrm{faecal\ N} - \mathrm{metabolic\ N})}{\mathrm{N\ intake}} \times 100$$

$$\mathrm{NPU} = \frac{\mathrm{TD.\ BV}}{100}$$

$$\mathrm{UN\ (Utilizable\ nitrogen)} = \frac{\mathrm{NPU.\ N\,(in\ \%\ of\ dry\ matter)}}{100}$$

$$\mathrm{RC\ (Retention\ coefficient)} = \frac{\mathrm{N\ intake} - (\mathrm{faecal\ N} + \mathrm{Urinary\ N})}{\mathrm{N\,intake}} \times 100$$

Equipment and Glassware

1. Rat cages complete with trolleys.
2. Balance for weighing rats.
3. Analytical balance for weighing diet.
4. Diet mixer.
5. Sieves (100 mesh).
6. Plastic containers (2 l).

Other Requirements

1. Rats (male, albino rats Wistar strain about 70.0 g in weight).
2. Casein.

3. Mineral mixture consists of the following:

Calcium carbonate ($CaCO_3$)	68.6 g
Calcium citrate ($Ca_3C_{12}H_{10}O_{14}\cdot 4H_2O$)	308.3 g
Calcium hydrogen phosphate ($CaHPO_4\cdot 2H_2O$)	112.8 g
Dipotassium hydrogen phosphate (K_2HPO_4)	218.8 g
Potassium chloride (KCl)	124.7 g
Sodium chloride (NaCl)	77.1 g
Magnesium sulphate ($MgSO_4$)	38.3 g
Magnesium carbonate ($MgCO_3$)	35.2 g
Ammonium ferric citrate (brown, 20.5–22.5% Fe)	15.3 g
Manganese sulphate ($MnSO_4\cdot H_2O$)	0.201 g
Copper sulphate ($CuSO_4\cdot 5H_2O$)	0.078 g
Potassium iodine (KI)	0.041 g
Aluminium ammonium sulphate ($AlNH_4\ (SO_4)_2\cdot 12H_2O$)	0.090 g
Sodium fluoride (NaF)	0.507 g

4. Vitamin mixture consists of the following:

1.000 g ~ 200.000 I.U. vitamin A
0.008 g ~ 15.000 I.U. vitamin D_3
0.040 g thiamine (vitamin B_1)
0.100 g riboflavin (vitamin B_2)
0.020 g α-tocopherol (vitamin E)
0.010 g pyridoxine (vitamin B_6)
0.400 g nicotinamide
0.100 g pantothenic acid
N – free starch to make up 1 kg. Mix thoroughly.

5. N – free diet consists of the following:

Sucrose	9.00%
Cellulose powder	5.20%
Soybean oil (or groundnut oil)	5.20%
N-free Starch (Potato starch autoclaved)	80.60%

Preparation of Diets

Five hundred gram diet (dry weight basis) is prepared for each treatment. Initially determine the protein and dry matter content in the sample, then calculate the amount of sample required to give 7.5 g Nitrogen on dry weight basis. To this add 20.0 g mineral mixture and 8.0 g vitamin mixture in order to make up 500 g diet (dry wt. basis) add rest of the amount from N-free diet after determining the moisture in N-free diet.

Procedure

1. Weigh 40.0 g diet on dry weight basis (actual weight will very depending upon the moisture in the diet) into boxes for each rat sufficient for preliminary

period of 4 days. Thus, each animal receives 150 mg N and 10 g dry matter per day throughout the preliminary period.

2. Put cages on the rack without funnel.
3. Weight the rats (approximately 75 g) in the beginning of the experiment, divide the rats into groups of five, such that average weight of the group differs by not more that ± 0.5 g and record the weight (lesser the difference better the standardization).
4. Transfer diet equivalent to 10 g dry weight to the feed box of respective cage. In all the experiments group 1 is always kept for casein diet.
5. Press the feed with suitable flat surface.
6. Put plastic bowls below each rat cage for the collection of urine and faeces.
7. Feed every rat once a day in the morning and check for the water in bottles and clean plastic bowls in which urine and faeces have been collected with water.
8. After transferring feed to cage feed boxes, reposition the wire netting properly and put the clips in the slot provided, clamping one of the wire of the gauge.
9. On the last day of the preliminary period, when all the diet from diet box had been transferred to the rat cage feed box, weigh 50 g equivalent of dry weight diet and transfer to respective aluminium boxes. Apply Vaseline grease to nylon net as well as Perspex funnel.
10. At the end of the preliminary period of 4 days, again weigh the rats and clean the rat cage feed box as well.
11. Put both the greased Perspex funnel and nylon net in proper position.
12. Clean inside and bottom lid of the rat cage with lukewarm water. Transfer the top lid to the bottom and then transfer the rat after weighing to the respective cage. Then close the cage with earlier bottom lid.
13. Secure the rat cages with the clips on both the sides and follow this procedure for all the rats.
14. Transfer 10.0 g dry weight equivalent diet from respective boxes to cage feed box and follow as in preliminary period for the 5 days of N-balance period.
15. Put flask containing 35 ml 5% H_2SO_4 and funnel with glass wool below the Perspex funnel to collect the urine and beaker containing 50 ml 5% H_2SO_4 below nylon net to collect faeces.
16. During 5 day of N-balance period, daily weighing of the feed from diet box and its transfer to cage feed boxes is followed. Any of the faeces remaining, at the neck of the nylon netting is also transferred to beaker with the help of forceps. If by chance faeces have fallen on the funnel, these are also removed and put in respective beakers.
17. Every morning spray the net in situ with small quantity of 20% citric acid from a plastic wash bottle to prevent N losses and then finally wash the glass wool with small quantity of 5% H_2SO for the same reason.
18. Thus following this procedure, during the N-balance period of 5 days, urine and faeces are collected respectively in flasks and beakers.
19. At the end of the N-balance period, remove all the water bottles and prevent the access of rats to feed box 1 h before the termination of the experiment.
20. Weigh the rats and transfer to bigger cage.

21. Transfer any remaining feed in the feed boxes and spill tray to the respective diet boxes.
22. Wash bottom lids and lower portion of the cage, funnel and nylon net with approximately 75.0 ml of lukewarm water, using a soft brush through a large glass funnel down the urine flask with funnel and glass wool. Further wash funnel with glass wool 3 times to ensure that all N has been washed out.
23. Transfer urine and washings quantitatively to a graduated 500 ml flask, make up the volume. Mix well.
24. To beaker containing faeces, add 4 times 25 ml conc. H_2SO_4 at hourly intervals. After each addition stir and mix thoroughly with spatula and allow it to cool.
25. This process if followed 4 times, the resultant faeces solution would become homogenous. Transfer this mixture to 500 ml volumetric flask and make up the volume.
26. Weigh the remaining feed in diet boxes and record in the note book.
27. Determine N in urine and in faeces by taking out 25 ml sample of urine and 50 ml sample of faeces by macro-Kjeldahl method. Samples are taken in duplicate.
28. Calculate the total amount of N excreted in urine and in faeces by each rat during balance period.
29. Calculate biological value using the following formula.

$$BV = \frac{\text{N intake} - (\text{faecal N} - \text{Metabolic N}) - (\text{Urinary N} - \text{Endogenous N})}{\text{N intake} - (\text{faecal N} - \text{Metabolic N})} \times 100.$$

Protein Efficiency Ratio (PER)

It is the gain in body weight of a growing animal for per gram of protein intake. PER is the simplest and the most obvious measure of the nutritive value. It varies with the level of protein in the diet, though the convention is to give the protein at a constant level (10%). The PER is determined mainly in feeding experiments with small animals like albino rats, mice etc. and has also been used in studies on infants.

In this determination feeding with test protein is carried out for 4 weeks and the ratio of weight gain to protein consumed is calculated as PER.

$$PER = \frac{\text{(g) gained in body weight}}{\text{(g) protein intake}}$$

Materials

1. Weaning male rats of Wistar strain 20–23 days old. Five to ten animals for each diet. Body weight 50–55 g.
2. Basal diet on an air-dried basis:

   Corn starch – 80%
   Corn oil or groundnut oil – 10%
   Non-nutritive cellulose – 5%

Salt mixture – 4%
Vitamin mixture – 1%

3. Individual cages provided with feeder.

Procedure

1. Randomize 5–10 rats for each diet and feed them for 4 weeks with diet and water at uniform level.
2. Record the food consumption and body weight at weekly intervals.
3. For PER determination: In addition to the test group, maintain reference groups of rats on a diet consisting of the basal ratio with casein to the level of 10% protein.
4. For NPR determination: In addition to the test group, maintain a control group of rats, matched with test animals with respect to weights, on a diet consisting of the unmodified basal ratio.

13.24 Determination of In Vitro Protein Digestibility in Food Grains

The nutritive value of a protein depends primarily on its capacity to supply needs of nitrogen and essential amino acids. Although, the chemically determined amino acid composition is used to measure the quality of a protein, the biological availability of these amino acids is the real measure of the quality of the proteins. The availability of amino acids depends upon the extent of digestibility of proteins by the proteolytic enzymes of the alimentary tract. Digestibility of a protein can be assessed using rats which is termed "in vivo" digestibility. It can also be measured using proteolytic enzymes and called "in vitro" protein digestibility.

1st Method (Satterlee et al. 1979)

Principle

Proteolytic enzymes are used to digest the protein and the pH change due to the release of amino acids at the fixed time interval is measured. By using the given formula:

$$\%\,\text{digestibility} = 234.84 - 22.56\,X,$$

where "X" is the pH after 20 min incubation the in vitro digestibility can be calculated.

Equipment and Glassware

1. pH metre
2. Water bath

Reagents

1. Powdered sample which passes through 80-mesh sieve.
2. Glass distilled water.
3. Three enzyme solution: 1.6 mg trypsin, 3.1 mg chymotrypsin and 1.3 mg peptidase per ml in glass distilled water.
4. Bacterial protease solution 7.95 mg protease (type IV from *Streptomyces griseus*) per ml in glass distilled water.

Procedure

1. Add 10 ml of glass distilled water to the powdered sample of the grain (amount of sample is adjusted so as to contain 6.25 mg protein/ml).
2. Allow the sample to hydrate for at least 1 h (not longer than 25 h) at 5°C.
3. Equilibrate the sample to pH 8.0 at 37°C.
4. Equilibrate the three enzyme solution pH 8.0 at 37°C.
5. Add 1 ml of three enzyme solution to the sample suspension and stir while being held at 37°C.
6. Exactly after 10 min from the time of adding the three enzyme solution (still stirring) add 1 ml of the bacterial protease solution.
7. Immediately transfer the solution to water bath maintained at 55°C.
8. Nine min after adding the bacterial enzyme, transfer the solution back at 37°C water bath (in total 19 min after the addition of the three enzyme solution).
9. Exactly after 10 min of adding the bacterial enzyme, the pH of the hydrolysate is measured (designated as 20 min pH).

Calculation

In vitro protein digestibility is calculated using the following equation:

$$\%\,\text{digestibility} = 234.84 - 22.56\ \text{X}$$

Where "X" is the pH after 20 min of incubation.

2nd Method (Saunders and Kohler 1972)

Equipment and Glassware

1. Shaking water bath
2. Centrifuge
3. Micro-Kjeldahl apparatus
4. Digestion unit
5. Water bath
6. Oven

Reagents

1. Sodium barbiturate buffer (0.02 M, pH 7.0)
2. Digestive enzyme solution: 5 mg pronase and 1 mg trypsin dissolved in 5 ml of 0.02 M barbiturate buffer (pH 7.0) containing 0.15 M NaCl.
3. Trichloroacetic acid (2.5%)

Procedure

1. Weigh 100 mg of finely ground (100 mesh) defatted sample in a glass centrifuge tube and add 5 ml water.
2. Heat on a boiling water bath for 5 min and cool it to room temperature.
3. After cooling add 0.5 ml of digestive enzyme solution.
4. Mix gently and shake for 18 h at 30°C.
5. Stop the reaction of adding 0.05 ml TCA solution.
6. After 30 min, centrifuge at 10,000 × *g* for 10 min.
7. Transfer the supernatant as well as the residue to the nitrogen digestion tubes separately and estimate the nitrogen in each of the fractions.
8. Estimate N in the sample also.

Calculation

1. $\text{Protein digestibility (\%)(Supernatant)} = \frac{\text{N in supernatant}}{\text{N in sample}} \times 100$

2. $\text{Protein digestibility (\%)(Residue)} = \frac{\text{N in sample} - \text{N in residue}}{\text{N in sample}} \times 100$

   The two values are averaged

13.25 Estimation of Blood Glucose and Cholesterol

Biological evaluation of nutritional quality of food grains is carried out by conducting feeding trials with diets composed of test materials or their dietary constituents (proteins, fat, carbohydrate, crude fibre) on animals especially albino rats. In order to get more insight into the metabolic changes that have been brought about, in addition to determining digestibility, biological value or animal growth efficiency, it is also important to record information of other parameters like blood glucose and cholesterol. The level of these two molecules is likely to get affected by the intake of test diets. Any change, if occurring, in these moieties, it is of special significance. The methods of estimation are given below:

1. Blood Glucose (Nelson 1994)

Principle

Blood proteins are precipitated by zinc hydroxide. The filtrate is heated with alkaline copper reagent and the reduced copper so formed is treated with arsenomolybdate reagent resulting in the formation of violet colour the intensity of which can be read at 500 nm.

Equipment and Glassware

1. Spectrophotometer
2. Water bath
3. Volumetric flasks (100 ml)
4. Micro pipettes

Reagents

1. 5% zinc sulphate solution.
2. 0.3 N Barium hydroxide: Dissolve 15 g of $Ba(OH)_2$ or 28 g of $Ba(OH)_2{\cdot}8H_2O$ in 500 ml of hot water, boil for few min, cool filter and keep it in stoppered bottle.

   These two solutions should be so adjusted that 5 ml zinc sulphate require 4.7–4.8 ml barium hydroxide for complete neutralization. Phenolphthalein can be used as an indicator.
3. Arsenomolybdate reagent (Nelson's reagent) – 25 g of ammonium molybdate is dissolved in 450 ml of water, and 21 ml concentrated sulphuric acid is slowly added with stirring. Then 3 g of sodium hydrogen arsenate dissolved in 25 ml distilled water, is poured slowly with constant stirring. The solution is incubated at 37°C for 24 h, filtered and stored in amber coloured reagent bottle.
4. Alkaline copper reagent

   (a) Copper reagent A – 25 g of anhydrous sodium carbonate, 25 g of sodium potassium tartrate (Rochelle salt) and 20 g sodium bicarbonate are dissolved in about 700 ml of water. Finally, add 200 g of anhydrous sodium sulphate and dilute the solution to one litre, and allow to keep for 2 days at room temperature. Collect the clean supernatant by filtering.
   (b) Copper Reagent B: 15 g $CuSO_4{\cdot}5H_2O$ is dissolved in 100 ml water. Add 1 drop of H_2SO_4.
   (c) Working copper reagent C (Alkaline Copper reagent): Prepare fresh before use by mixing 25 ml of copper reagent A and 1 ml of copper reagent B.

5. Standard glucose solution (Stock solution): 1.0 g of anhydrous pure glucose is dissolved in 10–15 ml 0.2% benzoic acid and diluted to 100 ml with benzoic acid solution.

Three working standards are prepared by diluting 0.5, 1.0 and 2.0 ml of the stock solution to 100 ml with benzoic acid. The solutions in benzoic acid can be kept indefinitely at room temperature.

Procedure

1. Procure 0.1 ml blood from the experimental animal into a clean dry centrifuge tube containing 3–5 ml of distilled water, mix well.
2. Add 0–2 ml of 0.3 N barium hydroxide solution to the diluted blood sample, after the mixture turned brown, add 0.2 ml zinc sulphate.

3. Shake well and centrifuge at 3,000 rpm for 20 min.
4. Take 1.0 ml of the above aliquot in a test tube and add 1 ml of alkaline copper reagent to it. Cover the tubes with glass marble. Keep it in a boiling water bath for 20 min.
5. After 20 min. cool the tubes under running water, add 1 ml of arsenomolybdate reagent and dilute with water to 10 ml.
6. Run simultaneously standard and reagent blank. The intensity of colour produced is read at 500 nm. The colour formed is stable.

Calculation

$$\text{mg glucose/100 ml} = \frac{\text{Reading of unknown}}{\text{Reading of standard}} \times \text{conc. of standard} \times 4 \times \frac{100}{0.1} \times \frac{1}{100}.$$

B. Cholesterol in Serum

Principle

Serum proteins can be dispersed by using 2,5 dimethylbenzene sulphonic acid (DMBSA) together with acetic anhydride and acetic acid, with the evolution of heat. On addition of sulphonic acid to the mixture blue green sulphuric acid derivative of cholesterol is produced, the absorbance of which is read at 560 nm.

Equipments and Glassware

1. Volumetric flasks 100 ml
2. Spectrophotometer
3. Water bath
4. Pipette

Reagents

1. Glacial acetic acid.
2. Dissolve 11.1 g of DMBSA in 400 ml of glacial acetic acid and make up the volume to 1 l with acetic anhydride.
3. Concentrated sulphuric acid.
4. Standard: Dissolve 200 mg of cholesterol in 70–80 ml of glacial acetic acid by warming. Make up the volume to 100 ml in a volumetric flask.

   Store all the reagents at room temperature. All are stable for 1 year.

Procedure

1. Take 0.1 ml each of serum, distilled water (blank) and standard in three test tubes.
2. Add 0.1 ml of acetic acid (Reagent 1) to the test sample and the blank. Add 0.1 ml of water to the standard.

3. Add 2.5 ml DMBSA solution (Reagent 2) to each tube. Mix well and keep all the tubes at 20–25°C, then add 0.5 ml of sulphuric acid with constant shaking and maintain the temperature below 25°C.
4. Mix the contents well to dissolve the precipitate that may be formed. Keep the tubes for exactly 20 min at 20–25°C.
5. Read immediately the intensity of the colour against blank at 560 nm and calculate the amount of cholesterol by comparing with standard.

Calculation

Concentration of cholesterol in test sample (mg/100 ml)

$$= \frac{\text{Absorbance of sample}}{\text{Absorbance of standard}} \times 100$$

13.26 Estimation of High Density Lipoprotein (HDL), Low Density Lipoprotein (LDL) and Very Low Density Lipoprotein (VLDL) Cholesterol in Serum

Principle

HDL and VLDL + LDL cholesterol are separated from serum by precipitation of VLDL + LDL with phosphotungstate in the presence of magnesium ions. HDL cholesterol is estimated from the supernatant and VLDL + LDL cholesterol is estimated from the precipitate.

The reagent ferric acetate/uranyl acetate in acetic acid is used to precipitate proteins and for the elimination of different interfering substances namely; bilirubin, haemoglobin. It also prevents turbidity caused by lipemia and provides a clear and colourless blank solution. Ferrous sulphate in sulphuric acid is used to develop the purple colour which is measured at 560 nm.

Equipment and Glassware

1. Spectrophotometer
2. Conical centrifuge tubes (50 ml)
3. Centrifuge

Reagents

1. Ferric acetate/uranyl acetate reagent: Dissolve 500 mg of ferric chloride hexahydrate in 40 ml water in a 50 ml conical centrifuge tube, add 3.0 ml 28% aqueous ammonia solution, mix and centrifuge. Remove supernatant, suspend the precipitate in 40 ml water and centrifuge again. Wash the precipitate 4–5 times until the supernatant gives a negative test of chloride with $AgNO_3$. After the last centrifugation aspirate the supernatant solution and transfer the precipitate in 1 l

flask containing glacial acetic acid, add 100 mg of uranyl acetate. Dissolve the precipitate of ferric hydroxide and uranyl acetate in about 600 ml glacial acetic acid. Make up the volume to 1.0 l with glacial acetic acid and mix well.

2. Ferrous acetate/sulphuric acid reagent: Weigh 100 mg of anhydrous ferrous sulphate in 1 l volumetric flask, add 100 ml of glacial acetic acid and 100 ml of concentration sulphuric acid. Mix thoroughly. Add 700 ml more of concentrated sulphuric acid and mix. Cool to room temperature and make up the volume. Mix thoroughly and store in a brown coloured bottle.
3. Sodium phosphotungstate reagent: Dissolve 40 g of phosphotungstic acid in 1 l of solution (160 ml of 1 M sodium hydroxide + 840 ml of distilled water) to get 4% of final concentration.
4. 2 M magnesium chloride solution: Dissolve 190 g of magnesium chloride in 1 l of water.
5. Standard cholesterol: Dissolve 200 mg cholesterol in 100 ml glacial acetic acid.
6. Precipitating mixture: It is prepared fresh. Mix 1 part of reagent 4 with 4 parts of reagent 3 (v/v).

HDL Cholesterol

Procedure

1. To 1 ml of serum in a centrifuge tube add 0.1 ml of precipitating mixture. Allow it to stand for 30 min. Centrifuge at 4,000 rpm for 30 min.
2. Separate the supernatant and to 0.2 ml supernatant add 5 ml of reagent 1.
3. Prepare a blank by taking 0.2 ml acetic acid and 5 ml of reagent 1.
4. Run a standard consisting of 0.1 ml of reagent 5, 0.1 ml of glacial acetic acid and 5 ml of reagent 1. Keep it for 10 min for complete precipitation and centrifuge.
5. In another set of clean and dry test tubes add 2 ml of reagent 2 and add to each tube 3 ml of the supernatant from the above reaction mixture. Mix the contents, thoroughly.
6. Cool the tubes to room temperature and read the intensity at 560 nm against blank. Calculate the amount of cholesterol by comparing the standing reading. Dilution factor should be taken into account

Calculation

$$\text{HDL cholesterol in serum (mg/100 ml)} = \frac{\text{Absorbance of test sample}}{\text{Absorbance of standard}} \times 100$$

VLDL + LDL Cholesterol

Procedure

1. To 1 ml of serum add 0.1 ml of precipitating mixture. Allow to stand for 20 min at room temperature of complete precipitation. Centrifuge at 4,000 rpm for 20 min and remove the supernatant completely by inverting the tubes. Remove the supernatant sticking on the rim of the tube with the help of filter paper.

2. Dissolve precipitate in 0.4 ml of glacial acetic acid.
3. Prepare a reagent blank by taking 0.2 ml of acetic acid and 5 ml of reagent 1.
4. Run a standard by taking 0.1 ml acetic acid, 0.1 ml reagent 5 and 5 ml of reagent 1. Take 0.2 ml of the dissolved precipitate and add 5 ml of reagent 1. Mix the contents thoroughly and allow it to stand at room temperature for 10 min. and centrifuge.
5. Take 2 ml of reagent 2 in another set of tubes; add 3 ml of the supernatant obtained from above. Mix the contents thoroughly; cool the tubes to room temperature.
6. Read the absorbance at 560 nm against a reagent blank.

   Normal Values
   Total cholesterol – 150–250 mg/dl
   HDL cholesterol – 30–70 mg/dl
   LDL + VLDL Cholesterol – 90–210 mg/dl

Calculation

$$\text{(VLDL + LDL) cholesterol in serum (mg/100 ml)} = \frac{\text{Absorbance of test sample}}{\text{Absorbance of standard}} \times 400.$$

Clinical Interpretations

HDL cholesterol is regarded as a risk factor for coronary heart diseases (CHD). The level of HDL cholesterol is found to be decreased in patients suffering from coronary heart disease, atherosclerosis, diabetes mellitus and in smokers also.

It is reported that HDL cholesterol increases during physical exercise, alcohol intake, oestrogen therapy and also insulin therapy in diabetics.

Women are found to have slightly higher values of HDL cholesterol.

Ratio of HDL cholesterol to total cholesterol decreases as in CHDs, diabetes mellitus, smokers, atherosclerosis and this ratio increases in physically active persons.

13.27 Estimation of Phenols

Phenols, the aromatic compounds with hydroxyl groups are widespread in plant kingdom. They occur in all parts of the plants. Phenols are said to offer resistance to diseases and pests in plants. Grains containing high amount of polyphenols are resistant to diseases. Phenols include an array of compounds like tannins, flavonols etc.

Principle

Phenols react with phosphomolybdic acid in Folin-Ciocalteau reagent in alkaline medium and produce blue coloured complex.

Materials

1. 80% Ethanol
2. Folin-Ciocalteau reagent
3. 20%Na_2CO_3
4. Standard: 100 mg catechol in 100 ml water, dilute 10 times for a working standard.

Procedure

1. Weigh exactly 0.5–1.0 g of the sample and grind it with pestle and mortar in 10-time volume of 80% ethanol.
2. Centrifuge the homogenate at 10,000 rpm for 20 min. Keep the supernatant and re-extract the residue with 5 times the volume of 80% ethanol, centrifuge and pool the supernatants.
3. Evaporate the supernatant to dryness.
4. Dissolve the residue in a known volume of distilled water (5 ml).
5. Pipette out different aliquots (0.2–2 ml) into test tubes.
6. Make up the volume in each tube to 3 ml with water.
7. Add 0.5 ml of Folin-Ciocalteau reagent.
8. After 3 min, add 2 ml of 20% Na_2CO_3 solution to each tube.
9. Mix thoroughly and place the tubes in boiling water for exactly 1 min, cool and measure the absorbance at 650 nm against a reagent blank.
10. Prepare a standard curve using different concentrations of catechol.

Calculation

From the standard curve find out the concentration of phenols in the test sample and express as mg phenols/100 g material.

Notes

1. If any white precipitate is observed on boiling, the colour may be developed at room temperature for 60 min.
2. Express the results in terms of catechol or any other phenol equivalents used as standard.

13.28 Chlorogenic Acid (Michael et al. 1978)

Phenolic compounds are widely distributed in plants. They contribute to the colour and flavour of plant parts. The presence of phenolic compounds in oilseeds and grains pose nutritional problems.

Principle

Chlorogenic acid is extracted with alcohol, dried and dissolved in acetone. It is reacted with titanium ion to form a coloured complex which is measured at 450 nm.

Materials

1. Titanium Reagent: 20% $TiCl_4$ in con. HCl
2. Standard 25–200 μg/ml Chlorogenic acid in Acetone
3. Acetone
4. 80% Ethanol
5. 2.5 N HCl

Method

1. Reflux twice a known quantity of defatted sunflower meal in 80% ethanol (adjusted to pH 4.0 with 2.5 N HCl) for 30 min (125 ml to 1 g meal).
2. Discard the precipitate and collect 250 ml of the extract.
3. Remove 0.5 ml samples and dry in a vacuum oven at 50°C and 700 mm pressure for 2 h.
4. Dissolve dried extract in 4.7 ml of acetone.
5. Add 0.25 ml of $TiCl_4$.
6. Read the colour at 450 nm against a reagent blank (acetone plus $TiCl_4$).
7. Similarly treat the standards with TiCl4 and read the colour intensity.
8. Draw a standard curve and find out the chlorogenic acid content in the sample.

Calculation

Express chlorogenic acid content as grams per 100 g sample.

13.29 Estimation of Tannins

Tannins and tannin-like substances are widespread in nature and are probably present in all plants. These are polyphenolic compounds and are divided into two main groups-hydrolysable and condensed.

1. *Hydrolysable tannins* contain a polyhydric alcohol usually, if not always, glucose esterified with gallic acid or with hexahydroxydiphenic acid.
2. *Condensed tannins* are mostly flavonols and are probably polymers of flavan 3-ol (catechin) and these cannot be hydrolyzed to simple components.

Although the seeds with high polyphenol content seeds are resistant to common diseases, they also display anti-nutritional attributes. Among the cereals, sorghum contain higher amounts of polyphenols

The tannins are estimated by the following two methods:

(i) *Folin-Denis Method*: This is based on the non-stoichiometric oxidation of the molecules containing a phenolic hydroxyl group.
(ii) *Vanillin Hydrochloride Method*: Vanillin method is specific for dihydroxyphenols and is particularly sensitive for meta-substituted, di and tri hydroxybenzene containing molecules.

(i) Folin-Denis Method

Principle

Tannin-like compounds reduce phosphotungstomolybdic acid in alkaline solution to produce a highly coloured blue solution, the intensity of which is proportional to the amount of tannins. The intensity is measured in a spectrophotometer at 700 nm.

Materials

1. *Folin-Denis Reagent*: Dissolve 100 g sodium tungstate and 20 g phosphomolybdic acid in 750 ml distilled water in a suitable flask and add 50 ml phosphoric acid. Reflux the mixture for 2 h and make up to 1 l with water. Protect the reagent from exposure to light.
2. *Sodium Carbonate Solution*: Dissolve 350 g sodium carbonate in 1 l of water at 70–80°C. Filter through glasswool after allowing it to stand overnight.
3. *Standard Tannic Acid Solution*: Dissolve 100 mg tannic acid in 100 ml of distilled water.
4. *Working Standard Solution*: Dilute 5 ml of the stock solution to 100 ml with distilled water. One ml of solution would contain 50 μg tannic acid.

Procedure

- *Extraction of Tannin*: Weigh 0.5 g of the powdered material and transfer to a 250 ml conical flask. Add 75 ml water. Heat the flask gently and boil for 30 min. Centrifuge at 2,000 rpm for 20 min and collect the supernatant in 100 ml volumetric flask and make up the volume.
- Transfer 1 ml of the sample extract to a 100-ml volumetric flask containing 75 ml water.
- Add 5 ml of Folin-Denis reagent, 10 ml of sodium carbonate solution and dilute to 100 ml with water. Shake well and read the absorbance at 700 nm after 30 min.
- Prepare a blank with water instead of the sample.
- If absorbance is greater than 0.7, make appropriate dilution.
- Prepare a standard graph by using 0–100 μg tannic acid.

Calculation

Calculate the tannin content of the samples as tannic acid equivalents from the standard graph.

(ii) Vanillin Hydrochloride Method

Principle

The vanillin reagent will react with any phenol that has an unsubstituted resorcinol or phloroglucinol nucleus and forms a coloured substituted product which is measured at 500 nm.

Materials

1. *Vanillin Hydrochloride Reagent*: Mix equal volumes of 8% hydrochloric acid in methanol and 4% vanillin in methanol. The solutions must be mixed just before use, and avoid using even if it is slightly coloured.
2. *Catechin-Stock Standard Solution*: Prepare a standard solution containing 1 mg catechin/ml methanol.
3. *Working Standard*: Dilute the stock solution 10 times from 10 to 100 ml (100 μg/mL).
4. *Preparation of the Extract*: Extract 1 g of ground seed in 50 ml methanol. Mix occasionally by swirling. After 20–28 h, centrifuge and collect the supernatant.

Procedure

1. Pipette out 1 ml of the supernatant and quickly add 5 ml of vanillin hydrochloride reagent. Take readings in a spectrophotometer at 500 nm after 20 min.
2. Prepare a blank with vanillin hydrochloride reagent alone.
3. Prepare a standard graph with 20–100 μg catechin using the diluted stock solution.

Calculation

From the standard graph, calculate the amount of catechin, i.e. tannin in the sample as per the absorbance values and express the results as catechin equivalents.

13.30 Estimation of Lignin (Goering and Van Soest 1975)

Lignins constitute around 30% of the organic matter of trees and besides other functions acts as a physical barrier against invading pathogens. It is present in the cell walls of plants which along with cellulose contribute rigidity and stiffness to plant stems.

Principle

Refluxing the sample material with acid detergent solution removes the water-soluble and materials other than the fibrous component. The left-out material is weighed after filtration, dried, treated with 72% H_2SO_4 and filtered, dried and ashed. The loss of weight on ignition gives the acid detergent lignin.

Reagents

1. *Acid detergent solution*: Dissolve 20 g of cetyl trimethyl ammonium bromide (CTAB) in 1 l of 1 N sulphuric acid.
2. Sintered glass crucible-G2
3. 72% H_2SO_4 (w/v)

4. Acetone
5. Round bottom flask
6. Refluxing Set
7. Muffle Furnace

Procedure

C. Acid Detergent Fibre (ADF)

1. Place 1 g of powdered sample in a round bottom flask and 100 ml of acid detergent solution. Heat to boil in 5–10 min. Reduce heat to avoid foaming as boiling begins. Reflux for 1 h after the onset of boiling. Adjust boiling to slow, even level.
2. Remove container, swirl and filter the contents through a pre-weighed sintered glass crucible (G-2) by suction and wash with hot water twice.
3. Wash with acetone breaks up the lumps. Repeat acetone washing until the filtrate is colourless.
4. Dry at 100°C for overnight.
5. Weigh after cooling in a desiccator.
6. Express ADF content in percentage i.e. W/S x 100, where W is the weight of the fibre and S is the weight of the sample.

D. Determination of Acid Detergent Lignin (ADL)

1. Transfer ADF to a 100-ml beaker with 25–50 ml of 72% sulphuric acid. Add 1 g asbestos. Allow it to stand for 3 h with intermittent stirring with a glass rod.
2. Dilute the acid with distilled water and filter with pre-weighed Whatman No. 1 filter paper. Wash the glass rod and the residue several times to get rid of the acid.
3. Dry the filter paper at 100°C and weigh after cooling in a desiccator.
4. Transfer the filter paper to a pre-weighed silica crucible and ash the filter paper with the content in a muffle furnace at 550°C for about 3 h.
5. Cool the crucible in a desiccator and weigh. Calculate the ash content.
6. For blank, take 1 g asbestos, add 72% H_2SO_4 and follow the steps 2–5.

- For preparing pre-weighed filter paper, see the note under pectic substance.
- NDF = Hemicellulose + Cellulose + Lignin + Minerals
  ADF = Cellulose + Lignin + Minerals
  Hemicellulose = NDF – ADF
  Cellulose = ADF – Residue after extraction with 72% H_2SO_4
  Lignin = Residue after extraction with 72% H_2SO_4 – ash
- Wet materials can also be used for lignin estimation, provided the wet sample is equivalent to 1 g of dry material.
- If the weight loss of asbestos blank on ashing is below 0.002 g/g, discontinue the determination of blank.

Calculation

Weight of 72% H_2SO_4

$$\% \text{ ADL} = \frac{\text{washed fibre (Test} - \text{Asbestos blank)} - \text{Ash (Test} - \text{Asbestos blank)} \times 100}{\text{Weight of sample}}.$$

13.31 Estimation of Capsaicin in Chillies

Capsaicin is a protoalkaloid which is responsible for the pungency of chillies. The quality of chilli fruits, extracts or oleoresins is determined by the capsaicin content.

(i) Colorimetric Method

Principle

The phenolic group in capsaicin reduces phosphomolybdic acid to lower acids of molybdenum. The resulting component is blue in colour and is read at 650 nm. The colour intensity is directly proportional to the concentration of capsaicin.

Reagents

1. 0.4% sodium hydroxide
2. 3% Phosphomolybdic Acid
3. Dry acetone (Add about 25 g anhydrous sodium sulphate to 500 ml acetone of analytical grade at least 1 day before use).
4. *Stock Standard Capsaicin Solution:* Dissolve exactly 50 ml of 0.4% sodium hydroxide solution (1,000 μg/ml).
5. *Working Standard*: Dilute 10 ml of the stock standard to 50 ml with 0.4% sodium hydroxide solution (200 μg/ml).

Procedure

- Weigh 0.5 g dry chilli powder into a glass-stoppered test tube or volumetric flask.
- Pipette out 10 ml dry acetone into the flask and shake it for 3 h in a mechanical shaker. Let the contents settle down or centrifuge (10,000 rpm for 10 min). Pipette out 1 ml of clear supernatant into a test tube and evaporate to dryness in a hot water bath. Dissolve the residue in 5 ml of 0.4% sodium hydroxide solution.
- Add 3 ml of 3% phosphomolybdic acid. Shake the contents and let stand for 1 h. Filter the solution quickly into centrifuge tubes to remove any floating debris. Centrifuge at about 5,000 rpm for 10–15 min. Transfer the clear blue coloured solution directly into the cuvette and read the absorbance at 650 nm.

- Run a reagent blank along with the test samples.
- Prepare a standard graph using 0–200 μg capsaicin simultaneously i.e. pipette out 0.2, 0.4, 0.6, 0.8 and 1 ml of working standard solution and proceed as above.

(ii) Spectrophotometric Method

Principle

Capsaicin is extracted from red pepper (chillies) with ethyl acetate and made to react with ethyl acetate solution of vanadium oxychloride. Then it is read at 720 nm. This method is sensitive and is useful to measure small quantities (less than 0.05%).

Reagents

1. Vanadium oxychloride (0.5%) in ethyl acetate.
2. Pure capsaicin (0.01%) in ethyl acetate (10 mg in 100 ml)

Procedure

1. Grind the sample well to pass through No. 40 sieve.
2. Place 2 g sample in a 100-ml volumetric flask.
3. Let it stand for 24 h to extract (otherwise reflux the contents for 2.5 h) and then make up to volume. Dilute 1 ml of extract to 5 ml with ethyl acetate. Add 0.5 ml vanadium oxychloride solution (just before reading) and shake.
4. Read at 720 nm in a spectrophotometer.
5. Subtract the absorbance value.
6. Prepare a standard curve using 0.5, 1.0, 1.5, 2.0 and 2.5 ml of standard capsaicin solution containing 50, 100, 150, 200 and 250 μg capsaicin, respectively.

Calculation

$$\% \text{ capsaicin} = \frac{\mu\text{g capsaicin}}{1{,}000 \times 1{,}000} \times \frac{100}{1} \times \frac{100}{2} = \frac{\mu\text{g capsaicin}}{200}.$$

13.32 Estimation of Glucosinolates Content in Oil Seeds (Brezezinski and Mendelewski 1984)

Rapeseed mustard meal obtained after oil extraction is a valuable component of animal feedstuff. However, its utilization is limited as its feeding causes undesirable physiological effects due to the presence of glucosinolates (GSL's). The glucosinolate content in defatted meal ranges from 7 to 10%. There is need to screen varieties containing low glucosinates produced by breeding programmes.

Principle

Sinigrin (a glucosinolate) on reaction with sulphuric acid is first hydrolyzed to thioglucose which then dehydrates to thiofurfural derivative. This furfural reacts with thymol to a coloured compound whose absorbance is measured at 505 nm against the sample blank.

Equipment and Glassware

1. Grinder
2. Fat extractor apparatus
3. Micro ion-exchange columns ; 100 × 18
4. Water bath for boiling.
5. Spectrophotometer

Reagents

1. Potassium sulphate solution (0.3 M): Weight 26.1 g potassium sulphate into 500 ml volumetric flask and make up the volume with water.
2. Petroleum ether (40–60°C boiling range)
3. Sulphuric acid solution (80%)
4. Thymol reagent: Weigh 1 g thymol into a 100-ml volumetric flask and make up the volume with ethanol.
5. Lead acetate solution: Weigh 22.75 g lead acetate trihydrate, dissolve and dilute to 100 ml with water.
6. Pyridine/acetic acid buffer (0.5 M): Add 30 ml glacial acetic acid and 40 ml pyridine to 930 ml water, mix and store in refrigerator.
7. Sodium hydroxide (0.5 N): Weigh 20 g NaOH and dissolve. Dilute it to 1,000 ml with water.
8. DEAE: Sephadex A-25
9. Formic acid (30%)
10. Allyl glucosinolate stock solution (5 μmoles/ml): Weigh out 519.5 mg sinigrin into 250 ml volumetric flask and make up the volume with water.
11. Allyl glucosinolate standard solution (0.3 μmoles/ml): Dilute 3 ml of the stock (5 μmoles/ml) sinigrin with 0.3 M K_2SO_4.

    Preparation of oil free meals: Weigh about 4 g raw mustard seed and extract with petroleum ether in a soxhlet extractor. Dry the extracted meals and store in screw-cap vials.

Extraction of Glucosinolates

1. Weigh 100 mg meal into a test tube.
2. Add 9 ml cold water and place the mixture for 15 min in a boiling water bath.
3. Cool and add 1 ml of lead acetate solution and allow the mixture to stand for 15 min.
4. Centrifuge the mixture (4 min, 1,400 × *g*).
5. Use the supernatant for purification on DEAE-Sephadex A-25 column.

Preparation of DEAE-Sephadex A-25 Pyridine Acetate Column

1. Stir suspension of sephadex in an excess of 0.5 M pyridine/acetic acid buffer such that the settled volume of the resin is one half of the total.
2. Fill half of the column with water and add 1 ml of sephadex suspension.
3. After setting of the column with water and insert the sieve.
4. Allow it to sink on the bed.
5. Wash the column twice with water and drain the water between each wash.

Isolation and Measurement of Glucosinolates

1. Add 3 ml of meal extract to the prepared columns without disturbing the surface.
2. Wash the column with 2 × 2 ml water, 2 × 2 ml 30% formic acid and 2 × 2 ml water.
3. Place clean graduated tubes or vials beneath the column.
4. Elute the column with 2 × 5 ml of 0.3 M potassium sulphate and dilute to 10 ml mark.
5. Take 1 ml aliquot in the glass tubes with screw caps.
6. Add 7 ml of 80% sulphuric acid and 1 ml of 1% thymol solution to each tube.
7. Thoroughly mix the contents and place at 100°C for 60 min.
8. Immediately cool the tubes under tap water and mix again.
9. Measure the absorbance at 505 nm against 0.3 M potassium sulphate blanks.

Calculation of Total Glucosinolate Content

1. Calculate the micromolar concentration of glucosinolates from the absorbance reading of the micromolar extinction coefficient of the sinigrin standard solution.
2. Prepare a set of (4 or 5) blanks and sinigrin standard with each batch of samples.
3. Heat them and read on the spectrophotometer.
4. Prepare the blank by mixing 1 ml 0.3 M potassium sulphate, 7 ml 80% sulphuric acid and 1 ml of 1% thymol solution.
5. Prepare sinigrin standard by taking 1 ml sinigrin standard (0.3 μmoles/ml), add to it 7 ml 80% sulphuric acid and 1 ml of 1% thymol solution.
6. Read the blanks before and after the sample batch.

Calculate the micromolar extinction (absorption) coefficient (K) from the absorption reading of the standard and blank as follows:

$$K = \frac{\text{(Mean absorbance of standards)} - \text{(Mean absorbance of blanks)}}{0.3 \text{ (the micromolar concentration of the standard)}}$$

Calculate the micromolar concentration of glucosinolates in the samples on a dry weight basis as follows:

$$\mu\text{M/g dry basis} = \frac{\text{Corrected absorbance} \times \text{Dilution Factor} \times 100}{\text{K} \times \text{Wt. of the sample (g)} \times (100 - \text{\% moisture of the sample})}.$$

13.33 Determination of Polyphenols in Pulse Grains (Swain and Hills 1959)

Polyphenols are tannin-like compounds and are termed as anti-nutritional constituents of pulse grains and some other food grains. Polyphenol combine with the proteins present in grains thereby adversely affecting their digestibility to a considerable extent. Although, polyphenols are nutritionally undesirable to a human being, however, they are useful to plants as they are involved in the defence mechanism of plants conferring them disease resistance.

Principle

Polyphenols reduce phosphotungstomolybdic acid in alkaline solution and produce a highly coloured blue solution, the intensity of which is proportional to the amount of polyphenols present in grain sample. The intensity is measured at 725 nm.

Reagents

1. Folin-Denis reagent: Dissolve 100 g sodium tungstate and 20 g phosphomolybdic acid in 750 ml distilled water in a suitable flask and add 50 ml phosphoric acid. Reflux the mixture for 2 h and make up to 1 l with water. Protect the reagent from exposure to light.
2. Sodium carbonate solution: Dissolve 350 g sodium carbonate in 1 l of water at 70–80°C. Filter through glasswool after allowing it to stand overnight.
3. Standard tannic acid solution: Dissolve 100 mg tannic acid in 100 ml of distilled water.
4. Working standard solution: Dilute 5 ml of the stock solution to 100 ml with distilled water. One ml contains 50 μg tannic acid.

Procedure

1. Take 500 mg of powdered sample and add 10 ml methanol containing 1 ml of 1% HCl and keep for 24 h at room temperature with occasional shaking or reflux for 1 h and cool.
2. Then centrifuge the contents at 2,000 rpm for 20 min of filter and transfer the supernatant into 25 ml volumetric flask.
3. Make up the volume up to mark by 80% methanol, take 1 ml of this solution into another 25 ml volumetric flask and add 5 ml distilled water followed by the addition of 1 ml Folin-Denis reagent and keep it for 30 min.
4. After this add 2 ml of sodium carbonate solution. Make up the volume up to 25 ml with the help of distilled water.
5. After this keep the whole contents for 1 h to develop blue colour.
6. Read the colour intensity at 725 nm.
7. Run a blank using distilled water in place of sample.
8. Prepare a standard graph by using 0–100 μg tannic acid.

Calculation

Calculate the polyphenol content of the samples as tannic acid equivalents from the standard graph.

Modified method for tannins (A.O.A.C 1995a, b)

Procedure

1. Weigh 1 g of the powdered material 4 times and transfer to 500 ml conical flasks. Add 150 ml water to each flask.
2. Add to three flasks 10, 15 and 20 ml of standard tannic acid solution, respectively.
3. Heat all the flasks gently and boil for 30 min. Then centrifuge at 2,000 rpm for 20 min.
4. Collect supernatant in 250 ml flasks and make up the volume.
5. Transfer 10 ml of the supernatant extract in 100 ml flasks and add 75 ml of water, 2.5 ml of Folin-Denis reagent and add 5 ml of sodium carbonate solution and make up the volume.
6. After 30 min take the reading at 740 nm.
7. Prepare a graph by plotting O.D. vs. tannic acid concentration. The value wherever it cuts *X*-axis was taken as new origin. The difference in O.D. in first and second origin is taken as a measure of the content of tannins in the samples.

13.34 Estimation of Aldehydes in Food Stuffs

The occurrence of aldehydes in food stuffs is highly undesirable. These compounds are usually formed *via* the process of auto-oxidation of oil/fats or oxidation of alcohols. In order to maintain the quality standards of food items, it is important to know the contents of these undesirable molecules whose regular intake might create serious health problem to the consumers.

Principle

Aldehydes possess a unique property of reacting with hydroxylamine hydrochloride. The liberated acid when reacts with alcoholic KOH, gives yellow colour. From the weight of material taken, volume of alkali used and the factor corresponding to a particular aldehyde, its percentage in food can be calculated with the help of formula derived for this purpose.

Equipment and Glassware

1. Tubes for weighing the sample
2. Volumetric flasks (100 ml)

Reagents

1. Benzene.
2. Ethanol (60% v/v)

3. Hydroxylamine hydrochloride (0.5 M). Dissolve 3.475 g of pure hydroxylamine hydrochloride in 95 ml ethanol (60% v/v). Add 10 drops of methyl orange and adjust, using 0.5 M alcoholic potassium hydroxide to a yellow colour. Make up the volume to 100 ml with 60% ethanol.
4. Alcoholic potassium hydroxide (0.5 M). Dissolve approximately 2.8 g of KOH in a few drops of water in a flask and then make up its volume to 100 ml with alcohol.

Procedure

1. Transfer 5 g of sample to a weighed tube, add ml benzene and 15 ml 0.5 M hydroxylamine hydrochloride solution.
2. Shake vigorously and titrate the liberated acid with 0.5 M alcoholic potassium hydroxylamine hydrochloride solution.

Calculation

$$\text{Percent aldehydes} = \frac{t \times f \times 1.008 \times 100}{W},$$

where, t = titre; W = weight of the sample (g)
The factor "f" are as follows:

Benzaldehyde – 0.053
Cinnamic aldehyde – 0.066
Citral – 0.076
Cuminaldehyde – 0.074
Decyclic aldehyde – 0.078

13.35 Estimation of ODAP (Rao 1978)

Lathyrus sativus L. (Khesari) contains a neurotoxin, β-*N*-Oxaly-α–β-diaminopropionic acid (ODAP) and prolonged consumption of this pulse causes neurolathyrism/lower limb paralysis.

Principle

ODAP on hydrolysis with KOH yield α,β-diaminopropionic acid (DAP) which reacts with o-Phthalaldehyde (OPT) under alkaline condition to give an intense yellow colour.

Equipments and Glassware

1. Spectrophotometer
2. Centrifuge

Reagents

1. 0.5 M Dipotassium tetraborate buffer (pH 9.9).
2. o-Phthalaldehyde (OPT) reagent: 100 mg of OPT in 1 ml of 95% ethanol and 0.2 ml of mercaptoethanol are added to 99 ml of potassium borate buffer. This reagent although freshly made and used, can also be used for as long as 3–4 days.
3. 3 N potassium hydroxide in distilled water.

Procedure

1. Boil 25 mg of finely ground powder of *Lathyrus sativus* seed in 5 ml of distilled water for 30 min.
2. Centrifuge at 4,000 rpm for 10 min. Collect the supernatant.
3. Take 0.1 ml of the supernatant in duplicate in test tubes and add 0.2 ml of 3 N KOH.
4. Keep one set of tubes in boiling water bath for 30 min for hydrolysis, while the other set is kept at room temperature.
5. Cool the tubes and make the volume to 100 ml with water and add 2.0 ml of 0.5 M Borate buffer (pH 9.9) along with 2 ml of reagent.
6. Keep the tubes at room temperature for 30 min for colour development and read the absorbance at 420 nm.
7. The difference between the absorbance readings with and without hydrolysis gives an estimation of DAP.
8. Run a set of control which includes all the reagents except the experimental material.
9. Make a standard curve using different concentration (10–100 n mol) of DAP.
10. Calculate DAP content using the standard curve.
11. For calculating ODAP content, multiply DAP content by a factor of 1.25

$$\text{ODAP in sample } (\%) = 1.25 \times \text{ DAP } (\%).$$

13.36 Assessment of Rancidity of Oil and Fats (A.O.A.C 1984)

Oils and fat undergo changes during storage which result in production of unpleasant taste and odour commonly referred to as "Rancidity". Similar changes also occur when oils/fats are subjected to heating processes during cooking. The rancidity caused by air is known as oxidative rancidity. When it is caused by micro-organisms, it is known as ketonic rancidity. Generally, oils with high unsaturated fatty acid content are prone to oxidative rancidity. The situation is caused in numerous ways involving features such as light, air, high temperature, enzymes, micro-organisms, metals and presence of free fatty acids but prime among them is enzyme-lipoxygenase during storage of oilseeds.

The principle toxic substance occurring in oxidized oils is lipid hydroperoxide, a primary product of rancid oils. The hydroxyl and carbonyl compounds are originated only by decomposition of hydroperoxides on prolonged aeration which make even greater contribution to the toxicity than hydroperoxides; these decomposition products are also injurious to health.

The two important parameters of oxidative type of rancidity are peroxide value and carbonyl value which are of nutritional significance. The procedures for their determination have been described below:

(i) Peroxide value

The presence of peroxide oxygen in fat resulting from auto-oxidation is determined by iodometric method. The peroxide value is expressed as the number of milliequivalents peroxide in 1 kg of fat.

Equipment and Glassware

1. Erlenmeyer flasks (250 ml)
2. Volumetric flasks (100 ml)
3. Pestle and mortar

Reagents

1. Acetic acid: Chloroform mixture (3:1 v/v)
2. Saturated KI solution (15% approximately).
3. Sodium thiosulphate solution (0.1 N)
4. Soluble starch solution (1%).

Procedure

A. Standardization of sodium thiosulphate (Hypo) solution

1. Mix 10 ml of 0.1 N $K_2Cr_2O_7$ and 5 ml of 1 N HCl and 10 ml of saturated KI in a 100-ml volumetric flask. Titrate against sodium thiosulphate solution.
2. When brownish yellow colour is formed, add 2–3 drops of 1% starch solution. On addition of starch, blue colour is formed.
3. Titrate it further till the disappearance of blue colour and the solution turns light green. Note down the titre value. The normality of Hypo is calculated by using the equation:
   N_1V_1 (for Hypo) $= N_2V_2$ (for $K_2Cr_2O_7$)

B. Procedure

1. Weigh 1.0 g oil/fat in 250 ml Erlenmeyer flask. Add 30 ml of acetic acid–chloroform mixture. Dissolve the fat completely by using wrist action shaker for 5–10 min.
2. Add 0.5 ml of saturated KI and allow to stand for 1 min with occasional shaking and then add 30 ml water.
3. Titrate against standardized Hypo solution with vigorous shaking until yellow colour disappears.

4. Add 2–3 drops of 1% starch and continue titration to release all the iodine from chloroform or until blue colour disappears.

Calculation

(i) Weight of fat = W g.
(ii) Volume of Hypo solution = V ml
(iii) Normality of Hypo solution = N

$$\text{Peroxide value} = \frac{V \times N \times 1{,}000}{W}.$$

(ii) Carbonyl value

When fat is made to react with TCA and 2,4-dinitrophenyl hydrazine, corresponding hydrozones are formed which on treatment with KOH give rise to development of colour, the intensity of which can be measured spectrophotometrically. Carbonyl value is expressed in terms of number of micromoles per gram of oil/fat.

Equipment and Glassware

1. Spectrophotometer
2. Water bath.
3. Volumetric flask (50 ml)

Reagents

1. Carbonyl-free benzene: Analytical reagent grade benzene is usually sufficiently carbonyl-free as received, but if the blank has an absorbancy greater than 0.35 against water at 43 nm, then the benzene should be purified as follows: To 1 l of benzene add 5 g of 2, 4-dinitrophenyl hydrazine and 1 g of trichloroacetic acid; reflux for 1 h and then distill through a short column.
2. Trichloroacetic acid solution (4.3%): Dissolve 43 g of trichloroacetic acid in carboxyl-free benzene and make to 1 l with water.
3. 2,4 Dinitrophenyl hydrazine solution (0.05% in water): Dissolve 0.5 g 2,4-dinitrophenyl hydrazine twice re-crystallized from carboxyl-free methanol (which can be prepared in same manner as carbonyl-free ethanol).
4. Potassium hydroxide solution (4%): Dissolve 4 g of potassium hydroxide in 100 ml of absolute carbonyl-free ethanol with the aid of gentle heating and shaking. Filter the solution through fine glasswool using suction.
5. Carbonyl-free ethanol: To 1 l of ethyl alcohol add 5–10 g of aluminium granules and 8–10 g KOH and reflux the mixture for 1 h. Distill it, discard the first 50 ml of distillate, and stop the distillation before the last 50 ml has distilled.

Procedure

1. Weigh 1 g fat in 50 ml volumetric flask, add 5 ml of benzene (The solution of fat should not have more than 250×10^{-6} molar carbonyls). Shake till the fat is dissolved.

2. Add 3.0 ml of 4.3% TCA and 5 ml of 0.05% 2,4-dinitrophenyl hydrazine solutions.
3. Stopper the flask and heat it in a water bath at 60°C for 30 min. Cool to room temperature. The solution is stable for several hours.
4. Develop the colour by adding 10 ml of 4% KOH solution. Dilute the solution with carbonyl-free absolute ethanol. Mix well.
5. After exactly 10 min. Read the absorbance at 430 and 460 nm against a blank prepared exactly in the same manner, substituting 5 ml of carbonyl-free benzene for the sample solution.

Calculation

When the measurements are made in 1 cm cuvettes with Beckman DU spectrophotometer, the analysis can be calculated using the following equations.

$$\text{Unsaturated} = \frac{3.861\,A_{460} - 3.012\,A_{430}}{0.854}$$

$$\text{Saturated} = 3.861 A_{460} - 2.170\,\text{Unsaturated}.$$

Important Points

1. In preparation of samples for analysis, fats and oils are easily dissolved in benzene and 5 ml aliquots used. Solid foods are ground in a mortar or a mill. Samples are weighed out into glass-stoppered centrifuge bottles, benzene is added, the bottle is stoppered, shaken and centrifuged. Five ml aliquots of these extracts are then taken for analysis.
2. Solutions and extracts of fat for this test should be protected from undue exposure to light and air before use to prevent further oxidation of fat and deterioration of existing carbonyl compounds.
3. Potassium hydroxide solution should be prepared fresh daily.
4. Using the present method of analysis the 2,4-dinitrophenyl hydrozones of saturated aldehydes exhibited maximum absorption at 432 nm, and the a_M is 16,670, while for this derivative of crotonaldehyde, an α,β-unsaturated aldehyde, maximum absorption is at 150 nm and a_M is 28,100. The most suitable wavelengths for this determination therefore, are 430 and 460 nm.

13.37 Determination of Phytin Phosphorus (Wheeler and Ferrel 1971)

Phytic acid (1, 2, 3, 4, 5, 6,-hexakis dihydrogen phosphate myoinositol) is a common storage form of phosphorus in seeds and is also considered as an anti-nutritional factor. The complexing of phytic acid with nutritionally essential elements and the possibility of interference with proteolytic digestion have been

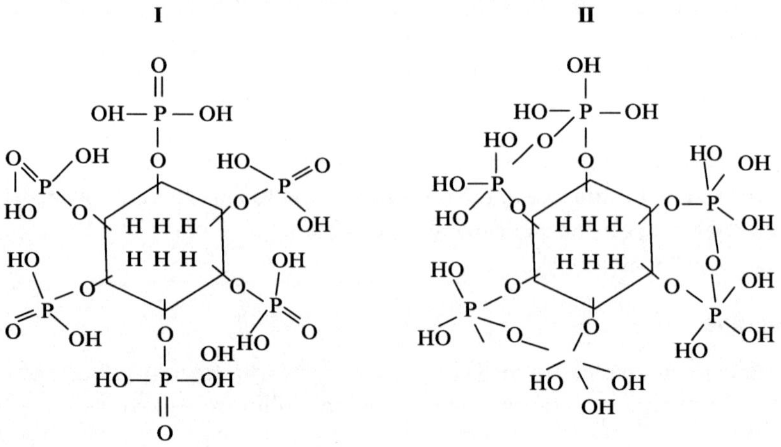

Fig. 13.3 Inositol hexaphate (phytic acid)

suggested as factors responsible for anti-nutritional activity. The phosphorus in phytic acid is not nutritionally available to the monogastric animals. Phytic acid also interferes with calcium and iron absorption. Hence, estimation of phytic acid in food grains becomes essential especially in cereals.

Principle

The phytate is extracted with trichloroacetic acid and precipitated as ferric salt. The iron content of the precipitate is determined colorimetrically and the phytate phosphorus content calculated from this value assuming a constant 4 Fe: 6P molecular ratio in the precipitate (Fig. 13.3).

Materials

1. 3% Trichloroacetic Acid
2. 3% Sodium Sulphate in 3% TCA
3. 1.5 N NaOH
4. 3.2 N HNO_3
5. $FeCl_3$ Solution (Dissolve 583 mg $FeCl_3$ in 100 ml of 3% TCA).
6. 1.5 M Potassium thiocynate (KSCN)(Dissolve 29.15 g in 200 ml water)
7. Standard $Fe(NO_3)$ solution

Procedure

1. Weigh a finely ground (40 mesh) sample estimated to contain 5–30 mg phytate P into a 125-ml Erlenmeyer flask.
2. Extract in 500 ml, 3% of TCA for 30 min with mechanical shaking or with occasional swirling by hand for 45 min.
3. Centrifuge the suspension and transfer a 10-ml aliquot of the supernatant to a 40 ml conical centrifuge tube.
4. Add 4 ml of $FeCl_3$ solution to the aliquot by blowing rapidly from the pipette.

5. Heat the contents in a boiling water bath for 45 min. If the supernatant is not clear after 30 min, add one or two drops of 3% sodium sulphate in 3%TCA and continue heating.
6. Centrifuge (10–15 min) and carefully decant the clear supernatant.
7. Wash the precipitate twice by dispersing well in 20–25 ml 3% TCA, heat in boiling water for 5–10 min and centrifuge.
8. Repeat washing with water.
9. Disperse the precipitate in a few ml of water and add 3 ml 1.5 N NaOH with mixing.
10. Bring volume to approximately 30 ml with water and heat in boiling water for 30 min.
11. Filter hot (quantitatively) through a moderately retentive paper Whatman No.2.
12. Wash the precipitate with 60–70 ml hot water and discard the filtrate.
13. Dissolve the precipitate from paper with 40 ml hot 3.2 N HNO_3 into a 100-ml volumetric flask.
14. Wash paper with several portions of water, collecting the washings in the same flask.
15. Cool flask and contents to room temperature and dilute to volume with water.
16. Transfer a 5-ml aliquot to another 100 ml volumetric flask and dilute to approximately 70 ml.
17. Add 20 ml of 1.5 M KSCN dilute to volume, and read colour immediately (within 1 min) at 480 nm.
18. Run a reagent blank with each set of samples.

Standard

Weigh accurately 433 mg $Fe(NO_3)_3$ and dissolve in 100 ml distilled water in a volumetric flask. Dilute 2.5 ml of this stock standard and make up to 250 ml in a volumetric flask. Pipette out 2.5, 5, 10, 15 and 20 ml of this working standard into a series of 100 ml volumetric flasks and proceed from step 16.

Calculation

Find out the μg iron present in the test from the standard curve, and calculate the phytate *P* as per the equation.

$$\text{Phytate } P \text{ mg/100 g sample} = \frac{\mu g\,\text{Fe} \times 15}{\text{Weight of sample (g)}}.$$

13.38 Estimation of Trypsin Inhibitor in Legumes (Kakade et al. 1974)

The food legumes are rich sources of proteins and their importance has been very well recognized in human nutrition, particularly in the countries where cereals form the staple diet of the people. Though grain legumes are rich in proteins, historically they are also known to possess a wide variety of chemical substance which interferes with the nutritive value when ingested. Among such anti-nutritional factors, enzyme inhibitors, the substances which have the ability to inhibit the activity of the enzymes, are present in an appreciable quantity in the grains. The protease inhibitors which are proteinaceous in nature are the inhibitors of the enzyme – trypsin, chymotrypsin, papain, elastase, carboxypeptidase A and B, pepsin etc.

Principle

The trypsin inhibitor activity is measured indirectly by inhibiting the activity of trypsin. A synthetic substrate (BApNA) is subjected to hydrolysis by trypsin to produce yellow coloured p-nitroanilide. The degree of inhibition by the extract of the yellow colour production is measured at 410 nm.

Reagents

1. 30% Glacial acetic acid (v/v)
2. Trypsin: Dissolve 6.25 mg lyophilized trypsin and make up to 25 ml with 0.001 M HCl. Dilute 2 ml of this solution to 25 ml for assay.
3. *Substrate*: Benzoyl-DL-Arginine-paranitroanilide (BApNA):Completely dissolve 40 mg BAPNA in 0.5 ml of dimethyl sulphoxide and then make up to 100 ml with Tris-HCl buffer pH 8.2.
4. *Tris-HCl Buffer* (pH 8.2): Weigh 6.05 g Tris (hydroxymethyl aminomethane) and 2.94 g $CaCl_2 \cdot H_2O$, dissolve in 90 ml water, adjust to pH 8.2 with dilute HCl and make up to 1,000 ml with distilled water.
5. Trypsin inhibitor source: Extract 0.5 g sample in 25 ml water by grinding in prechilled mortar and pestle. Extract the ground sample in a refrigerator for 2–3 h with occasional shaking for complete extraction of TI. Centrifuge the homogenate at 12,000 rpm for 20 min at 4–6°C. Dilute 1 ml of the supernatant to 10 ml with distilled water and use as TI source.

Procedure

1. Pipette out 0–1 ml of extract in duplicate sets of test tubes, one to serve as endogenous (E) and the other test (T).
2. Make up the volume to 2 ml with buffer in the endogenous set.
3. Make up the volume to 1 ml in the test set.
4. Add 1 ml of trypsin solution (20 μg) to each tube in the test set. Pipette out into a separate test tube 1 ml of buffer and 1 ml of trypsin solution for standard (S).
5. Incubate all the tubes in a water bath at 37°C.
6. After a few minutes, add 2.5 ml of substrate (1 mg BApNA) to each tube.
7. Allow the reaction to proceed for 10–60 min at 37°C.
8. Stop the reaction by adding 0.5 ml of 30% glacial acetic acid.
9. Read the absorbance at 410 nm in a spectrophotometer.
10. Determine the protein content in the extract by Lowry's method.

Calculation

Find out T and S absorbance. Plot the absorbance against the volume of extract. Determine the aliquot size of the extract required to inhibit 50% of the trypsin activity (S/2). That aliquot size is considered to be one unit of trypsin inhibitor.

One unit of activity corresponds to that amount of trypsin inhibitor in μg protein which gives 50% inhibition of enzyme activity under experimental conditions. The trypsin inhibitor activity is expressed as trypsin inhibitor units (TIU) per gram sample or per mg protein. (The dilutions of trypsin inhibitor source are made in such a way that 0.5 ml produces 50% inhibition).

13.39 Estimation of Cyanogens

Cyanogens are cyanogenic glucosides and are widespread in the plant kingdom in trace amounts. These glucosides release hydrogen cyanide by an endogenous enzymatic reaction. Relatively high concentration is found in certain grasses, root crops and fruit kernels. Young sorghum plants contain high amounts of cyanogens which could result in adverse effects on the livestock.

Principle

Hydrocyanic acid which is evolved from the sample forms a red coloured compound with sodium picrate and the intensity of developed colour is measured at 625 nm.

Materials

1. Chloroform.
2. Whatman No.1 filter Paper: Cut filter paper into strips 10–12 cm long and 0.5 cm wide, and saturate them with alkaline picrate solution.
3. Alkaline picrate solution: Dissolve 25 g sodium carbonate and 5 g picric acid in 1 l of water.
4. Standard hydrogen cyanide solution: Dissolve 0.241 g of KCN/litre of water. This gives a solution containing 100 μg hydrogen cyanide/ml.

Procedure

1. Homogenize 1 g of the sample in 25 ml water with 3–4 drops of chloroform.
2. Place the homogenate in 500 ml conical flask.
3. Place the saturated filter paper in the hanging position with the help of a cork stopper inside the conical flask.
4. Incubate the mixture at room temperature (20°C) for 20–24 h.
5. The sodium picrate present in the filter paper is reduced to reddish compound in proportion to amount of hydrocyanic acid evolved.

6. Elute the colour by placing the paper in a clean test tube containing 10 ml distilled water and compare it with standards at 625 nm.
7. *Preparation of the Standard Curve*: Place 5 ml of the alkaline picrate solution and 5 ml of the potassium cyanide solution in a test tube. Heat for 5 min in boiling water. Deliver the following volumes from the above KCN alkaline picrate solution of six test tubes: 0.1, 0.2, 0.4, 0.6, 0.8 and 1 ml. Bring the volume of each test tube to 10 ml by adding distilled water. Press rubber stopper to the tubes and keep them in a cool place. Measure the absorbance at 625 nm. The amount of hydrogen cyanide in the above tubes is equal to 5, 10, 20, 30, 40 and 50 μg, respectively. Prepare a blank with 10 ml distilled water (as KCN is poisonous, therefore use automatic pipette for the transfer of cyanide solution).

Calculation

Calculate the hydrogen cyanide content of the sample from the standard graph.

13.40 Estimation of Indoleactic Acid (Knegt and Bruinsma 1973)

Indole-3-acetic acid (IAA) is an important hormone involved in plant growth and development.

Principle

IAA is reacted with trifluoroacetic acid and acetic anhydride to convert it into indole-α-pyrone which is measured fluorimetrically.

Materials

1. Methanol (redistilled)
2. *Trifluoroacetic Acid-Acetic. Anhydride Reagent:* Mix equal volume of each liquid, pre-cool to 0°C. Freshly prepare and store on ice until use.
3. *Spectrophotofluorimeter:* Measure primary filter at 440 nm and secondary filter at 490 nm (for low concentration measurement).

Extraction of IAA

Put 5 g of plant material in liquid nitrogen and grind to a fine powder using a pestle and mortar. Continue grinding with 10 ml methanol (redistilled) to a fine suspension. Filter the homogenate through a G4 glass filter under suction into a 100-ml flask. Extract the material on the filter twice by adding 10 ml methanol and then once with 5 ml. Evaporate the filtrate in a rotary evaporator at 30°C to an aqueous residue. To the aqueous residue add 10 ml of cold 0.5 M K_2HPO_4 solution so that pH reaches to about 8.5. Transfer to a suitable separating funnel and shake with 10 ml light petroleum ether. Repeat this step again and then shake with 10 ml diethyl ether each time the lipid fraction is discarded. Adjust the aqueous layer to pH 3 by adding about 3 ml of 2.8 M phosphoric acid. Extract IAA with 10 ml diethyl ether.

The 10 ml diethyl ether is then extracted with 10 ml cold 50 mM, K_2HPO_4 solution. The pH of the solution is adjusted to 3 with phosphoric acid (0.28 M) and the IAA is passed into a final 10 ml diethyl ether. The ether is then evaporated in a few minutes under reduced pressure. Dissolve the residue in a known volume (5 ml) of cold redistilled methanol.

Procedure

1. Pipette out 1 ml of the above methanolic extract each in four different test tubes.
2. To each tube add 1 ml of methanol containing 0, 10, 20 or30 ng of IAA, respectively.
3. Dry the contents in each tube completely under reduced pressure and cool to 0°C.
4. To each flask add 0.2 ml ice-cold trifluoroacetic acid-acetic anhydride reagent and mix.
5. Place the tubes on ice for exactly 15 min to ensure the complete conversion of IAA into indole-α-pyrone. Stop the reaction adding 3 ml water.
6. A blank may be prepared occasionally by adding first 3 ml water to one of four aliquots and 0.2 ml reagent after 15 min.
7. Take the readings in a spectrophotofluorimeter with excitation at 440 nm and emission at 490 nm for low concentration samples.
8. Calculate the amount in unknown sample.

13.41 Estimation of Ethylene

Ethylene, the ripening hormone is usually estimated for different physiological studies.

Principle

The ethylene evolved is measured in a gas chromatography based on the adsorption principle on activated silica gel or poropak.

Materials

1. Conical Flasks with facility to seal with rubber gaskets
2. Air-tight syringes
3. Ethylene gas
4. GLC

Procedure

1. Place the fruits in conical flasks.
2. Seal the mouth with rubber septum or gasket.
3. Incubate for 1 h at 20°C.
4. Withdraw gas samples with hypodermic syringes and inject into GLC.
5. For standard, inject pure ethylene into empty conical flasks or cylinder of same volume and satisfy identical assay conditions. Remove the same volume of internal atmosphere as that of sample from the flask, inject into GLC and measure ethylene peak height.

Calculation

The quantity of ethylene produced is expressed as μl ethylene per hour per kg material.

Chapter 14
Nutritional Evaluation of Forages

14.1 Preparation of Plant Extract for Analysis

The general problem in the study of natural plant products is that their nature and amount are dependent on various factors, which must be controlled as far as possible. Some of these factors are (i) stress; the metabolic state of the plant may change when it is stressed in any manner. This can be a problem before as well as after harvesting a plant part for analysis. As cells die (the senescent process), the cellular integrity is lost and as a result the enzymes come in contact with substrates to which they are not normally exposed in living cells. In addition, it also increases the oxidation process, which is a problem with phenolics since these are prone to oxidation. On oxidation, phenolics oxidize to quinones and then polymerization reaction could follow. If a plant is cut and dried under "near ambient" conditions, which generally requires a large time to dry, the nature and content of phenolic compounds can change. In order to avoid these changes, the metabolic activities of the cells need to be curbed immediately. The next important step is to bring the chemical constituent into solutions for their measurement.

1. Collection, drying and storage of plant material
   Leaf age and stage of development affect levels and nature of phenolics. Therefore, it is important to define the stage of maturity of plant and leaf as close as possible before collecting leaf material for analysis. When the collection site is close to the laboratory, the material can be transported to the laboratory in fresh state. The fresh material should be kept on ice and transported under dark conditions. Transportation of large amount of leaves in plastic bags should be avoided, since temperature in the bag could rise leading to sweating and wilting, which can change the nature and level of phenolics. If liquid nitrogen is available, the better option is to freeze the sample and then freeze-dry the material without thawing it. Thawing can rupture cell membranes leading to changes in phenolics. If the material is frozen using a freezer, make sure that the material is not thawed during transport. Solid carbon dioxide should be used to transport such material. Once the material is dried, it should be kept in a dry

R. Katoch, *Analytical Techniques in Biochemistry and Molecular Biology*,
DOI 10.1007/978-1-4419-9785-2_14, © Springer Science+Business Media, LLC 2011

place (preferably in a desiccator) in the dark. The freeze-dried material generally is hygroscopic. Light is also known to change the nature of phenolics. After freeze-drying, the cell structure is broken and the enzymes are in the native state. With the absorption of water, enzymes and phenolics can react, which can produce drastic changes in phenolics. The freeze-drying, though considered to be one of the safest method for preservation of phenolics, can lead to drastic changes if the storage conditions are not appropriate.

If a lyophilizer is not available, the plant material has to be dried under far from ideal conditions. The sample can be dried at about 50–52°C using a forced air oven. This will hasten the process of drying, and the enzymes present in the plant sample will not have much time to react with phenolics. Drying at temperatures higher than 55°C should be avoided, since it can lead to inactivation of phenolics or could decrease their extractability in solvent and effect the quantification.

2. Grinding of sample

   Fresh material, when frozen using liquid nitrogen, can be ground using "Polytron" homogenizers. Phenolics are generally extracted in aqueous organic solvents. The moisture present in the fresh material needs to be taken into account while preparing organic solvents for extraction.

   It is suggested to grind the sample after drying the sample. About 500 g of the plant material should be ground first to pass a 2-mm screen. All the ground material including those parts remaining inside the mill should be taken, mixed well and approximately 100 g of this sample is again ground to pass through a 0.5-mm screen. For fibre analysis, and in situ nylon bag and in vitro gas techniques, the sample passed through 2-mm screen should be taken. At any stage of the grinding, the sample temperature does not rise above 40°C.

14.2 Proximate Analysis of Forages (Van Soest 1963)

Proximate analysis was given by Henneberg and Stohmann in 1865 who were then working at Weende experiment station in Germany. This scheme of analysis is commonly referred to as the "Weende analysis" and forms the basis for description of quality composition of feed and forages. This scheme partitions biological materials into fractions which are referred for the nourishment of animal body. The Weende proximate analysis as a chemical procedure is shown in the following Fig. 14.1.

This scheme groups together a variety of substances having common chemical characteristics and hence not an analysis of a lone nutrient of the feed. Except for water, each component represents a combination of substances, some of which are nutrition or a combination of nutrients, while others may not be of any nutrition value to the animal. Most prominent points of significance and drawbacks of the scheme are enumerated below.

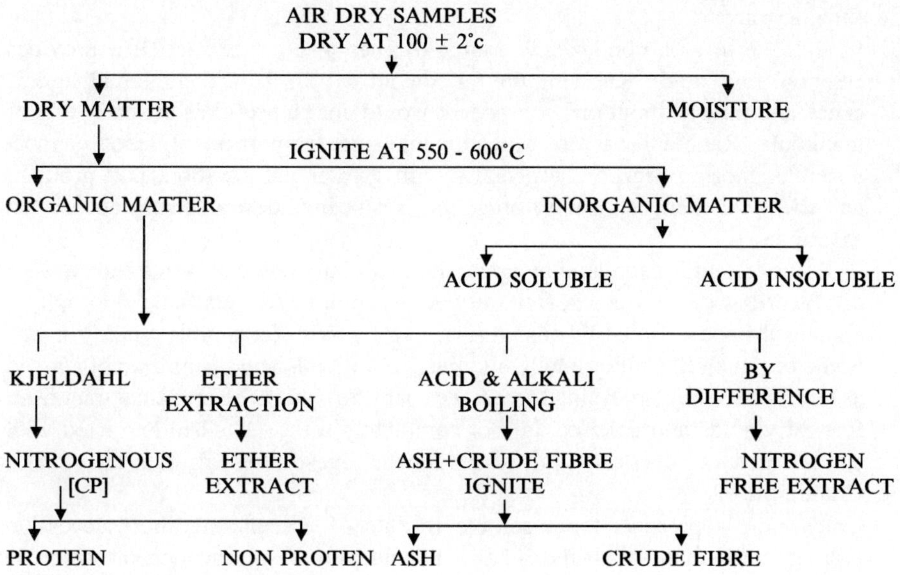

Fig. 14.1 Weende proximate analysis

1. Moisture
   Feeds containing more than 14% moisture cannot be stored in bulk for the danger of mould formation as well as spontaneous combustion. Calculation of relative cost of nutritional value involves consideration of moisture content. While ensiling, the moisture of forages has to be brought nearer to 70%. Feeds with high moisture may restrict the intake of other nutrients. Feeding standards do not include water requirements and therefore need correction for this omission. Moisture is usually determined as loss in weight by oven drying to a constant weight just above the boiling point of water at atmospheric pressure. Such drying sometimes results in a loss of heat labile substances such as volatile fatty acids. In addition, some sugars decompose at above 70°C. The loss of these during oven drying is also taken as moisture and inflates its value.
2. Crude protein
   Knowing the protein (CP) content of a feed gives an idea about the class of feed to which it belongs. Crude protein determination involves multiplication of estimated nitrogen value usually by a factor 6.25 basing on the assumption that most of the feeds contain 16% nitrogen. It is an estimate of total protein without any consideration of its quality. It does not distinguish the nitrogen contribution from true protein and non-protein nitrogenous substances such as urea, uric acid and ammonium salts. While analysing faecal nitrogen, conversion factor for feed protein may not apply for portion of nitrogen of metabolic origin.

3. Ether extract

   In addition to as a non-specific source of energy, ether extract (EE) provides essential fatty acids. Knowing the fat content in milk is of commercial significance and milk without this component would not be useful as the early diet of mammals. Rancid feeds are objectionable and fat portion of feeds is most unstable. Storing high fat, especially with greater unsaturation, is a problem and chemical changes during storage may result in undesirable substances such as amines.

   Ether extract is estimated by extraction with fat-soluble solvents and consists of glycerides of fatty acids, free fatty acids, cholesterol, lecithins, chlorophyll, alkali substances, volatile oils, resins, carotenoids, fat-soluble vitamins, etc. Some of which like chlorophyll, alkalies, volatile oils and resins are not classed as nutrients. While analysing faeces, the soaps formed in the intestinal tract from free fatty acids and calcium are not completely removed when extracted with ether. It gives erroneously high values for the digestibility of ration fat.

4. Crude fibre

   Crude fibre represents the insoluble residue of a feed left after successive boiling with dilute acid and alkali, the nutritional significance of which is not clear and precise. As per original supposition of Weende analysis, crude fibre is the indigestible portion of total carbohydrate of the feed. Sometimes it is misleading index of overall digestibility of a feed as in a number of cases the crude fibre is digested as high as soluble carbohydrates, normally referred to as nitrogen-free extract, especially in case of ruminants. Moreover, crude fibre estimation simulates monogastric digestion than that of the ruminants.

5. Ash

   The inorganic residue left after ignition at 550–600°C represents total ash. It does not discriminate the proportion of mineral matter and sand or silica due to either contamination or adulteration. Some volatile mineral elements, such as iodine and selenium, are lost on ashing.

   Despite various limitations, Weende analysis thus forms a basis for chemical description of feeds, body tissues and biological excreta to find digestibility and utilization and in the feeding standards of different categories of animal species.

6. Nitrogen-free extract

   Nitrogen-free extract (NFE) is a non-cellulose portion of feed carbohydrates and is a non-specific source of energy to the animal. This fraction makes up about 40% of dry weight of forage feeds and 70% of basal feeds. The proportion of NFE will be inversely related to protein content in concentration feeds.

   NFE is the difference between actual sample weight and sum of weights of water, ether extract, crude protein, crude fibre and ash. Hence, its numerical value is affected by analytical errors of these five as well as by the lack of precission of crude fibre determination in separating functional categories of carbohydrates as it is a mixture of all the starches and sugars plus some hemicellulose and much of the lignin.

14.3 Estimation of Dry Matter (DM)

Principle

A known quantity of sample is dried in hot air oven for 8–12 h at 100 ± 2°C to a constant weight and the loss of weight is expressed as moisture percentage from which dry matter percent is calculated.

14.3.1 DM in Bulk Samples

The representative samples collected as per described standard procedure are brought to the laboratory in closed containers such as polyethylene bags and are immediately subjected to moisture estimation.

Equipment

1. Physical balance
2. Hot air oven
3. Aluminium tray or Glass Petri dish

Procedure

1. Weigh representative sample of approximately required quantity in a clean dry preweighed aluminium tray or glass Petri dish.
2. Dry the weighed sample in hot air oven at 100 ± 2°C for 8–12 h (overnight) to a constant weight.
3. Weigh quickly after cooling at the room temperature without undue exposure to atmosphere for prolonged period to avoid absorption of moisture.
4. The difference in weight is expressed as moisture percent from which DM percent can be calculated.
5. Store the oven-dried sample, after grinding in laboratory grinder (1-mm size), in air tight containers for further estimation of various proximate principles.

Calculations

$$\text{Moisture (\%)} = \frac{(b-a)-(c-a)}{(b-a)} \times 100$$

$$\text{Dry matter (\%)} = 100 - \text{Moisture percent (or)}$$

$$\text{Dry mater (\%)} = \frac{(c-a)}{(b-a)} \times 100$$

where

a = empty weight (g) of aluminium tray or glass Petri dish
b = weight (g) of tray or Petri dish with sample before oven drying

c = weight (g) of tray or Petri dish with sample after oven drying

Note

1. In case of long fodders, they may be cut into small pieces for convenience of handling with least disturbance to its structure.
2. Keep half open the cover of the ventilator on the top of oven, especially in case of samples of high moisture, to avoid water accumulation inside the oven due to condensation.
3. If extraneous moisture is present in case of green forages, care should be taken to overcome its loss during handling to assess actual dry matter truly representing the situation of bulk or lot from which it is drawn.

14.3.2 Dry Matter in Laboratory Samples

In order to get uniform idea about the chemical composition, it is customary to express it on moisture-free basis which can be estimated in ground and stored bulk samples preserved after DM estimation.

Equipment

1. Moisture cup (aluminium) with lid (2 cm depth and 5 cm diameter)
2. Hot air oven
3. Chinametric balance
4. Desiccator

Procedure

1. Dry a clean moisture cup with lid in a hot air oven at 100 ± 2°C for 10–15 min and cool in a desiccator and not the empty weight.
2. Spread uniformly the ground sample to be analysed on a sheet of white paper for sufficient time to attain room temperature.
3. Transfer representative sample of suitable quantity with the help of spatula from different places into the moisture cup and note the weight with lid.
4. Dry the sample in hot air oven at 100 ± 2°C for 8–12 h with partially opened lid.
5. Cool in a desiccator with closed lid and weigh.
6. The process of drying, cooling and weighing is repeated till the difference between two successive weighings is less than 1 mg. The moisture-free sample is retained in desiccator for the estimation of ether extract.

Calculations

$$\text{Moisture }(\%) = \frac{(b-a)-(c-a)}{(b-a)} \times 100$$

$$\text{Dry matter }(\%) = 100 - \text{Moisture percent (or)}$$

$$\text{Dry mater }(\%) = \frac{(c-a)}{(b-a)} \times 100$$

where

a = empty weight (g) of moisture cup with lid
b = weight (g) of moisture cup with lid and sample before oven drying
c = weight (g) of moisture cup with lid and sample after oven drying

14.3.3 DM in Silage, Haylage and Molasses

Oven drying of materials having heat labile compounds may underestimate the moisture content. Alternatively, the moisture in such feedstuffs can be estimated by toluene distillation method.

Principle

The actual volume of water is measured by distillation in the presence of excess toluene after correcting the volume for total acidity that is occupied by volatile acids.

Equipments and Glassware

1. Toluene distillation set with condenser
2. Graduated flask
3. Pipette

Reagents

1. Toluene
2. Ethanol (80% alcoholic ammonia solution neutralized to phenolphthalein)
3. 0.1 N NaOH
4. Phenolphthalein indicator

Procedure

1. Add 400 mL of toluene to the distillation flask containing 70–80 g of sample.
2. Distil the contents for 6–8 h at the rate of 2–3 drops per second.
3. Stop distillation after observing two equal consecutive volumes at 15 min interval.
4. Wash the condenser with toluene to remove traces of water present in it and continue distillation for another 15 min. Keep the graduated receiver flask in water bath at 20°C for 20 min and note the volume of water at room temperature.
5. Transfer the aqueous layer to 100-mL volumetric flask and make up the volume with CO_2-free distilled water.
6. Add 40 mL neutralized ethanol to 10 mL of diluted distillate and titrate against 0.1 N NaOH in the presence of phenolphthalein.

Calculation

$$DM = 100 - \frac{99.8v}{w} \times (t - f/10)$$

where

v = volume (mL) of aqueous phase
w = weight (g) of sample
t = volume (mL) of 0.1 N NaOH utilized
f = factor depending on the constituents of acids of the sample (normally 0.00555 is taken for all practical purposes)

Note: No acid correction is needed in case of molasses.

14.4 Crude Protein (CP)

Principle

The Kjeldahl Nitrogen method is the most frequently used procedure for measuring nitrogen and in turn the protein content in biological materials. In this method, the amino ($-NH_2$) nitrogen is oxidized by sulphuric acid in the presence of catalyst of $(NH_4)_2SO_4$. The ammonium ion is converted to NH_3 by NaOH and collected by distillation. The NH_3 is then quantitatively titrated against standard acid (HCl or H_2SO_4) of known strength and nitrogen in the sample is computed. The crude protein is obtained by multiplying the nitrogen content with factor 6.25 (16% nitrogen in protein for most of the feeds in general).

Equipment and Glassware

1. Digestion bench placed in digestion chamber
2. Distillation unit (micro-Kjeldahl distillation apparatus)
3. Kjeldahl flask (500–800 mL)
4. Burette
5. Conical flask
6. Volumetric flask

Reagents

1. Commercial sulphuric acid
2. Digestion mixture [1 part copper sulphate (acts as catalyst) : 10 parts sodium or potassium sulphate (raises boiling point)]
3. 40% Sodium hydroxide
4. Tashiro's indicator
5. N/7 H_2SO_4

Procedure

The method of estimation of nitrogen/crude protein encompasses digestion, distillation and titration.

Digestion

1. Transfer exactly weighed suitable quantity (as described under sampling) of sample into Kjeldahl flask.
2. Add 20–50 mL commercial sulphuric acid depending upon the type of sample.
3. Add 5–10 g of digestion mixture.
4. Boil the contents for 2–3 h on a digestion bench placed in a digestion chamber till the solution is clear without leaving any undigested black particles. Adhering materials inside walls of the flask needs one or two washings in between after cooling. To avoid bumping, a few glass beeds may be placed inside the Kjeldahl flask.
5. Transfer the digested material present after cooling by dissolving with nitrogen-free tap water into volumetric flask followed by 5–6 repeated washings. Make up the final volume after cooling.
6. Similarly run a blank without sample.

Distillation

1. Place a conical flask containing 10 mL of Tashiro's indicator at the end of condenser of micro-Kjeldahl distillation apparatus. Care should be taken to see the tip of the condenser is completely dipped inside the indicator to avoid escape of released ammonia during distillation.
2. Pipette out 5–10 mL of aliquot of digested sample from the volumetric flask into the distillation unit.
3. Add 10–20 mL of 40% NaOH sufficient to make the contents alkaline (i.e. till the contents turn blue or black). Wash with a small quantity of distilled water and close the receiving end immediately with a pinch cock. Seal the funnel with a little amount of distilled water to avoid escape of ammonia.
4. Steam distil the contents of the distillation unit by boiling the water in a round bottom flask (avoid bumping by addition of glass beads) connected to the distillation unit.
5. Collect around 30–50 mL distillate (or at least 2 times the quantity of indicator taken) to ensure all nitrogen in the form of ammonia is distilled. Red colour turns to green.
6. Remove the conical flask with distillate after washing the tip of the condenser with a few millilitres of distilled water.
7. Wash the distillation unit 2 or 3 times with distilled water with the help of back suction developed by vacuum due to displacement of boiling flask from the heater to make the apparatus ready for distillation of next sample.

Note: In place of Tashiro's indicator, standard N/7 H_2SO_4 may also be taken with methyl red indicator in the receiver flask (conical flask). In which case, the distillate is back titrated against N/7 NaOH to know the actual volume of N/7 H_2SO_4 consumed.

Titration

1. Titrate the distillate in the conical flask against standard N/7 H_2SO_4 solution taken in a burette till the red colour just reappears.
2. Note the volume of N/7 H_2SO_4 consumed.

Calculations

$$\text{Crude protein (\%)} = 100 \times \frac{Y \times (B - B1) \times 0.002 \times 6.25}{X \times W}$$

where

Y = volume (mL) made out of digested sample
X = volume (mL) of aliquot taken for distillation
B = volume (mL) of N/7 H_2SO_4 consumed for titration of sample
$B1$ = volume (mL) of N/7 H_2SO_4 consumed for titration of blank distillate
W = weight (g) of oven-dried sample taken for digestion
1 mL of N/7 H_2SO_4 = 0.002 g N
6.25 = factor for converting nitrogen into protein of feedstuffs

It is assumed that most of the proteins in feedstuffs contain 16% nitrogen. Therefore, for conversion of estimated nitrogen, a factor 6.25 (100/16) is usually adopted to arrive at crude protein content of feed samples.

14.5 Determination of True Protein and Non-Protein Nitrogen

The insoluble protein left by precipitation with a suitable precipitating agent after filtering NPN is termed as true protein. The NPN is calculated by difference between the total crude protein nitrogen and the value of the precipitated true protein nitrogen. The various precipitating agents commonly used are tungstic acid, trichloroacetic acid (TCA), copper hydroxide, zinc-barium hydroxide, etc. The choice of method depends on the kind of procedure and objective being followed.

Apparatus

Erlenmeyer flask (125 mL), Whatman no. 54 or 541 filter paper, analytical balance, filter funnel and Kjeldahl apparatus.

Reagents

TCA 10% (w/v) in water (keep refrigerated). The remaining materials are described in nitrogen estimation.

Procedure

1. Weigh 0.5 g ground dry sample into a 125-mL Erlenmeyer flask.
2. Add 50 mL of distilled water. Allow to stand 30 min.
3. Add 10 mL of 10% TCA. Let it stand for 20–30 min.
4. Filter on Whatman no. 54 or 541 filter paper by gravity.

5. Wash twice with TCA solution.
6. Transfer paper to digestion tube and determine residual nitrogen by using Kjeltec Auto Analyser as described earlier and arrive at true protein value.
7. Calculate NPN by subtracting residual nitrogen from total nitrogen. Value of NPN may be expressed as crude protein (N × 6.25) or as percent of total feed nitrogen.

14.6 Estimation of Ammonia Nitrogen

Equipment and Glassware

1. Micro-Kjeldahl distillation unit
2. Burette
3. Conical flask
4. Volumetric flask
5. Funnel and funnel stand

Reagents

1. Tashiro's indicator
2. N/7 H_2SO_4
3. Carbonate-free magnesium oxide
4. 40% Sodium hydroxide

Procedure

1. Shake 2–4 g of ground sample with water and filter. Wash the residue with water and make up the volume.
2. Distil as described earlier by taking 2–5 mL filtrate after adding magnesium oxide at 25 mg/mL.
3. Distil a blank without taking the filtrate.
4. Calculate the amount of ammonia nitrogen present in the sample

$$\text{Ammonia nitrogen (\%)} = \frac{100 \times Y(B - B_1) \times 0.002}{X \times W}$$

where

Y = volume (mL) of filtrate made up to
X = volume (mL) of aliquot taken for distillation
B = volume (mL) of N/7 H_2SO_4 consumed for the titration of test sample
$B1$ = volume (mL) of N/7 H_2SO_4 consumed for the titration of blank distillate
W = weight (g) of oven-dried sample taken for filtration 0.002 g = 1 mL of N/7 H_2SO_4

14.7 Estimation of Urea Nitrogen

Equipment and Glassware

1. Macro-Kjeldahl distillation unit
2. Conical flask
3. Burette

Reagents

1. Standard sulphuric acid solution (0.1–0.5 N, depending on nitrogen concentration)
2. Standard sodium hydroxide solution (0.1–0.25 N as per expected nitrogen concentration)
3. Methyl red indicator.
4. Calcium chloride solution (25 g in 1,000 mL water)
5. Defoaming solution (dissolve 50 g diglycol stearate in 375 mL benzene, 75 mL alcohol and 250 mL dibutyl phthalate; warm, if necessary).
6. Freshly ignited carbonate-free magnesium oxide.
7. Urease solution (dissolve standardized urease afresh).

Standardization of Urease Solution

1. Dissolve 0.1 g of commercial urease in 50 mL of distilled water
2. Find the alkalinity by titrating against 0.1 N HCl in the presence of methyl red.
3. Prepare neutralized urease solution as standardized above by dissolving equal volumes of 0.1 N HCl and urease solution of 0.1 g in water so that each 10 mL of neutralized solution may convert more than 0.1 g of urea into nitrogen.

Determination of Enzyme Activity

Determine the enzyme activity by preparing 50 mL of 1% neutralized solution. Add different quantities of neutralized urease to each 0.1 g pure urea. Distil after enzyme digestion as per procedure described later. Calculate the activity of urease solution.

Procedure

1. Shake 2 g of ground sample placed in a Kjeldahl flask with 250 mL of distilled water, and add 10 mL of urease solution (more urease solution may be needed if urea is more than 5% to a level 12% protein equivalence).
2. Stopper tightly and let it react for 1 h at room temperature or warm it to 40°C for 20 min and cool to room temperature.
3. Rinse the stopper and neck with a few millilitres of distilled water.
4. Add 2 g magnesium oxide, 1 mL calcium chloride and 5 mL defoaming solution.
5. Distil about 100 mL into a receiver flask containing standard sulphuric acid and methyl red indicator by directly connecting the Kjeldahl flask to a condenser.
6. Titrate the contents against standard sodium hydroxide solution.
7. Run a blank simultaneously without test material.

Calculation

$$\begin{array}{l}\text{Urea nitrogen (\%)} \\ \text{(as total ammonical N)}\end{array} = \frac{X(B - B_1)N}{W} \times 100$$

where

B = mL of standard sodium hydroxide required to neutralize excess acid in the test
B_1 = mL of standard sodium hydroxide required to neutralize excess acid in blank
N = normality of standard sodium hydroxide
W = weight of moisture-free sample
X = factor for converting 1 mL of standard sulphuric acid (actually utilized) into nitrogen

14.8 Estimation of Ether Extract (Crude Fat)

Principle

Ether extract is determined by extracting known amount of moisture-free sample with fat solvents such as petroleum ether in soxhlet ether extraction apparatus. The collected extract is dried to a constant weight at 100 ± 2°C in hot air oven and expressed as percentage on dry matter basis. The ether extract may, in addition to glycerides of fatty acid, contain other ether-soluble compounds such as free fatty acids, cholesterol, lecithins, chlorophyll, carotenoids, waxes, sterols, alkali substances, volatile oils, resins and fat-soluble vitamins and thus also known as crude fat.

Equipment and Glassware

1. Soxhlet apparatus (condenser, extractor and flask fitted to a thermostatically controlled heating mantle bench)
2. Thimble (made out of Whatman no. 1 filter paper with cotton plug)
3. Hot air oven
4. Balance (Chinametric or electronic with facility to weight up to 4th decimal)
5. Desiccator

Reagents

Petroleum ether (60–80°C) (or any fat solvents such as benzene, chloroform and diethyl ether).

14.8.1 Preparation of Thimble

Wrap around Whatman no. 1 filter paper (8 cm diameter) on a 2 cm diameter wooden stick or glass test tube and tie with cotton thread.

Procedure

1. Carefully transfer the moisture-free sample after dry matter estimation into the thimble and plug with cotton.
2. Place the sample containing thimble in the extractor having small cotton plug at its bottom to avoid passage of any spilled over particles of test material. Care should be taken to see the height of thimble to be below siphon level of extractor.
3. Fix the extractor to a clean dry preweighed oil flask.
4. Insert extractor with oil flask to the condenser.
5. Pour required quantity of solvent from the top of condenser with the help of a funnel to about 1.5 times capacity of extractor and plug the condenser with cotton to avoid any evaporation loss of solvent.
6. Start cold tap water circulation through outer jacket of condenser after checking for any water leakage.
7. Depending upon the condensing rate of solvent from 5–6 to 2–3 drops per second, extract for at least 4–16 h by setting the thermostat of heating mantles of the apparatus (roughly 8 h extraction with a minimum total 250 condensations may be sufficient for complete extraction).
8. Collect back all the solvent for reuse, by removing it from extractor when the level of solvent is well below the siphon, till no solvent is left in the extractor and oil flask. Care should be taken to avoid charring of contents of oil flask due to over heating while collecting the solvent. Always put off the heaters at the time of separating extractor from the unit for transferring solvent to the container.
9. Weigh the flask with contents after oven drying to a constant weight.
10. Dry and store the thimble having moisture- and fat-free sample for the estimation of crude fibre.

Calculations

$$\text{Ether extract (\%)} = \frac{(y - x)}{w} \times 100$$

where

y = weight of oil flask after extraction
x = weight of oil flask before extraction
w = weight of oven-dried sample

14.9 Determination of Crude Fibre (CF) (Goering and Van Soest 1970)

Principle

The moisture- and fat-free sample is successively refluxed with weak sulphuric acid and alkali (NaOH), each followed by filtration and repeated hot water washings. The remaining residue comprises lignocellulose and cellulose is oven-dried and ashed. The loss in weight is expressed as crude fibre.

Equipment and Glassware

1. Hot plate
2. Hot air oven
3. Muffle furnace
4. Spoutless beakers (1 L)
5. Round bottom flask (as condenser)
6. Muslin cloth (24 threads each in warf and weft per sq. inch)
7. Wash bottle
8. Steel spatula
9. Buchner funnel with suction arrangements (suction pump attached to swan neck tap at one end and the other end to side neck of conical flask having funnel).

Reagents

1. 2.04 N H_2SO_4
2. 2.50 N NaOH
3. Ethyl alcohol (as antifoaming agent, if necessary)

Procedure

1. Transfer the moisture- and fat-free sample (after ether extraction) from the thimble, by untying the thread, into a spoutless beaker of 1 L capacity previously marked to 200 mL.
2. Add 25 mL of 2.04 N H_2SO_4 and make up the volume with water up to 200 mL marking (0.255 N) from the sides of the beaker.
3. Place the beaker, sealed with a round bottom flask filled with cool water (ordinary) to act as a condenser for maintaining the volume of the contents of the beaker, on a previously warmed hot plate. Care should be taken to wipe water droplets on outside the beaker to avoid its breakage due to expansion of moistened glass surface.
4. Reflux the contents for 30 min from the start of boiling (add a few drops of antifoaming agent, if necessary)
5. Remove the beaker and cool by adding around 200 mL of water to avoid charring of muslin cloth during washing, which results passage of finer particles during filtration and washing.
6. Filter the beaker contents with the help of Buchner funnel with suction arrangements and wash beaker and residue left on muslin cloth with repeated hot water washings to make it completely acid free.
7. Transfer the residue to same spoutless beaker carefully with smooth steel spatula followed by a little washing of muslin cloth, if necessary.
8. Add 25 mL of 2.50 N NaOH to the beaker and make up the volume to 200 mL with ordinary water (0.312 N).
9. Reflux the contents on hot plate for 30 min from the point of boiling, after placing round bottom flask with cold water as condenser on the top of the beaker (add a few millilitre of antifoaming agent, if necessary).

10. Filter the contents of beaker after addition of around 200 mL water for cooling and make alkali free by repeated hot water washings to the beaker and residue on muslin cloth.
11. Transfer the residue to a clean silica basin of suitable capacity with the help of smooth steel spatula and a little water washing, if required.
12. Dry the contents of silica basin in hot air oven at $100 \pm 2°C$ to a constant weight. Cool it in a desiccator and note the weight of dried residue along with silica basin.
13. Ash the dried residue in silica basin, after decarbonization on a heater or flame, in muffle furnace at 550–600°C for 1–2 h. Weigh the silica basin with left back ash (including sand and silica, if any) after cooling in a desiccator.
14. The difference in oven-dried and ashed weights gives the weight of crude fibre.

Calculations

$$\text{Crude fibre } (\%) = \frac{(a - b)}{w} \times 100$$

where

a = weight (g) of silica basin plus oven-dried residue left after aid and alkali digestion
b = weight (g) of silica basin plus ash
w = weight (g) of oven-dried sample

Precautions

- Do not scratch the muslin cloth with spatula as it may spoil the texture of cloth.
- Secure muslin cloth placed on the funnel tightly for efficient suction.
- Avoid spillage of residue from muslin cloth while filtering and washing.
- If residue is not acid and alkali free, it may inflate the values.
- Transfer of residue to the beaker after acid digestion and to the silica basin after alkali digestion should be done carefully without losing any fibre particles.

14.10 Estimation of Total Ash

Principle

Ash is the inorganic residue left after ignition of a decarbonized material in a muffle furnace at 550–600°C for 2–3 h, which essentially can be partitioned into soluble portion comprising of mineral matter and insoluble ash consisting silica or sand. It can be used for preparing mineral extract in the estimation of various minerals as well as for finding out organic matter and NFE by difference.

Equipment and Glassware

1. Muffle furnace
2. Tongs

3. Silica basin
4. Desiccator

Procedure

1. Take approximately required quantity of ground material, as suggested under sampling depending upon the type of sample, into a clean and dry preweighed silica basin and note the weight of silica basin along with the sample.
2. Decarbonize on a heater or flame till no smoke in emitted.
3. Transfer the silica basin with decarbonized sample into a muffle furnace and ignite at 550–600°C for 2–3 h until no black particles are left.
4. Cool the silica basin with ash in a desiccator and not the weight quickly.

Calculations

$$\text{Total ash (\% on as such basis)} = \frac{(c - a)}{(b - a)} \times 100$$

or

$$\text{Total ash (\% in DM)} = \frac{(c - a)}{w} \times 100$$

where

a = empty weight (g) of silica basin
b = weight (g) of silica basin with sample
c = weight (g) of silica basin with ash
w = weight (g) of moisture-free sample

Precautions

- Switch off the fans and put on the exhaust while decarbonizing (smoking) the sample.
- Transfer decarbonized sample into muffle by carrying in a closed container to avoid loss of sample due to air movement.
- Weigh cooled ash sample as quickly as possible since it is highly hygroscopic.
- Though silica basin can with stand reasonably high temperature, monitor to maintain temperature of muffle furnace between 550 and 600°C to avoid breakage.

14.11 Estimation of Acid Insoluble Ash

Principle

The residue left after dissolving inorganic portion of the total ash represents acid insoluble ash (AIA), majorly containing sand and silica. It is useful to estimate the extent of contamination while handling feedstuffs.

Equipment and Glassware

1. Muffle furnace
2. Hot air oven
3. Silica basin
4. Tongs
5. Desiccator
6. Volumetric flask
7. Funnel
8. Beaker (250 mL)

Reagents

1:2 Hydrochloric acid

Procedure

1. Add about 25 mL of dilute (1:2) HCl to the total ash obtained in the previous experiment and quantitatively transfer to a 250-mL beaker with repeated hot distilled water washings to the silica basin (chipping due to scratching of silica basin with glass rod should be avoided, which otherwise may inflate the values).
2. Boil the contents for 5–10 min.
3. Filter through Whatman no. 1 filter paper into a volumetric flask of around 250 mL capacities. Make it acid free through repeated hot water washings to the beaker and to the residue on filter paper.
4. Retain the filtrate collected in volumetric flask containing dissolving inorganic salts for further estimation of minerals after making up the volume when cooled.
5. Transfer the filter paper with retained residue to a preweighed silica basin.
6. Dry in hot air oven at 100 ± 2°C to a constant weight.
7. Transfer the silica basin to muffle furnace after decarbonization and ignite at 550–600°C for 1–2 h.
8. Cool in a desiccator and weigh the silica basin with left back AIA.

Calculations

$$\text{Acid insoluble ash (\%)} = \frac{b-a}{w} \times 100$$

where

b = weight (g) of silica basin with acid insoluble ash
a = empty weight (g) of silica basin
w = weight (g) of moisture-free sample taken for ashing

- Weight of filter paper need not be considered as it is usually made of ash-free organic cellulose which is lost during ignition.

14.12 Nitrogen-Free Extract

Principle

The NFE is determined by subtracting the percentage of crude protein, ether extract, crude fibre and total ash on dry matter basis and including moisture on as-fed basis from 100, respectively.

Calculation

$$\text{Nitrogen free extract (\% as fed basis)} = 100 - [\text{CP\%} + \text{EE\%} + \text{Total ash \%} + \text{Moisture \%}]$$

or

$$\text{Nitrogen free extract (\% on DM basis)} = 100 - [\text{CP\%} + \text{EE\%} + \text{CF\%} + \text{Total ash \% on DM basis}]$$

- NFE can also be calculated by subtracting the percentages of CP, EE and CF from percentage of organic matter (OM).
- Since determination of NFE is by difference and not by direct analysis, the cumulative errors during analysis of other principles are reflected in the NFE content.
- Organic matter is derived by subtracting percentage of total ash from 100 or through the addition of percentages of CP, EE, CF and NFE.
- Total carbohydrate content of feed can be arrived by addition of CF and NFE or by deletion of CP and EE from OM or by deletion of CP, EE and total ash from DM.

14.13 Determination of Free Fatty Acids (Duncombe 1963)

Principle

Free fatty acids are extracted from lipids of feed sample by using Dolls extraction mixture (40:10:1, heptane: isopropanol: acetic acid). Then the heptane phase is taken and evaporated. The free fatty acids form a complex with cupric ions when mixed with copper reagent; the coloured complex formed with copper with copper is soluble in chloroform and diethyldithiocarbamate is used as a colour developer.

Reagents

1. Copper reagent: a volume of aqueous 1 M triethanolamine, 1 volume of 1 N acetic acid and 10 volumes of 6.45% $Cu(NO_3)_2{\cdot}3H_2O$.
2. Sodium diethyl dithiocarbamate: The diethyldithiocarbamate reagent is 0.1% (w/v) solution of sodium diethyl dithiocarbamate in redistilled secondary butanol. (Both the reagents are stored in the refrigerator and use within 7 days.)
3. Standard fatty acids solution: Make up in chloroform in the range 10–100 μM myristic, palmitic, stearic and linoleic acids.

Equipment

Centrifuge, centrifuge tubes, hypodermic needle and spectronic-20.

Procedure

Take about 5 mL of the chloroform solution of fatty acids of feed sample and place in centrifuge tube. Add 2.5 mL of copper reagent. The tubes are stoppered and shaken vigorously for 1 min. Centrifuge the tubes for a few minutes. Remove the supernatant aqueous phase by suction with a few hypodermic needles. The surface of the chloroform phase can easily be left clean with only traces of aqueous phase adhering to the wall of the tube. A portion (2.5 mL) of chloroform layer is taken into a clean dry tube; care should be taken that the pipette does not touch the inner wall of the either tube, as traces of copper containing aqueous phase might be transferred. Then 0.5 mL of diethyl dithiocarbamate reagent is added to the chloroform solution and mixed. The extinction is read at 440 nm. All measurements are also made against a blank solution that had been subjected to the same procedure.

Calculation

A standard curve is made with various concentrations ranging from 10 to 100 μM of myristic, palmitic, stearic and linoleic acids. The sample reading is matched with the curve and the amount of free fatty acid is calculated.

14.14 Estimation of Fibre Fractions (Van Soest 1963)

P. J. Van Soest in 1963 partitioned the carbohydrates into various fractions by a system of analysis using detergents. The method makes use of the concept that the dry matter of plant origin consists of two principal parts:

1. Cell wall
2. Cell contents

Cell contents are soluble in neutral detergent, whereas cell wall (fibre) is insoluble in neutral detergent but is soluble in acid detergent.

14.14.1 Estimation of Neutral-Detergent Fibre (Determination of Cell Contents and Cell Wall Constituents)

Principle

The neutral-detergent procedure for cell wall constituents is a rapid method for analysing the total fibre in vegetable feedstuffs. It appears to divide the dry matter of feeds very near the point that separates the nutritively available (98%) and soluble constituents from those that are incompletely available and dependent on a microbial fermentation. The cell contents are determined as the difference between the percent cell wall and 100%.

Apparatus

1. Refluxing apparatus:
   (a) Tall beakers (spoutless) of about 1,000 mL capacity.
   (b) Round bottom flask as condenser.
2. Sintered glass crucibles with coarse porosity (Grade 1) of 50 mL capacity.
3. Electronic balance
4. Vacuum pump
5. Hot plate
6. Wash bottle
7. Hot air oven and muffle furnace

Reagents

1. Neutral-detergent solution: Put EDTA and $Na_2B_40_7.10H_2O$ together in a large beaker, add some of the distilled water, and heat until dissolve, then add to solution containing disodium lauryl sulphate and 2-ethoxy ethanol. Put Na_2HPO_4 in a beaker, add some of the distilled water and heat until dissolved and then add to solution containing other ingredients. Check pH range 6.9–7.1. If solution is made accurately pH adjustment is rarely required (Table 14.1).
2. Decahydronaphthalene (Decalin) reagent grade.
3. Acetone
4. Sodium sulphite (anhydrous)

Table 14.1 Composition of neutral detergent solution

Distilled water	1 L
Sodium lauryl sulphate	30 g
Disodium ethylene diamino tetra acetate (EDTA) dehydrate	18.61 g
Sodium borate decahydrate	6.81 g
Disodium hydrogen phosphate (anhydrous)	4.56 g
2-ethoxyethanol (ethylene glycol monoethyl ether)	10 mL

Procedure

1. Take 0.5–1.0 g air dry sample ground to pass 20–30 mesh (1 mm) into a beaker of the refluxing apparatus.
2. Add in order 100 mL (preheated) neutral-detergent solution, 2 mL of decalin and 0.5 g sodium sulphite with a calibrated scoop and reflux for 60 min, time starting from the onset of boiling.
3. Filter off the regent, wash thrice with hot distilled water under vacuum, remove vacuum, break up mat and wash crucible with hot water.
4. Wash twice with acetone in the same manner and suck dry. Dry crucible at 100°C for 8 h or overnight and weigh it.
5. Report yield of recovered NDF as percent of cell wall constituents. Estimate cell-soluble material by subtracting this value from 100.
6. Ash residues in the crucible for 3 h at 500–550°C and weigh. Report ash content as ash insoluble in neutral detergent.

Observations

Empty wt. of crucible = ——— g
Wt. of dry sample = ——— g
Wt. of crucible + cell wall constituent = ——— g
Wt. of crucible + Ash = ——— g

Calculation

1. NDF%(cell wall constituents)

$$= \frac{(\text{Wt. of crucible} + \text{cell wall constituents}) - \text{Wt. of crucible}}{\text{Wt. of dry sample}} \times 100$$

2. Cell contents(%) = 100 – cell wall constituents

3. Insoluble ash in neutral detergent(%)

$$= \frac{\text{Wt. of crucible} + \text{Ash} - \text{Wt. of crucible}}{\text{Wt. of dry sample}} \times 100$$

14.14.2 Determination of Acid-Detergent Fibre

Principle

The acid-detergent fibre procedure provides a rapid method for lignocellulose determination in feeds stuffs. The residue also includes silica. The difference between cell walls and ADF is an estimate of hemicellulose; however, this difference does include some protein attached to cell walls. The acid-detergent fibre (ADF) is used as a preparatory step for lignin determination.

Apparatus

Same as given in NDF estimation.

Regents

(a) Sulphuric acid (H_2SO_4): Regent grade, standardized to 1 N (100% assay) is 49.04 g dissolved in 1,000 mL.
(b) Cetyl trimethyl ammonium bromide (CTAB) = 20 g: Weigh sulphuric acid and make up to volume with distilled water. Check normality by titration before addition of detergent. Then add CTAB and Stir.
(c) Decalin
(d) Acetone – Use grade that is free from colour and leaves no residue upon evaporation.
(e) *n*-Hexane (technical grade)

Procedure

(a) Weigh 1 g air dry sample ground to pass 20–30 mesh (1 mm) screen or approximate equivalent of wet material in a beaker suitable for refluxing.
(b) Add 100 mL cold (room temperature) acid-detergent solution and 2 mL decalin. Heat to boiling in 5–10 min. Reduce heat as boiling begins, to avoid foaming. Reflux 60 min from on set of boiling; adjust boiling to a slow, even level.
(c) Filter on a previously weighed crucible. Wash with hot distilled water 3–4 times breaking the mat. Repeat wash with acetone twice or until it removes no more colour and suck dry.
(d) Optional wash with hexane. Hexane should be added while crucible still contains some acetone (hexane can be omitted if lumping is not a problem in lignin analysis). Suck the ADF free of hexane and dry at 100°C for 8 h or overnight and weight after cooling of crucible in desiccator.

Observations

Empty wt. of crucible = ——— g
Wt. of dry sample = ——— g
Wt. of crucible + fibre = ——— g

Calculation

Acid detergent fibre %(on dry matter basis)

$$= \frac{(\text{wt. of crucible} + \text{fibre}) - \text{empty weight of crucible}}{\text{wt. of dry sample}} \times 100$$

14.15 Determination of Acid-Detergent Lignin

Principle

In the acid-detergent lignin (ADL) procedure, the ADF procedure is used as a preparatory step. The detergent removes the protein and other acid-soluble material that would interfere with the lignin determination. The ADF residue consists of cellulose, lignin, cutin and acid-insoluble ash (mainly silica); treatment with 72% sulphuric acid dissolves cellulose. Ashing of the residue will determine the crude lignin fraction including cutin.

Apparatus

1. Same as given for NDF estimation
2. Glass tray
3. Muffle furnace

Reagents

H_2SO_4, 72% by weight.

Take 417 mL water in a volumetric flask and add 583 mL pure H_2SO_4 slowly with occasional swirling. The flask must be cooled in water bath (Sink) in order to add the required weight of acid.

Procedure

1. Prepare the ADF.
2. Place the crucible in the glass tray. Have one end of the tray 2 cm higher so acid will drain away from the crucibles.
3. Cover the contents of crucible with cooled (15°C) 72% H_2SO_4 and stir with a glass rod to a smooth paste, breaking all lumps. Fill crucible about half way with acid and stir. Let glass rod remain in crucible, refill with 72% H_2SO_4 and stir at hourly intervals as acid drain away. Crucible does not need to be kept full at all times. Three additions suffice. Keep crucible at 20–23°C. After 3 h, filter off as much acid as possible with vacuum and then wash contents with hot water until free from acid. Rinse and remove stirring rod.
4. Dry crucible overnight at 100°C and weigh.
5. Ignite crucible in a muffle furnace at 500–550°C for 3 h and then cool to 100°C and weigh.

Observations

Wt. of oven dry sample = ——— g
Wt. of crucible and lignin = ——— g
Wt. of crucible and ash = ——— g

Calculation

$$\text{Acid detergent lignin(ADL)\%} = \frac{\text{Wt. of crucible and lignin} - \text{Wt. of crucible and ash}}{\text{Wt. of sample on dry matter basis}} \times 100$$

14.16 Permanganate Lignin, Cellulose and Silica (Van Soest and Wine 1968)

Principle

An indirect method to determine lignin was developed that makes possible the preparation of cellulose and insoluble ash in the same sample. The insoluble ash is an estimate of total silica content, which in many grasses is a primary factor in reducing digestibility. The permanganate lignin method is an alternative procedure to the 72% sulphuric acid method; each has its own advantages. The choice of methods depends on materials analysed and on the purpose for which the values are to be used.

The permanganate reagents are much less corrosive and require no standardization. The residue requires no filter aids, and lignin values are not subject to some interferences that affect 72% sulphuric acid lignin. Values are less affected by heat-damage artefacts and are closer to a true lignin figure.

Principal

Interfering matter is removed by preparing acid-detergent fibre, which is chiefly composed of lignin, cellulose and insoluble minerals. Lignin is oxidized with an excess of acetic acid buffered potassium permanganate solution, containing trivalent iron and monovalent silver as catalysts. Deposited manganate and iron oxides are dissolved with an alcoholic solution of oxalic and hydrochloric acids, which leaves cellulose and insoluble minerals. Lignin is measured as the weight loss by these treatments, whereas cellulose is determined as the weight loss upon ashing. The ash residue is mainly silica and much of the non-silica matter can be removed by leaching with concentrated hydrobromic acid.

Apparatus

Same as used for ADL.

Reagents

Saturated potassium permanganate

Distilled water: 1 L
$KMnO_4$ reagent grade: 50 g
Ag_2SO_4, reagent grade: 0.05 g

Dissolve $KMno_4$ and Ag_2SO_4 in distilled water. Keep out of direct sunlight. Add silver sulphate to dehalogenate the reagent.

1. Lignin buffer solution: 1 L

   Ferric nitrate, monohydrate: 6.0 g
   Silver nitrate: 0.15 g
   Acetic acid, glacial: 500 mL
   Potassium acetate: 5 g
   Tertiary butyl alcohol: 400 mL
   Distilled water: 100 mL

   Dissolve ferric nitrate nonhydrate ($Fe(NO_3)_3.9H_2O$ and silver nitrate in distilled water. Combine with acetic acid and potassium acetate. Add tertiary butyl alcohol and mix.
2. Combined permanganate solution:

   Combine and mix saturated potassium permanganate and lignin buffer solution in the ratio of 2:1 by volume, before use. Unused mixed solution that may be kept about a week in a refrigerator in the absence of light solution is usable if purple and containing no precipitate. Old solutions that assume a reddish colour should be discarded.
3. Demineralizing solution: 1 L

   Oxalic acid dehydrate: 50 g
   95% Ethanol: 700 mL
   Concentrate HCl (12 N): 50 mL
   Distilled water: 250 mL

   Dissolve oxalic acid dehydrate in 95% ethanol. Add concentrate HCl and distilled water and mix.
4. Ethanol 80%: 1 L

   95% Ethanol: 845 mL
   Distilled water: 155 mL

5. Hydrobromic acid, reagent grade

Procedure

1. Dry sample at less than 65°C and grind through 20–30 mesh (1 mm) serene. Prepare ADF according to standard procedure.
2. Use a 1.0 g sample; except on samples containing a high amount of lignin (15% or more) use 0.5 g sample.
3. Place previously weighed crucibles in a shallow enamel pan containing cold water to a depth of about 1 cm. Fibre in crucible should not be wet.
4. Add about 25 mL of combined saturated potassium permanganate and lignin buffer solution (2:1 by volume) to the crucible in the enamel pan containing cold water. Adjust level (2–3 cm) of water in pan to reduce flow of solution out of crucible. Place a short glass rod in each crucible to stir contents, to break lumps and to draw permanganate solution up on sides of crucible to wet all particles.

5. Allow crucible to stand at 20–25°C for 90 ± 10 min; add more mixed permanganate solution if necessary. Purple colour must be present at all times.
6. Remove crucible to filtering apparatus. Suck dry. Do not wash. Place in a clean enamel pan and fill crucible no more than half full with demineralizing solution. Demineralizing solution may be added directly to crucible in case filtering is difficult. Care must be taken to avoid spillage by foaming. After 5 min, suck dry on filter and refill half full with demineralizing solution. Repeat after second interval if solution is very brown. Rinse sides of crucible with solution from a wash bottle with a fine stream. Treat until fibre is white. Total time required 20–30 min.
7. Fill and thoroughly wash crucible and contents with 80% ethanol. Suck dry and repeat two times. Wash twice in similar manner with acetone. Suck dry.
8. Dry at 100°C overnight and weigh. Calculate lignin content as loss in weight from acid-detergent fibre.
9. Ash at 500°C for 3 h, coal and weigh calculate residual ash as the difference between this weight and original tare of crucible. Calculate cellulose by weight loss upon ashing.
10. A presumptive analysis for silica may be obtained by hydrobromic acid treatment of the ashed permanganate lignin of ADF residue. This determination has its greatest value when the residual ash is greater than 2%. Ash and weigh and then add enough drops of 48% HBr (not more than 4 mL) to moisten all particles. Allow to stand 1–2 h. Add more drops of HBr if much red colour forms. Suck off excess acid on vacuum and wash once with acetone. Use no water. Dry and ash briefly at 500°C, cool and weigh. Report silica as the difference between this weight and the original tare.

Observations

Empty wt. of crucible = ——— g
Wt. of dried sample = ——— g
Wt. of acid-detergent fibre = ——— g
Wt. of permanganate fibre residue = ——— g
Wt. of crucible and ash = ——— g
Wt. of crucible + ash, after hydrogen bromide washing = ——— g

Calculations

1. Lignin percent on DM basis

$$= \frac{\text{wt. of acid detergent fibre} - \text{wt. of permanganate fibre residue}}{\text{wt. of oven dried sample taken for ADF analysis}} \times 100$$

2. Cellulose percent on DM basis

$$= \frac{\text{wt. of crucible and permanganate fibre} - \text{wt. of crucible and ash}}{\text{wt. of oven dried sample taken for ADF analysis}} \times 100$$

3. Silica percent on DM basis

$$= \frac{(\text{wt. of crucible} + \text{ash after HBr washing}) - \text{Empty wt. of crucible}}{\text{wt. of oven dried sample taken for ADF analysis}} \times 100$$

14.17 Estimation of Non-Starch Polysaccharides

Non-structural carbohydrate (NSC) minus starch and sugars equals non-starch polysaccharide (NSP). Starch and sugars can be measured directly. The net fraction can reasonably be calculated by difference using one of two formulae:

1. NSC = 100 − (NDF + protein + fat + ash)
2. NSC = 100 − (NDF − NDF protein) + protein + fat + ash

The second equation deletes neutral-detergent insoluble protein, which is the slowest to be degraded and should therefore be excluded. The NSPs do not generally include native hemicelluloses and celluloses that are ordinarily a part of the lignified cell wall matrix, which recovers hemicellulose and cellulose. So the determination of organic matter, starch, protein, fat, NDF and lignin will help to estimate NSP indirectly or by directly estimating different components of NSP in the animal feedstuffs. It includes cellulose, hemicellulose, β-glucans, pentosans, pectins, galactomannans and L-galactosides.

14.17.1 Determination of Starch

Principle

Powdered feed sample is treated with alcohol to solubilize free sugars, lipids, most pigments and cuticular waxes. The residue rich in starch is solubilized with perchloric acid and the extract is treated with anthrone–sulphuric acid to determine glucose.

Reagents and Chemicals

1. Anthrone–sulphuric acid: Add 0.2 g anthrone in 100 mL cold 95% H_2SO_4 stored at 4°C (prepare fresh)
2. 80% ethyl alcohol–water : 95% ethyl alcohol = 1.68 L (make final volume 2 L with distilled water)
3. 52% perchloric acid: 270 mL 72% perchloric acid + 10 mL distilled water.
4. Standard glucose: Stock solution (1 mg/mL): 0.1 g glucose in 100 mL distilled water, 0.1 g benzoic acid as preservative and working solution (10–100 μg); dilute 2 mL stock to 100 mL with distilled water.

Procedure

Extraction

1. To 0.2 g sample add a few drops of 80% ethyl alcohol and 5 mL of distilled water.
2. Stir thoroughly and add 25 mL hot 80% ethyl alcohol.
3. Mix well and centrifuge, discard the supernatant and add 30 mL hot 80% ethyl alcohol.
4. Mix well and then centrifuge and discard the supernatant.

Table 14.2 Protocol for starch estimation

	Test tube no.					
Reagents	1 (blank)	2	3	4	5	6
Distilled water	2.5	2.0	1.5	1.0	0.5	0.0
Std. glucose	0.0	0.5	1.0	1.5	2.0	2.5
Conc. of glucose (μg)	0	10	20	30	40	50
Reagent 1	5.0	5.0	5.0	5.0	5.0	5.0

5. Repeat washing twice more or until a test with anthrone is negative.
6. Collect all the washings in a volumetric flask for the determination of free sugars.
7. To the residue add 5 mL of distilled water; cool in ice water bath and then add 6.5 mL diluted perchloric acid reagent while stirring. Stir and keep for 20 min with occasional stirring and then again add 20 mL of water.
8. Centrifuge and transfer aqueous solution to a 100 mL volumetric flask.
9. Repeat solubilization with perchloric acid reagent for 30 min at 0°C and transfer the content to the volumetric flask and make final volume with water.
10. Filter the content and store the filtrate at 0°C.

Estimation

1. Dilute 5–10 mL filtered starch solution to 250 mL with water (10–40 μg starch, mL extract)
2. Take 2.5 mL diluted extract in a test tube.
3. Cool in water bath and add 5 mL anthrone reagent.
4. Mix thoroughly and heat in boiling water bath for 7.5 min.
5. Cool and read OD at 630 nm.
6. Arrange tubes in triplicate for standard as given below and draw the standard curve (Table 14.2).

Calculations

$$\text{Starch(mg\%)} = \frac{\text{Concentration of glucose } (\mu\text{g/mL}) \times \text{dilution factor} \times 100 \times 0.9}{\text{g sample taken (DM basis)} \times 1{,}000}$$

where 0.9 = conversion factor of glucose to starch.

14.18 Determination of Digestibility by In Vitro and In Sacco Techniques

The in vitro digestibility of feeds for ruminants can be measured by fermenting them with rumen liquor and then treating with pepsin. This is also known as two-stage in vitro method. During the first stage, finely ground sample of feed is incubated for 48 h with buffered rumen liquor in a tube under anaerobic conditions. In the second stage, the microbial activities are stopped by acidifying with hydrochloric acid to pH 2.0 and then digested by incubating with pepsin for another 24 h. The insoluble residue is filtered off, dried and ignited, and its organic matter is subtracted from the feed to obtain digestible organic matter.

14.18.1 In Vitro Dry Matter Digestibility (Tilley and Terry 1963)

Apparatus

1. Conical flask/tube – 100 mL
2. Water bath
3. Cork with Bunsen valve
4. Oven
5. Carbon dioxide gas cylinder

Reagents

1. Phosphate–carbonate buffer

   Na_2HCO_3: 9.80 g
   $Na_2HPO_4.2H_2O$: 7.00 g
   KCl: 0.57 g
   NaCl: 0.47 g
   $MgSO_4.2H_2O$: 0.12 g
   $CaCl_2$: 0.04 g

   Mix the above chemical except $CaCl_2$ in 800 mL distilled water in 1 L volumetric flask, stir to dissolve and make the volume to 1 L. Just before use, add $CaCl_2$, keep at 39°C and pass CO_2 through the solution.
2. 6 N HCl: Add 530.3 mL conc. HCl in 400 mL distilled water and make the volume to 1 L after cooling the solution.
3. Pepsin powder (1:3,000).

Procedure

1. Take 0.50 g finely ground (particle size < 1 mm) sample in 100-mL conical flask/tube.
2. Add 40 mL CO_2 saturated phosphate–carbonate buffer and 10 mL strained rumen liquor.
3. Pass CO_2 through the contents for 10 s and put a stopper (cork fitted with Bunsen valve) on the flask/tube immediately.
4. Incubate the flask/tube at 39°C with periodic shaking.
5. After 48 h of incubation, add 2 mL 6 N HCl and 0.1 g pepsin powder.
6. Incubate the tubes for another 24 h.
7. Filter the contents through filter paper (no. 54) or sintered crucible (G1).

8. Dry the residue at 100°C overnight and weigh.
9. Run parallel blank with phosphate–carbonate buffer and rumen liquor without feed sample.

Calculation

$$\text{DM disappearance} = \text{Wt. of sample} - (\text{wt. of residue of test} - \text{wt. of residue of blank})$$

$$\text{DM digestibility (\%)} = \frac{\text{DM disappearance}}{\text{Wt. of sample (DM basis)}} \times 100$$

Precaution

1. Collection of rumen liquor
   (a) Feed normal diet twice daily for about 7 days to the donor animal.
   (b) Draw homogenous rumen liquor samples from different parts of rumen by suction using plastic tube about 30 mm in diameter with several 9 mm holes in the lower 2 in. portion.
   (c) Filter rumen liquor through four layers of cheese cloth.
   (d) Transport rumen liquor to the laboratory in an insulated jug with temperature maintained at 39°C.
2. Rumen liquor should be collected 3 h of post-feeding.
3. The animal should not be given water between 1 and 3 h post-feeding.

14.18.2 Modified Method of In Vitro Dry Matter Digestibility

This is the modification of the two-stage in vitro method developed by Tilley and Terry (1963) and is completed in two stages

(a) Fermentation with rumen liquor and
(b) Extraction with neutral-detergent solution

Apparatus

1. Concial flask/tube – 100 mL
2. Water bath
3. Cork with Bunsen valve
4. Filter paper/crucible
5. Oven
6. Carbon dioxide gas
7. Spoutless beaker
8. Hot plate
9. Toulene

Reagents

1. Mc Dougall's buffer
2. Sodium hydroxide (1 N)
3. Casein hydrolysate
4. Reducing agent: 625 mg of cystine–hydrochloride dissolved in 95 mL distilled water and 4 mL 1 N NaOH. Add 625 mg anhydrous sodium sulphite.
5. Resazurin solution (0.1%, w/v)
6. Neutral-detergent solution

   Sodium lauryl sulphate: 30.00 g
   Disodium ethylene diamine tetra acetate (EDTA) dihyrate: 18.61 g
   Sodium borate decahydrate: 6.81 g
   Disodium hydrogen phosphate: 4.56 g
   2-Ethoxy-ethanol (Ethylene glycol): 10.00 mL

   Make up volume 1 L with distilled water

Put EDTA and sodium borate decahydrate ($Na_2BO_7.10H_2O$) in a beaker of 2 L capacity. Add 400 mL distilled water, shake and heat until dissolved. Add sodium lauryl sulphate and 2-ethyoxy-ethanol. Take disodium hydrogen phosphate (Na_2HPO_4) in a separate beaker and add 400 mL distilled water and heat until dissolved. Mix both the reagents and make the volume to 1 L. The pH of the solution should be in between 6.9 and 7.1.

Procedure

1. Take 0.5 g finely ground sample in 100-mL Erlenmeyer flask.
2. Prepare fermentation medium by adding in the order: 2.5 g casein hydrolysate and 1.5 mL of resazurin solution in 1,000 mL of McDougall's buffer. Keep the medium on magnetic stirrer and bubble CO_2 for 30–40 min. Add reducing agent (50 mL) and pass CO_2 gas till it becomes colourless.
3. Transfer 40 mL medium in each flask and add 10 mL rumen liquor.
4. Bubble CO_2 for 5 min.
5. Seal the flasks with Bunsen valve and incubate at 39°C for 48 h with shaking.
6. Wash the flask with 100 mL NDS in 500 mL spoutless beaker to make total volume to 150 mL. Reflux the sample for 1 h at 100°C and then filter on previously weighed sintered glass crucible. Wash the sample with hot water to remove the detergent completely.
7. Dry the crucible at 100°C for 24 h and record weight.
8. If samples are not to be processed immediately for NDS extraction, preserve the contents of the flask by adding 1 mL toluene and store in refrigerator till analysed.

Calculation

Wt. of crucible $= W$
Wt. of sample on DM basis $= W1$

Wt. of residue $\pm$ crucible $= W2$
Wt. of residue left $= W2 - W = W3$
Net digestible dry matter $= W1 - W3 = W4$

$$\% \text{ IVDMD} = \frac{W4}{W1} \times 100$$

14.19 Quantification of Tanins in Foliage

Tannins are generally defined as naturally occurring polyphenolic compounds of high enough molecular weight to form complexes with proteins. These are classified into two groups based on their structural types: (a) hydrolysable tannins and (b) condensed tannins. Methods for quantification of tannins may be based on the chemical properties of tannins or their capability to bind substrates, particularly proteins. The methods for the quantification of tannins could be described in three categories.

1. Chemical methods

   (a) *Determination of total phenolics*: The method is based on the fact that phenolics are reducing agents. It may be noted that all tannins are phenolics, but not all phenolics are tannins.
   (b) *Determination of total tannins*: It is partly chemical, based on reducing property of tannins and partly physical because tannins are measured as the reduction in phenolics that occur when a binding agent (polyvinyl polypyrrolidone, PVPP) is added to the extract.
   (c) *Determination of condensed tannins (proanthocyanidins)*: It is based on oxidative depolymerization of condensed tannins in butanol–HCl reagent. The presence of iron is considered to increase the reproducibility and sensitivity of the assay.
   (d) *Determination of gallotannins*: It is based on hydrolysis of gallotannins to gallic acid under acidic conditions and measurement of the released gallic acid by reaction with rhodanine or by HPLC.

2. Protein precipitation/binding methods

   (a) *Determination of protein-precipitable phenolics*: It is based on the formation of tannin–protein complexes (tannins in the plant extract and the protein, bovine serum albumin [BSA]). Tannins present in the complex are determined using ferric chloride assay for total phenolics. Iron forms a complex with phenols to give a pink chromatophore which is measured spectrophotometrically.
   (b) *Filter paper–protein Ponceau S dye assay*: The protein–tannin complexes are formed on a sheet of filter paper, and the protein bound to the complex is measured by dyeing it with Ponceau S dye. This dye is specific for proteins

and does not bind tannins. The colour of the dye bound to proteins is eluted and measured spectrophotometrically.

(c) *Radial diffusion assay*: In this method, tannin molecules migrate through agarose gel which is impregnated with the protein, BSA. The tannin–protein complex is formed in the gel which appears as a ring. The diameter of the ring is a measure of protein precipitation/binding capacity of tannins.

3. Tannin bioassay
   The samples under investigation (or tannins) are incubated in the absence and presence of a tannin-complexing agent, polyethylene glycol (PEG), in the *in vitro* rumen fermentation system containing rumen microbes. The affinity of PEG for binding to tannins is very high; it even breaks the already formed tannin–protein complexes and releases proteins from the complex. In the in vitro rumen fermentation system, release of gas and production of microbial mass are measured. The difference between these parameters observed in the absence and presence of PEG is a measure of tannin activity in relation to their effects on rumen fermentation.

14.20 Method for Extraction of Tannins

The aim is to quantitatively diffuse phenolics present in the plant material to liquid phase. For the extraction process, a suitable solvent is required. Generally, aqueous methanol (50%) and aqueous acetone (70%) are common choices. The latter has been reported by various workers to be better in extracting phenolics from tree leaves. One can try both these solvents for extraction, and then based on the efficiency of extraction of phenolics (using Folin–Ciocalteu method) and/or condensed tannins (using butanol-HCl method), one can decide the solvent to be used for a particular plant material.

Dried (finely ground) plant material (200 mg) is taken in a glass beaker of approximately 25 mL capacity. Ten millilitres of aqueous acetone (70%) is added and the beaker is suspended in an ultrasonic water bath and subjected to ultrasonic treatment for 20 min at room temperature. The contents of the beaker are then transferred to centrifuge tubes and subjected to centrifugation for 10 min at approximately 3,000 × *g* at 4°C (if refrigerated centrifuge is not available, cool the contents by keeping the centrifuge tube on ice and then centrifuge at 3,000 × *g* using an ordinary centrifuge). Collect the supernatant and keep it on ice. Transfer the pellet left in the centrifuge tube to the beaker using two portions of 5 mL each of 70% aqueous acetone and again subject the contents to ultrasonic treatment for 20 min centrifuge and collect supernatant as described above. Folin–Ciocalteu method is used for determination of total phenols in the two supernatants.

- Very long extraction at too high a temperature may lead to degradation and loss of phenolics.

- Pigments and fat can be removed from the dried leaf sample by extracting with diethyl ether containing 1% acetic acid before extracting tannins. Freshly prepared extract should be used for tannin analysis.
- Tubes/container containing the extract should be kept on ice till the analysis is complete.

14.21 Measurement of Total Phenolics and Tannins Using Folin–Ciocalteu Method

The method for total phenol is useful in order to know the efficiency of extraction of phenolics in solvents. This method can be coupled with the use of insoluble matrix, PVPP (binds tannin-phenolics), for measurement of tannins. The results can be expressed as tannic acid equivalent. The nature of tannic acid varies from one commercial source to the other.

Reagents

1. *Folin–Ciocalteu reagent* (1 N): Dilute commercially available Folin–Ciocalteu reagent (2 N) with an equal volume of distilled water. Transfer it in a brown bottle and store in a refrigerator (4°C). It should be golden colour. Do not use it if it turns olive green.
2. *Sodium carbonate* (20%): Weigh 40 g sodium carbonate (×10 H_2O), dissolve it in about 150 mL distilled water and make up to 200 mL with distilled water.
3. *Insoluble polyvinyl pyrrolidone* (PVPP)
4. *Standard tannic acid solution* (0.1 mg/mL): Dissolve 25 mg tannic acid (TA) obtained from 25 mL distilled water and then dilute 1:10 in distilled water (always use a freshly prepared solution).

Analysis of Total Phenols

Take suitable aliquots of the tannin-containing extract (initially try 0.02, 0.05 and 0.1 mL) in test tubes, make up the volume to 0.5 mL with distilled water and add 0.25 mL of the Folin–Ciocalteu reagent and then 1.25 mL of the sodium carbonate solution. Vortex the tubes and record absorbance at 725 nm after 40 min. Calculate the amount of total phenols as tannic acid equivalent from the above calibration curve. Express total phenolic content on a dry matter basis (x%).

Estimation of Tanins

Removal of tannin from the tannin-containing extract: PVPP binds tannins. Weigh 100 mg PVPP in a 100 × 12-mm test tube. Add to it 1.0 mL distilled water and then 1.0 mL of the tannin-containing extract (100 mg PVPP is sufficient to bind 2 mg of total phenols; if total phenolic content of feed is more than 10% on a dry matter basis, dilute the extract appropriately). Vortex and keep the tube at 4°C for 15 min,

Table 14.3 Calibration curve for estimation of tanins

Tube	Tannic acid solution (0.1 mg/mL), mL	Distilled water (mL)	Folin reagent (mL)	Sodium carbonate solution (mL)	Absorbance at 725 nm	Tannic acid (μg)
Blank	0.00	0.50	0.25	1.25	0.000	0
T1	0.02	0.48	0.25	1.25	0.112	2
T2	0.04	0.46	0.25	1.25	0.218	4
T3	0.06	0.44	0.25	1.25	0.327	6
T4	0.08	0.42	0.25	1.25	0.432	8
T5	0.10	0.40	0.25	1.25	0.538	10

vortex it again, then centrifuge (3,000 × *g* for 10 min) and collect the supernatant. This supernatant has only simple phenolics other than tannins (the tannins would have been precipitated along with the PVPP); the procedure for binding of tannins to PVPP is presently being modified, and the modification is to bind tannins to PVPP at pH 3 since PVPP binds maximally to tannins at this pH. Measure the phenolic content of the supernatant as mentioned above (take at least double the volume (preferably three times) used for total phenol estimation, because extract has already been diluted twofold and expect to lose tannin-phenols though binding with PVPP). Express the content of non-tannin phenols on a dry matter basis (*y*%).

Preparation of calibration curve

See Table 14.3.

14.22 Determination of Condensed Tannins (Proanthocyanidins)

- *Butanol–HCl reagent* (*butanol–HCl 95:5; v/v*): Mix 950 mL *n*-butanol with 50 mL conc. HCl (37%).
- *Ferric reagent* (2% ferric ammonium sulphate in 2 N HCl): Make 16.6 mL of conc. HCl up to 100 mL with distilled water to make 2 N HCl. Dissolve 2.0 g of ferric ammonium sulphate in this volume of 2 N HCl. This reagent should be stored in a dark bottle.

Procedure

In a 100 nm × 12 mm glass test tube, pipette 0.50 mL of the tannin extract diluted with 70% acetone. The quantity of acetone should be large enough to prevent the absorbance (550 nm) in the assay from exceeding 0.6. It will depend on the quantity of condensed tannin expected in the sample, and occasionally will need to be determined by trial and error. To the tubes add 3.0 mL of the butanol–HCl reagent and 0.1 mL of the ferric reagent. Vortex the tubes. Cover the mouth of each tube with a glass marble and put the tubes in a heating block adjusted at 97–100°C (or in a boiling water bath) for 60 min. Cool the tubes and record absorbance at 550 nm. Subtract at suitable blank, which is usually the absorbance of the unheated mixture. If the extract has flavan-4-ols, a pink colour develops without heating. If this happens, use one heated blank for each sample, comprising 0.5 mL of the extract, 3 mL of butanol and 0.1 mL of the ferric reagent. Condensed tannins (% in dry matter) as leucocyanidin equivalent are calculated by the formula:

$$(\text{A550 nm} \times 78.26 \times \text{Dilution factor}^{*})/(\%\ \text{dry matter})$$

This formula assumes that the effective $E^{1\%,\ 1\ \text{cm},\ 550\ \text{nm}}$ of leucocyanidin is 460

- The dilution factor is equal to 1 if no 70% acetone was added and the extract was made from 200 mg sample in 10 mL solvent. Where 70% acetone is added (for example to prevent the absorbance from exceeding 0.6) the dilution factor is:

  0.5 mL/(volume of extract taken)

- The presence of pigments may interfere in this method. The pigments can be removed by extracting the dried leaves with petroleum ether containing 1% acetic acid.
- Ascorbic acid (generally added to prevent oxidation of phenols) does not interfere in this assay.

14.23 Gallotannin Determination by Rhodanine Assay (Inoue and Hangerman 1988)

The plant material was extracted as described previously.

Reagents

- *Rhodanine solution* (0.667%, w/v, in methanol): Weigh 667 mg rhodanine and dissolve it in 100 mL methanol. Stable for at least 2 weeks when stored in a refrigerator.
- *KOH* (0.5 N): Dissolve 2.8 g potassium hydroxide in 100 mL distilled water.
- *Sulphuric acid solutions*: Prepare 0.3, 0.4, 22 and 26 N sulphuric acid solutions by appropriately diluting 98% commercially available sulphuric acid (36 N).

Determination of free gallic acid

Pipette 200 μL supernatant A in a 160 × 12 mm culture test tube (4 tubes per sample). Remove acetone from the sample using a vacuum drying oven adjusted at 40°C (pressure 300 mbar) for about 2 h or by flushing with nitrogen gas (bring nitrogen gas beam as close as possible to liquid surface). Using nitrogen gas, it takes about 8–10 min to dry the sample. One can construct a multi-tube (8–10 outlets) assembly for flushing nitrogen gas, which allows simultaneous drying of 8–10 samples.

Add to 200 μL of 0.2 N sulphuric acid to the tubes containing dried supernatant A. To three tubes add 300 μL of the rhodanine solution and to the fourth tube 300 μL methanol. This fourth tube acts as a proper blank. After 5 min add 200 μL of 0.5 N potassium hydroxide solution to all the tubes. Wait for 2.5 min and then add 4.3 mL distilled water. After 15 min measure absorbance at 520 nm against proper blank.

Determination of gallic acid present in free and in gallotannin forms

Pipette 3.34 mL of supernatant A, in duplicate, in the above-mentioned culture test tube. Remove acetone by flushing the tubes with nitrogen gas. After removal of acetone, 1 mL of the supernatant remains (or make to 1 mL with distilled water). To it add 0.1 mL of 22 N sulphuric acid so that the final sulphuric acid concentration is 2 N. Freeze the contents and remove air and tightening of the cap of the culture tube. Keep these tubes at 100°C for 26 h to hydrolyse gallotannins to gallic acid. After hydrolysis, make up the volume to 11 mL by adding 9.9 mL distilled water. Sulphuric acid concentration in this solution is 0.2 N. This solution is addressed as "hydrolysed supernatant A".

Pipette 200 μL of the hydrolysed supernatant A (4 tubes per sample; 1 blank and the rest "test" samples). Now add 300 μL of the rhodanine solution to the "test" tubes and 300 μL methanol to the blank tube. Wait for 5 min and then add 200 μL of 0.5 N KOH solution. Again wait for 2.5 min and add 4.3 mL distilled water. After 10 min, measure absorbance at 520 nm.

Calibration curve

- *Stock gallic acid solution* (1 mg/mL in 0.2 N sulphuric acid): Weigh 100 mg gallic acid and dissolve in approximately 80 mL of 0.2 N sulphuric acid and then make up the volume to 100 mL with 0.2 N sulphuric acid. It can be kept frozen for at least 1 month.
- *Working gallic acid solution* (0.1 mg/mL): Dilute the gallic acid stock solution 1:10 with 0.2 N sulphuric acid (1 mL of the stock solution plus 9 mL distilled water). It can be stored in a refrigerator for at least 2 weeks (Table 14.4).

Wait for approximately 10 min and then measure absorbance at 520 nm.

Total gallic acid minus free gallic acid = gallotannins as gallic acid equivalent.

Table 14.4 Calibration curve

Gallic acid (μg)	Working solution (μL)	0.2 N Sulphuric acid (μL)	Rhodanine (μL)	Wait for at least 5 min	0.5 N KOH (μL)	Wait for at least 2.5 min	Distilled water (mL)
4	40	160	300		200		4.3
8	80	120	300		200		4.3
12	120	80	300		200		4.3
16	160	40	300		200		4.3
20	200	0	300		200		4.3
Blank	0	200	300		200		4.3

14.24 Gallotannin Determination Using HPLC

1. Preparation of sample for determination of free gallic acid
   Pipette 1 mL of supernatant A into the above-mentioned culture tubes and remove acetone in a vacuum oven adjusted at 40°C (pressure 300 mbar). It takes about 3 h to remove acetone. Acetone can also be removed by flushing with nitrogen gas. Dry the contents completely in a heating block at 40°C by flushing with nitrogen. It takes about 25 min to dry the contents completely. It can be stored under nitrogen in a freezer.

   Just before the analysis, add 750 μL distilled water to the residue and dissolve the contents using an ultrasonic water bath (keep the tubes for about 5 min). Filter the contents through a 0.45-μm membrane filter (cellulose acetate or cellulose nitrate) before loading to the HPLC.
2. Preparation of sample for determination of gallic acid present in free and in gallotannin forms
   Pipette 2 mL of the hydrolysed supernatant A or B (see above) into a 25–50 mL capacity tube/beaker and then add 2 mL buffer (2.304 g $NH_4H_2PO_4$ dissolved in 1,000 mL distilled water) to it. Adjust the pH between 6.3 and 6.8 using 8 M KOH solution (approximately 120 μL will be required) and record the exact amount required for the pH adjustment. Do not exceed pH 7.0 because the sample will immediately get oxidized. In case the pH rises beyond 7, discard it and start afresh with 2 mL of the hydrolysed supernatant. Freeze overnight, thaw it and then filter through 0.45-μm membrane filter before loading to the HPLC.

HPLC conditions for analysis

Column: 250 × 4.6 mm filled with Nucleosil 120–5 C18
Flow rate: 1.2 mL/min.
Column temperature: Ambient (ca 22°C).
Injection volume: 20 μL.
Gallic acid stock solution: Dissolve 400 mg gallic acid in 1,000 mL distilled water.
Gallic acid working solution (6 mg/1,000 mL): Pipette 150 μL of the stock solution and make the volume to 10 mL with distilled water. Linearity between area and concentration was observed in the range of 2–10 mg gallic acid/mL on injection of 20 μL of gallic acid solution.
Elution time of gallic acid: Between 14 and 15.5 min.
Solvents: Buffer A: H_2O–methanol–H_3PO_4 (975.5:19.5.1; v/v/v), and Buffer B: methanol–H_2O (700:300; v/v) (Table 14.5).

Table 14.5 Gradient used

Time (min)	Solvent A (%)	Solvent B (%)
0.0	100	0
15.0	100	0
22.0	0	100
25.0	0	100
30.0	100	0
33.0	100	0

Detection wavelength: 280 nm.

- Ascorbic acid does not interfere in this assay and therefore can be added to the solvent used for extraction of phenols/tannins.

14.25 Determination of Protein-Precipitable Phenolics (Makkar et al. 1988)

Reagents

- *Acetate buffer* (pH 4.8–4.9, 0.2 M): Pipette 11.40 mL glacial acetic acid to about 800 mL distilled water. Adjust pH of this solution to 4.8–4.9 with 4 N sodium hydroxide solution and bring the final volume to 1 L. To it add 9.86 g NaCl to make its concentration 0.17 M.
- *Sodium dodecyl sulphate solution* (SDS) (1%, w/v): Dissolve 1 g SDS in 100 mL of distilled water.
- *SDS-triethanolamine* (TEA) (1% SDS (w/v) and 7% (v/v) triethanolamine in distilled water) solution: To 7 mL of triethanolamine add 93 mL distilled water and dissolve 1 g SDS in this solution.
- *Ferric chloride reagent* (0.01 M ferric chloride in 0.1 M HCl): For making 0.1 M HCl, dilute 4.2 mL conc. HCl (37%) to 500 mL with distilled water. Dissolve 0.81 g ferric chloride in 500 mL of 0.1 M HCl. Filter and store the contents in a brown bottle.
- Glacial acetic acid.
- *BSA solution*: Dissolve 100 mg BSA (fraction V) in 100 of the acetate buffer.

1. Formation of the tannin–protein complex
   To 2 mL of the BSA solution (containing 1 mg BSA/mL acetate buffer), add 50% methanol and increasing levels of the tannin-containing extract to make 3 mL. For example, use 0.95, 0.90, 0.85, 0.80, 0.75 and 0.70 mL of 50% methanol with 0.05, 0.10, 0.15, 0.20, 0.25 and 0.30 mL of the extract; this may vary depending on the amount of tannins in the sample. Vortex the contents. Allow the mixture to stand in a refrigerator (4°C) overnight. Centrifuge at about 3,000 × *g* for about 10 min. Remove the supernatant carefully without disturbing the precipitate. Add 1.5 mL of 1% SDS solution to the precipitate and vortex it to dissolve the precipitate.

   - It is essential to use 50% aqueous methanol for extraction of tannins form the plant material since acetone interferes in the protein-precipitation assay.

2. Determination of tannins (phenolics) in tannin–protein complex
   Take an aliquot (1 mL) of the above-dissolved complex. Add 3 mL of SDS-triethanolamine solution. Then add a 1-mL portion of the ferric chloride reagent. Record absorbance at 510 nm after 15–30 min. Convert the absorbance to tannic

Table 14.6 Estimation of tannic acid from extract

Tube	Extract (μL)	Leaf (mg)[a]	Absorbance at 510 nm	TA (mg)[b]	TA in complex (mg)[c]
1	100	2	0.121	0.054	0.081
2	150	3	0.167	0.077	0.116
3	200	4	0.234	0.109	0.164
4	250	5	0.292	0.136	0.204
5	300	6	0.341	0.160	0.240
6	350	7	0.422	0.199	0.299
7	400	8	0.472	0.222	0.333
8	500	10	0.591	0.280	0.420

TA tannic acid

[a] 200 mg leaf is extracted in 10 mL 50% aqueous methanol

[b] Conversion of absorbance at 510 nm to mg tannic acid by the standard curve

[c] Obtained by multiplying values in the previous column (which correspond to 1 mL of the soluble tannin–protein complex) by 1.5, because the tannin–protein complex is dissolved in 1.5 mL of 1% SDS

acid equivalent, using a standard curve. Multiply the values obtained by 1.5 to obtain tannins in the complex. Draw a linear regression between tannins precipitated as tannic acid equivalent and mg leaf (in aliquot taken for the assay). The slope of the curve (mg tannic acid precipitated/mg leaf; let it be x) represents the protein-precipitable phenolics in the sample (Table 14.6).

- Protein-precipitable phenolics (×; mg tannic acid equivalents precipitated/mg leaf dry matter) for the above example = 0.043/0.953 = 0.045, since dry matter of the leaves was 95.3%.

3. Protein-precipitable phenolics as percentage of total phenolics

Determination of total phenolics as percentage of total phenolics

Take different aliquots (0.05, 0.10, 0.15, 0.2, 0.25 and 0.30 mL, but this may vary depending on the amount of phenolics in the sample) of the extract (200 mg in 10 mL of 50% methanol), make up to 1 mL with 1% of SDS, and add 3 mL of the SDS-triethanolamine solution and 1 mL of the ferric chloride reagent. Record absorbance at 510 nm as described above. Convert the absorbance to tannic acid equivalent using the standard curve. Draw a linear regression between tannic acid equivalent and mg leaf (in the aliquot taken). The slope of the curve (mg tannic acid equivalent/mg leaf; let it be y) represent total phenolics.

Protein-precipitable phenolics have already been measured as x
The percentage of total phenolics which precipitate protein = $(x/y) \times 100$ (Table 14.7).

Calibration curve for the above example

See Table 14.8.

Table 14.7 Estimation of tannic acid from extract

Tube	Extract (μL)	Leaf[a] (mg)	Absorbance at 510 nm	TA[b]
1	50	1	0.145	0.066
2	100	2	0.280	0.131
3	150	3	0.404	0.190
4	200	4	0.532	0.251
5	250	5	0.674	0.319
6	300	6	0.824	0.391

TA tannic acid

[a] 200 mg leaf is extracted in 10 mL 50% aqueous methanol

[b] Calculated for the calibration curve below

Total phenolics (*y*; mg tannic acid equivalent/mg leaf dry matter) = 0.064/0.9535 = 0.067, since dry matter of the leaves was 95.35%

Therefore, protein-precipitable phenolics as percentage of total phenolics = $(x/y) \times 100$ = $(0.045/0.067) \times 100 = 67.2$

Table 14.8 Calibration curve

Tube	TA solution[a] (mL)	SDS, 1% (mL)	SDS-TEA (mL)	Ferric chloride (mL)	Absorbance at 510 nm	TA (mg)
Blank	0	1.0	3.0	1.0	0.000	0.00
T1	0.1	0.9	3.0	1.0	0.107	0.05
T2	0.2	0.8	3.0	1.0	0.225	0.10
T3	0.3	0.7	3.0	1.0	0.319	0.15
T4	0.4	0.6	3.0	1.0	0.426	0.20
T5	0.5	0.5	3.0	1.0	0.527	0.25

TA tannic acid

[a] TA solution: 0.5 mg/mL in 1% SDS.

14.26 Protein-Binding Capacity by Filter Paper Assay (Dawra et al. 1988)

Reagents

- *Tannic acid solution*: Dissolve 100 mg tannic acid in 100 mL of 50% aqueous methanol. Add 100 mg ascorbic acid to it to minimize oxidation of tannic acid during handling.
- *Acetate buffer* (pH 5, 0.05 M): Pipette 2.85 mL glacial acetic acid to about 800 mL distilled water. Adjust pH of this solution to 5.0 with 4 N sodium hydroxide solution and bring the final volume to 1 L.
- *Bovine serum albumin* (BSA) solution: Dissolve 200 mg of bovine serum albumin (fraction V) in 100 mL of 0.05 M of the acetate buffer (pH5, 0.05 M).
- *Dye solution*: Prepare 3% (w/v) TCA solution in distilled water. Dissolve 0.2 g of Ponceau S dye in 100 mL of the TCA solution.
- *Acetic acid solution* (0.2%, v/v): Add 2 mL of glacial acetic acid to 998 mL distilled water.

- *Sodium hydroxide solution* (0.1 N): Weight 4 g sodium hydroxide and dissolve it in approximately 500 mL distilled water and then make up the volume to 1 L with distilled water.
- *Acetic acid solution* (10%, v/v): Add 10 mL of glacial acetic acid to 990 mL distilled water.

Preparation of plant extract: In 70% acetone. Acetone does not interfere in this method.

Procedure

Take 1 mm Whatman paper chromatography sheet and cut it into an appropriate size (depends on the number of samples to be analysed). Draw squares of approximately 1.5–2 cm using a light lead pencil on the chromatography sheet. Apply different aliquots (5–25 μL containing 5–25 μg tannic acid) on the sheet; each aliquot, at least in triplicate and on three different squares. Similarly, apply appropriate aliquots (10–50 μL; depending on the amount of tannins present) of the plant extract on the middle of squares on the chromatography sheet. Allow the spots to dry and spray immediately with the BSA solution until the paper is wet. After 30 min, wash the paper with the acetate buffer (pH 5, 0.05 M) with three 10-min changes with slight shaking to remove the unbound BSA. The paper was stained with 0.2% Ponceau S dye solution by keeping the strips dipped for 10 min in the stain solution. The staining solution should not be used in successive experiments. Wash the stained strips in 0.2% acetic acid solution until no more colour is eluted from the strips. Normally, this requires three washings to make the background clear. Air-dry the strips and cut the stained areas. To prepare a corresponding blank, stain simultaneously a chromatography sheet and wash in a manner similar to other samples. Cut the stained area into small pieces and elute the colour by adding 3 mL of 0.1 N sodium hydroxide solution and vortexing and followed by addition of 0.3 mL of 10% acetic acid and centrifugation at approximately 2,500 × *g*. Measure the absorbance of the colour at 525 nm against corresponding blank. Convert these absorbance values to protein content by using a standard curve. For preparing a standard curve, apply different concentrations of BSA (5–50 μg; 5–50 μL of 1 mg/mL BSA solution in the acetate buffer) as separate spots (at least in triplicate for each concentration) on a chromatography sheet and cut into strips. Stain these strips with the dye solution, wash and read the colour as described above for the samples.

14.27 Characterization of Phenolic Compounds by Thin Layer Chromatography (TLC)

Characterization of condensed tannins

Preparation of plant material

About 500 g of the shade-dried plant material should be ground first to pass a 2-mm screen. All the ground material including those parts remaining inside the mill should be taken, mixed well and approximately 100 g of this sample is again ground to pass through a 0.5-mm screen. Take care that at any stage of the grinding, the sample temperature does not rise above 40°C.

Reagents

- 70% acetone: Mix 70 mL acetone and 30 mL distilled water.
- Butanol/HCl reagent: Thoroughly mix 95 mL butan-1-ol and 5 mL HCl (12 M).
- Anthocyanidin standards: Chlorides of cyanidin, delphinidin, pelargonidin, etc.

Material for TLC

- Cellulose MN300 plates
- Disposable micro-pipettes
- TLC sprayers which produce a fine mist (need a fine nozzle) and are resistant to 12 M HCl
- TLC solvents:

Solvent 1 (for first direction): Mix 100 mL concentrated formic (85%), 10 mL 12 M HCl and 30 mL water.

Solvent 2 (for second direction): Mix 20 mL pentan-1-ol, 10 mL glacial acetic acid and 10 mL water.

Extraction

Leaf samples (200 mg of dried plant material passed through a 0.5-mm screen) are taken in a glass beaker of approximately 25 mL capacity. To it is added 10 mL of aqueous acetone (70%) and the beaker is suspended in an ultrasonic water bath and subjected to ultrasonic treatment for 20 min at room temperature. The contents of the beaker are then transferred to centrifuge tubes and subjected to centrifugation for 10 min at approximately 3,000 × *g* at 4°C (if refrigerated centrifuge is not available, cool the contents by keeping the centrifuge tube on ice and then centrifuge at 3,000 × *g*). Collect the supernatant and keep it on ice.

Take 1 mL of the aqueous acetone extract and evaporate the solvent to less than 200 μL in a stream of nitrogen (not oxygen, as this would oxidize the tannins). Dilute the concentrate to 400 μL with water and mix well. Take an aliquot of 80 μL from this aqueous solution and add 1 mL of the HCl/butanol reagent, cover the tubes (marbles or loose Teflon lined Pyrex tube screw tops) and heat at 100°C for 60 min. Cool the solution and spot aliquots onto the TLC plates.

Procedure

In a fume cupboard, fill two small TLC tanks with solvents 1 or 2 to a height of about 3 mm and ensure that the atmosphere in the tanks is well saturated with the solvent (a sheet of filter paper can be dipped in the solvent and surround around the tank walls inside to speed up the saturation process and "seal" the lids with grease to prevent evaporation of the solvent; the TLC plate should not be run before 30 min after filling the tanks with the solvents).

Spot the sample carefully at the bottom left corner (7 mm from the edges). The diameter of the spot should exceed 5–7 mm. The volume of sample to be spotted depends on the concentration of the anthocyanidins. It is best to try out a range of volumes (e.g. 5, 10 and 20 μL) by repeatedly applying 5 μL onto the same spot and letting the spot dry between applications. Do not use hot air blower/hair drier to dry the spots.

Gently lower the plate into the TLC tank. Switch off the fume cupboard to prevent drafts crossing the tank and causing temperature gradients. Remove the plate, when the solvent front has just reached the top of the TLC plate (approximately 3 mm below the top); but switch the fume cupboard on again for this. Dry the plate in the draft of the fume cupboard (do not use hot air). When all solvent has evaporated, turn the plate by 90° and repeat the separation with solvent 2. The colours of the anthocyanidins should be clearly visible. They are identified from their position (Rf-values) on the plate and by their characteristic colours.

Analysis of condensed and hydrolysable tannins by TLC

Solvents

Solvents 1: Mix 2 mL glacial acetic and 98 mL water.
Solvents 2: Mix 60 mL butan-1-ol, 15 mL glacial acetic acid and 25 mL water.

Spray reagents to detect different classes of tannins

- *Vanillin/HCl reagent*: Prepare solution containing 1 g vanillin in 10 mL 12 M HCl. It detects flavan-3-ols, e.g. catechin and epi-catechin, plus condensed tannins. When these are present, pink spots are obtained. Keep a record of these red spots. This can be done by

    i. Using a sharp tool (edge of thin spatula, scalpel or razor blade) to surround each spot with a series of small holes.
    ii. Photocopying the TLC plate (but be careful, the acid might damage the photocopier; consider covering it with a transparent film first).
    iii. Placing some tracing paper over the TLC plate and copying the spots (but TLC surface is quite fragile).

- *Ferric ion reagent*: Prepare daily a fresh solution containing 1 g of $K_3Fe(CN)_6$ and 1 g $FeCl_3$ in 50 mL water. Then add five tiny crystals of $KMnO_4$. Lightly spray the TLC plate with a fine mist of this reagent (avoid spluttering large drops on the plate). The background of the plate tends to turn dark blue more or less quickly. This can be reduced by laying the TLC plate (Cellulose surface point up) into a glass basin containing 2–3 M HCl soon after applying the reagent.
- *Potassium iodate reagent*: Prepare a saturate solution KIO_3 (potassium iodate) in distilled water (i.e. add enough KIO_3 crystals until some of them no longer dissolve in the water).

- *Sodium nitrite reagent*: Cool 10 mL of water to near 0°C and add 20 mg of $NaNO_2$ (sodium nitrite) plus 1–2 drops of glacial acetic acid.
- This reagent detects ellagic acid and its esters. Orange-brown spots are obtained. This is the most tricky reagent and it will require several attempts to become confident with it. Again, try it out first using ellagic acid as the standard.

Procedure

Apply the aqueous acetone extracts (5–30 μL) to several cellulose MN300 TLC plates (10 × 10 cm) and try to keep the spot at the origin as small as possible without damaging the TLC surface (approximately 7 mm diameter). Place the TLC plate in "solvent 1" first and remove it when the solvent front has just reached the top of the plate. Dry in a cool stream of air, turn the plate by 90° and place the plate into a tank containing "solvent 2".

Then subject each TLC plate to 1 of the 4 spray reagents separately.

14.28 Estimation of Nitrates and Nitrites (Wiseman and Jacobson 1965)

Principle

Nitrate is reduced to nitrite by zinc and manganese sulphate. The reaction proceeds with diazotization of sulphonic acid by nitrite ion and subsequent coupling with 1-naphthylamine to form a red dye. Copper aids the reaction at low ppm level (0.2), but over 1.2 ppm it interferes. Interfering iron is complexed with citrate. Reaction is optimum at pH range 1.7–3.0.

Reagents and chemicals

1. 20% acetic acid reagent: (200 mL acetic acid + 5 mL copper sulphate, final volume 1 L with distilled water)
2. Bray's indicator:

   (a) 100 g barium sulphate ($BaSO_4$)
   (b) 10 g manganese sulphate ($MnSO_4.H_2O$)
   (c) 2 g zinc (metallic Zn)
   (d) 75 g citric acid
   (e) 4 g sulphanilic acid
   (f) 2 g 1-naphthylamine

   Zinc and citric acid are powdered; manganese sulphate, zinc, sulphanilic acid and 1-naphthylamine are mixed separately with a part of barium sulphate and then mixed with citric acid and remaining part of barium sulphate. The indicator is stored in dark bottle and stored away from light.

3. Hydrochloric acid (0.1 N) : 8.18 mL. 37–38% HCl in 1 L distilled water
4. Activated charcoal.

5. Standard nitrate solution:
   Stock solution (1 mg/mL): 1.37 g sodium nitrate in 1 L distilled water and working solution (10 μg/mL); dilute 1 mL of stock solution to 100 mL with distilled water.

Procedure

Sampling

Collect samples from different places in the field. Moist sample (silage) should be kept frozen until analysed. Green may be dried immediately in an oven at 60°C.

Analysis

1. Take 1 g air dry material or 3–5 g moist sample (nitrate content ranging between 1 and 8 mg) in a conical flask of 150 mL capacity.
2. To above add 100 mL of 0.1 N HCl and place for 1 h with occasional shaking.
3. If the extract is strongly coloured, decolourize with activated charcoal and then filter through Whatman filter paper (no. 1).
4. The sample extract, blank and standard are run as given below (Table 14.9).
5. Shake the above solutions avoiding exposure from light.
6. Centrifuge at 3,000 rpm for 5 min.
7. Remove any film on top and collect the clear red supernatant.
8. Read at 520 nm and find out concentration of nitrate in sample from the standard curve drawn taking OD against concentration.

Calculation

$$\text{Total nitrate (nitrate + nitrite)} = \frac{\text{Conc. of nitrate } (\mu\text{g/mL}) \text{ in final solution from standard curve}}{\text{g sample taken (DM basis)} \times 10}$$

$$\text{Nitrite concentration} (\mu\text{g/mL}) = \frac{\text{Conc. in final solution obtained from standard curve}}{\text{g sample taken (DM basis)} \times 10}$$

Table 14.9 Protocol

			Standard			
Reagents/chemicals	Sample	Blank	1	2	3	4
Extract (mL)	1.0	0.0	0.0	0.0	0.0	0.0
Distilled water (mL)	0.0	1.0	0.0	0.25	0.5	0.75
Working standard (mL)	0.0	0.0	1.0	0.75	0.5	0.25
Conc. Of standard (μg)	–	–	10.0	7.5	5.0	2.5
Reagent (1) (mL)	9.0	9.0	9.0	9.0	9.0	9.0
Bray's indicator (g)	0.5	0.5	0.5	0.5	0.5	0.5

14.29 Estimation of Total Glucosinolates (McGhee et al. 1965)

Principle

The glucosinolates (GSL) of sample are collected in water. It is then treated with 0.1 N $AgNO_3$ and 95% ethanol. The solution with ferric ammonium sulphate in acidic medium is titrated against N/100 potassium thiocyanate.

Reagents

(a) 0.1 N $AgNO_3$
(b) 95% Ethanol
(c) 8% (w/v) Ferric ammonium sulphate
(d) 6 N-HCl
(e) N Potassium thiocyanate solution

Equipments

Grinder, heater, filter paper, Buchner funnel, wash bottle, beakers, and burette.

Procedure

Put 10 g of ground sample into 250 mL boiling water for enzyme deactivation for 5 min. The enzyme deactivation step is normally not allowed to prolong 5 min to avoid the breakdown of glucosinolates. Filter the contents on a Buchner funnel. Wash the residue with 50 mL hot water and filter again. Make the volume of filtrate to 500 mL. Take 25 mL aliquot in a beaker and add 10 mL of 0.1 N $AgNO_3$ solution and 25 mL ethanol (95%). Reflux the contents on a water bath for 45 min. Cool to room temperature and make volume to 100 mL with distilled water and filtered through ordinary filter paper. Take 25 mL of supernatant in a 125-mL flask containing 2 mL of 6 N-NHO_3 and 6 mL of 8% (w/v) ferric ammonium sulphate solution. Titrate this homogenous mixture against 0.01 N potassium thiocynate till a pale salmon colour is obtained. Run parallel blank with each determination.

Calculation

$$\%\text{Glucosinolates} = \frac{(\text{Blank} - \text{titration}) \times 4.0 \times 0.01 \times \text{Mol. Wt. of GSL} \times \text{Total volume}}{1{,}000 \times 2.0 \times \text{Sample Wt.} \times 25.0}$$

Total volume = 500 mL
Mol. wt. of GSL = 411

14.30 Estimation of Oxalic Acid (Abaza et al. 1968)

Oxalates are present in free and salt forms in the feed, fodder and forages. Beet and spinach are rich in oxalates, and among straw/stover, paddy straw contains substantial amount of oxalic acid. Conversely, its low levels are present in peas, beans and brassicas, etc. In monogastric animals such as pigs, poultry and rabbits, oxalates containing diets reduce growth as well as calcium retention. However, ruminants are able to decompose oxalic acid with the help of rumen micro-organisms.

Principle

The sample is extracted in hydrochloric acid (HCl) and precipitated in the form of calcium oxalate by adding calcium chloride. The precipitate is then washed and titrated with N/20 potassium permanganet ($KMnO_4$) in the presence of dilute sulphuric acid (H_2SO_4) at 70°C. 1 mL of N/20 potassium permanganet is equivalent to 2.25 mg of oxalate.

Reagents

1. 6 N hydrochloric acid – Add equal amounts of conc. HCl and distilled water to prepare 6 N HCl.
2. Calcium chloride solution (5%) – Dissolve 5 g of calcium chloride anhydrous ($CaCl_2$) in 100 mL distilled water.
3. Sulphuric acid (H_2SO_4) in water (1:4 ratio) – A volume of sulphuric acid is slowly poured into four volumes of distilled water.
4. Concentrated ammonia (NH_4OH).
5. Methyl red indicator.

Procedure

1. To weigh 2 g of sample in a 250-mL volumetric flask, add 190 mL of distilled water and 10 mL of 6 N HCl. Boil the contents on a water bath. Make up the volume after cooling and filter the supernatant through Whatman (no. 41) filter paper.
2. Pipette out 50 mL of supernatant in a beaker and add 20 mL of 6 N HCl. Evaporate to about half of its volume and filter. The precipitate is washed in between to make the volume to about 125 mL.
3. Add 3–4 drops of methyl red to the filtrate followed by concentration ammonia till the solution turns faint yellow. Heat the contents to 90–100°C and filter (using Whatman no. 41 filter) after cooling to remove the precipitated impurities, if any. Add 10 mL of 5% $CaCl_2$ with constant stirring and allow to stand for 24 h.
4. Filter through Whatman no. 41 filter paper and wash the precipitate several times with hot water to make it free from calcium ions. Transfer the precipitate to the original beaker by washing with distilled water. Then add dilute sulphuric acid solution (1:4) till the precipitate is completely dissolved.
5. Warm the contents to 70°C and titrate with N/20 $KMnO_4$ near to end point and stir the filter paper to remove to the side contents of the beaker, wash with hot distilled water and complete the titration.

Calculation

$$\text{Oxalate}(g/100g) = N/20\ KMnO_4 \text{ used (mL)} \times 0.00225 \times 250/50 \times 100/2$$

Or

$$\text{Oxalate}(\%) = N/20\ KMnO_4 \text{ used (mL)} \times 0.5625$$

14.31 Estimation of Trypsin Inhibitor in Forages (Roy and Rao 1971)

Principle

The casein is used as a substrate for assaying the activity of the trypsin enzyme. The inhibition of the trypsin enzyme activity is measured in the extract prepared from the sample.

Reagents

1. 5% Trichloroacetic acid (TCA) solution
2. 2% Casein solution in phosphate buffer
3. 0.1 M Sodium phosphate buffer (pH – 7.5)
4. Trypsin solution (5 mg/mL)
5. 0.001 M Hydrochloric acid

Procedure

Put 4 g of the finely ground deffated material feed in stoppered 250-mL conical flask and treat with 40 mL of 0.10 M sodium phosphate buffer (pH 7.5) and 40 mL of distilled water. Shake the suspension in the flask for 3 h on shaker and centrifuge at 700 × *g* for 30 min at 15°C. Dilute the supernatant so that 40–60% activity of control enzyme is inhibited. Prepare incubation mixture containing 0.5 mL trypsin solution, 2.0 mL casein solution (2%), 1.0 mL 0.1 M sodium phosphate buffer, 0.4 mL HCl (0.001 M) and 0.1 mL extract or supernatant. Make the volume of incubation mixture to 4.0 mL. Incubate the mixture at 37°C for 20 min. Add 6.0 mL 5% TCA to stop the reaction. Run parallel blank. Record the absorbance on the spectrophotometer at 280 nm.

Computations

One trypsin unit is defined as an increase of 0.01 absorbance unit at 280 nm in 20 min for 10 mL reaction mixture and the trypsin inhibitory activity as the number of trypsin unit inhibits (TUI) is expressed per mg of protein. To express the value of specific activities, the total TUI expressed per mg of protein. The value for trypsin inhibitory activity must have about 40–60% inhibition to have reproducible and accurate results.

14.32 Estimation of Cyanogenic Glycosides (AOAC 1995b)

14.32.1 Qualitative Test

Reagents

1. Filter paper strips
2. Picric acid solution (1%)
3. 10% Sodium carbonate (Na_2CO_3)

Procedure

- Dip filter paper strips in 1% picric acid solution and after drying, further dip into 10% Na_2CO_3 solution and dry again. Store the strips in stoppered bottle.
- Place the sample of plant material in test tube. Insert a piece of moistened sodium picrate paper in tube while taking care that it does not come in contact with the sample. Add few drops of chloroform and stopper tube hermetically. The sodium picrate paper gradually turns orange and then brick red if plant tissue contains cyanogenic glycosides. The rapidity of change in colour depends upon amount of free HCN present.
- This test works well with fresh plant materials, but relatively dry substances particularly seeds of various plants should be ground and moistened with H_2O and allowed to hydrolyse in stoppered test tube containing sodium picrate paper.

14.32.2 Titrimetric Method for Quantitative Test

Acid titration method

Apparatus and Glassware

1. Micro-Kjeldahl distillation apparatus
2. Kjeldahl flasks
3. Conical flasks
4. Burette
5. Pipette
6. Gooch crucible

Reagents

1. Silver nitrate (0.01 N)
2. Nitric acid concentrate
3. 0.02 N Potassium cyanide (KHCN)
4. Ferric alum indicator

Place about 10–20 g finely ground sample (sieve no. 20) in 800-mL Kjeldahl flask. Add 100 mL H_2O and macerate at room temperature for 2 h. Further add

100 mL H_2O and steam distil for collecting distillate in 20 mL 0.01 N $AgNO_3$ acidified with 1 mL HNO_3. Adjust the distillation apparatus so that tip condenser dips below surface or liquid in receiver. When 150 mL distillate is collected then it is passed through Gooch crucible. Wash the receiver and Gooch with little H_2O and titrate excess $AgNO_3$ in combined filtrate, and washings with 0.02 N KH CN, using ferric alum indicator.

Calculations

1 mL 0.02 N $AgNO_3$ = 0.54 mg HCN

Alkaline titration method

Apparatus

1. Kjeldahl flask
2. Distillation apparatus (micro)
3. Conical flask
4. Burette
5. Pipette

Reagents

1. 0.5 g sodium hydroxide (in 20 mL H_2O)
2. Ammonium hydroxide (6 N)
3. Potassium iodide (5%)
4. Silver nitrate (0.02 N)

Procedure

- Place about 10–20 g of finely ground (sieve no. 20) sample in 800-mL Kjeldahl flask and add 200 mL H_2O and allow to stand for 2–4 h (analysis is done with apparatus completely connected for distillation). Now steam distil and collect 150–160 mL distillate in NaOH solution (0.5 g in 20 mL H_2O) and dilute to a definite volume.
- To 100 mL distillate (preferably dilute 250 mL and titrate 100 mL aliquot) add 8 mL 6 N NH_4OH and 2 mL 5% potassium iodide solution and titrate with 0.02 N $AgNO_3$ using micro burette. The end point is faint but permanent turbidity, which can be easily recognized, especially against black background.

Calculations

1 mL 0.02 N $AgNO_3$ = 1.08 mg HCN (1 Ag equivalent to 2 CN)

14.33 Qualitative Estimation of Ricin (Olsnes et al. 1974)

Principle

The ricin present in saline extract of castor (*Ricinus communis*) bean meal (CBM) exerts haemagglutinating activity and can be utilized for detecting the presence of CBM in feeds.

Reagents

1. Normal saline (0.9%, NaCl)
2. EDTA
3. Solution containing 0.14 M sodium chloride, 20 mM sodium phosphate per 100 μL (pH 7.1)
4. Bovine serum albumin

Procedure

Preparation of crude extract
Stir 50 g of sample mechanically in 150 mL normal saline (0.9%, NaCl) for 1 h. Filter the contents through muslin cloth and centrifuge the filtrate at 2,000 rpm for 5 min. An aliquot of clear supernatant is further tested for haemagglutinating activity of ricin by comparative qualitative test using plate agglutination technique.

Plate agglutination test
Collect 10 mL of blood from experimental animal in equal volume of normal saline having EDTA (1 mg/mL) blood and centrifuge at 2,500 rpm for 20 min to sediment red blood cells (RBC). Wash RBCs thrice with normal saline through centrifugation and dilute further with normal saline to a final suspension of 1%. Carry out microhaemagglutination test in Lambro plate.

To each well add 100 μL of a solution containing 0.14 M sodium chloride, 20 mM sodium phosphate (pH 7.1) and 100 μg BSA per mL. Add 100 μL of the crude extract to the first well and make serial dilutions by transferring after thorough mixing, 100 μL to the next well and so on. Subsequently, add 25 μL of normal saline washed 1% erythrocytes to each well and mix gently. Incubate the plate at 37°C and record the reading after 10 min. In case of agglutination, the erythrocytes will be sticky and cover the bottom of the well as a thin film or matrix, whereas nonagglutinated RBCs settle at bottom in the centre of the well. Express haemagglutinating activity as HA unit (the reciprocal of the end point dilution).

Chapter 15
Techniques in Molecular Biology

15.1 Polymerase Chain Reaction (PCR) (Mullis et al. 1986; Palumbi 1996)

PCR has been the most important invention of the past decade which has revolutionized the field of molecular biology. Beginning with a single molecular of DNA, the PCR can generate billions of copies of DNA in few hours, i.e. Nano gram(ng) of DNA can be amplified to get μg of DNA by using this technique. PCR technique is based on in vitro enzymatic amplification of a specific target DNA sequence in a cyclic process using two oligonucleotides. These oligos used as primers have different sequences and are complementary to the sequences on the opposite strands of the template DNA and flank the segment of target DNA that is to be amplified. Thus, given a particular target DNA, large amounts of that product and only that product are produced in sufficient quantities for subsequent experimental analysis.

Solutions and Reagents

1. Template DNA
2. Upstream and downstream oligonucleotide primers
3. Taq DNA polymerase (5 U/μl) and 10× PCR buffer
4. $MgCl_2$, 25 mM
5. dNTP mix (10 mM of each dNTP)
6. Nuclease-free water

Protocol (Basic)

1. Combine the first five reaction components in the order listed below in a thin-walled (0.2 or 0.5 ml) reaction tube and vortex for ~10 s and briefly centrifuge in a microcentrifuge. Initiate the reaction by adding the template and primers (Table 15.1).

R. Katoch, *Analytical Techniques in Biochemistry and Molecular Biology*, 377
DOI 10.1007/978-1-4419-9785-2_15, © Springer Science+Business Media, LLC 2011

Table 15.1 Reaction mixture

	Volume	Final concentration
1. Nuclease-free water (to a final volume of 50 μl)	× μl	
2. 10× PCR buffer	5 μl	1×
3. dNTP mix (10 mM of each dNTP)	1 μl	0.2-mM each
4. Taq DNA polymerase (5 U/μl)	0.25 μl	0.0025 U/μl
5. 25 mM $MgCl_2$	3 μl	1.5 mM
6. Downstream primer	50 p mol*	1 μM
7. Upstream primer	50 p mol*	1 μM
8. Template DNA	y μl**	

* The general formula for calculating nanograms of primer equivalent to 50 pmol is: 50 pmol = 16.3 ng × no of bases in promer
**Keep the final DNA concentration of reaction <10ng/μl

2. Place the tubes in a controlled temperature heat block and protocol with thermal cycling profile chosen for the reaction.
3. Analyse the PCR reaction products by agarose gel electrophoresis (1.2–1.5%) by loading a part of the aliquot.
4. Store the reaction products at −20°C until needed.

PCR Hygiene (Precautions)

Because PCR products are so concentrated and easily volatilized (by opening a microfuge tube or pipetting, for instance), cross-contamination of samples is potentially a serious problem. Certain simple precautions can be taken to avoid contamination or at least minimize it if it occurs.

- Aliquoting solutions makes it possible to contain and help resolve contamination problems that do arise. Each person working in the lab should have his or her own set of solutions. PCR reagents prepared in large amounts should be distributed in 1.5 ml microfuge tubes and stored at −20°C.
- Water used for PCR reagents, DNA, and primers should be double-distilled, sterilized, and then distributed in 1.5 ml microfuge tubes and stored at −20°C.
- When primers are made, the stock solutions usually are highly concentrated. From this highly concentrated stock solution, it is desirable to make a 100 μM stock solution which can then be used in making 10 μM solutions for individual use. The different stock solutions are stored separately. In this way, massive, laboratory-wide contamination problems can be avoided and any contamination problems that do arise can be contained.
- Different sets of pipettes should be designated for different procedures. One set of pipettes should be designated for preparing PCR reactions. These pipettes should never come in contact with any amplified DNA. Another set of pipettes can be designated for post-PCR use. One pipette should be designated to be used only in loading samples in agarose gels. Another set of pipettes should be designated for use with radiation only.

Common Problems with PCR

Problem: No PCR product, not even in positive controls.

Solution

- Repeat the experiment.
- Check buffer, dNTPs, and primer recipes and concentrations. Remake any questionable solutions.
- Try a different set of primer or a different positive control.
- Try a new batch of enzyme (this is seldom the problem unless the enzyme is very old).
- Was oil added to the reactions?
- Check the thermal cycler by watching it go through 2–3 cycles.

Problem: Positive control works, but there is no product

Solution

- Run 5 μl of the stock DNA solution on 1% agarose gel. If there is a large amount of high-molecular-weight DNA, try diluting the starting template DNA (try dilutions of 1:10 or 1:100). If there is no high-molecular-weight DNA, increase the amount of starting material or switch to better samples of genomic DNA.
- Try lowering the annealing temperature in the PCR cycle.
- Try a step-up cycle.
- Try using more cycles on the PCR machine (increase from 40 cycles to 50 cycles). This is effective only when the product is present but in small quantity.
- It is possible that something in the DNA temperature is interfering with the PCR reaction. This can be determined by setting up a single reaction with two templates (the added template should be known to work well with the primers being used). If the problem template prevents the added template form amplifying, then there is something in the problem template solution that is inhibiting the reaction. To solve this problem, try diluting the problem template, or try one of the rescue procedures outlined above.
- Switch primers and try again.

Problem: Bright bands in well of agarose gel following electrophoresis.

Solution

Such bands usually result from overamplification of the PCR product or from insufficient dilution of the product prior to electrophoresis. This is also a common result of amplifications from too much genomic DNA. Try diluting the template 100–1,000-fold.

Problem: Smearing of double-stranded PCR products or multiple bands following electrophoresis.

Solution

- Try less template. The most common cause seems to be too much template.

- Try annealing temperature 2–3°C higher. A lot of smearing, or multiple bands, may indicate that the primer is annealing to other parts of the template DNA.
- Try varying $MgCl_2$ concentration results in the best bands.
- Try fewer cycles. This is often recommended, but is probably not the best solution. While there may be less evidence of non-specific amplification, subsequent amplification from this PCR reaction will amplify even minute quantities of non-target DNA to visible levels (unless gel slices are used). A better solution is to optimize conditions to reduce mis-priming (e.g. temperature and salt concentration in buffer).
- Try gel purifying the double strands (only take the brightest part of the band) and then reamplify (with stringent conditions) the purified double-stranded product.

Problem: Bands in the negative controls.

Solution

- Often, in spite of all precautions, contamination problems occur. Once contamination becomes a visible problem, the contamination is more than one solution, so altering one solution may not be informative. Fresh preparation of all stock solutions is desirable.
- Wash the pipettes well, expose the tips to 10 min of UV light.
- Treat the solutions, including the primers, with UV light. Place the solutions in plastic tubes on a UV light source and illuminate them for 10 min (less if the UV source is a short wavelength source). This tends to break up contaminating DNA, making it less attractive as a PCR template.

15.2 Isolation of Plant DNA (Murray and Thompson 1980)

A number of methods are available for the isolation of high-molecular-weight DNA from plants. Generally, all methods involve removal of cell wall and nuclear membrane from around the DNA and the separation of DNA from other cell components such as cell wall debris, proteins, lipids, or RNA without affecting the integrity of DNA.

One of the most widely followed extraction procedures involves the use of a nonionic detergent cetyltrimethylammonium bromide (CTAB), which complexes with carbohydrates and can be phenol-extracted. It is a relatively simple procedure and is useful for the preparation of small samples of DNA needed for various experimentation.

Reagents and Materials

1. 2× CTAB

   CTAB – 10 g
   5 M NaCl – 140 ml
   2 M Tris-Cl, pH 8.0 – 25 ml
   0.5 M EDTA – 20 ml

2. *Chloroform and isoamyl alcohol* solution in the ratio of 24:1
3. *3 M Sodium acetate* (pH 5.2):

Sodium acetate = 408.1 g + Sterile H_2O = 800 ml

Adjust pH to 5.2 with glacial acetic acid and make up the volume to 1 l and autoclave the solution before use.

4. DNase-free RNase A:

RNase A (10 mg/ml) in 10 mM Tris (pH 7.5) + 15 mM NaCl

Heat to 100°C for 15–20 min to make it DNase free and cool slowly to room temperature (RT). Store in small aliquots at −20°C.

5. Proteinase K: 20 mg/ml proteinase K (store at −20°C)
6. 25% SDS

SDS – 25 g
Sterile H_2O – 100 ml (warm to dissolve)

7. TE buffer (10 mM Tris, 1 mM EDTA), pH 8.0

Tris – 1.211 g
EDTA – 0.372 g
Sterile H_2O – ~800 ml

Adjust pH to 8.0 and make up the volume to 1 l with sterile H_2O

Procedure

1. Material for DNA extraction
   Procure the seed material which will be the source of DNA. Wash the seeds thoroughly with distilled water and then with 50% ethanol for 10 min. Soak the seeds in 0.001% mercuric chloride for 10 min and then wash them several times with distilled water and soak them overnight. Keep the seeds for germination on wet germination towels at the desired temperature and humidity till they grow 3–6 in. in height. Cut the seedling 1 in. above the surface to minimize the bacterial contamination and cut them into smaller pieces. Weigh them and store them after a dip in liquid N_2 at −20 or −70°C for further use.
2. Isolation of plant DNA
   1. Add β-mercaptoethanol (β-ME) to the required amount of 2× CTAB extraction buffer to a final concentration of 0.2%. Heat β-ME/CTAB solution to 65°C in a waterbath for 5 min.
   2. Grind 10 g of etiolated seedlings in liquid N_2 to a fine powder with a pestle and mortar. Be very careful while powdering the tissue as the mortar and pestle can shatter due to the extreme cold.

3. Transfer the ground tissue (frozen powdered form) to a 500-ml conical flask containing the preheated β-ME/CTAB extraction solution (50 ml) and incubate the mixture for 45–60 min at 65°C with occasional mixing (Tissue:Extraction buffer : 1:5).
4. Extract the homogenate with equal volume of chloroform solution by gentle mixing (10 min).
5. Centrifuge for 5 min at 12,000 rpm at 4°C.
6. Transfer the upper (aqueous) phase to a fresh tube with a wide bore pipette tip and re-extract with chloroform solution.
7. Centrifuge as above and transfer the aqueous phase to a fresh tube.
8. Precipitate the DNA with 0.6–1 vol. of isopropanol (Mix well) at −20°C for atleast 30 min (precipitating overnight substantially increases the yield).
9. Centrifuge at ~12,000 rpm for 15 min at 4°C.
10. Wash the pellet (at least twice – necessary to remove traces of CTAB and chloroform) with 70% ethanol and dry it until all visible traces of ethanol are gone. Re-suspend the pellet in minimal volume of TE buffer (0.1–0.5 ml per gram of the starting material). Do not overdry the pellet.
11. For removal of RNA, to the re-suspended DNA add 20 μg/ml DNAse-free RNAse and incubate in a waterbath at 37°C for 1 h.
12. For removal of proteins, to the above mixture add 100 μg/ml of proteinase K solution, 1/10 volume 3 M sodium acetate (pH 5.2), and 1/100 volume of 25% SDS. Mix well and incubate at 60°C for 1 h.
13. Extract once with saturated phenol, then twice with 24:1, chloroform: isoamyl alcohol.
14. Precipitate the DNA using 2 volumes of 95% chilled ethanol. Wash the pellet twice with 70% ethanol and dry it partially. Re-dissolve it in TE buffer and check the purity of DNA.

3. Purity test of DNA

(i) Spectrophotometric test

In a spectrophotometer, check the optical density (OD) of a dilution of the DNA preparation at 260 and 280 nm. Pure DNA has an A_{260}/A_{280} ratio of 1.8–2.0 in 10 mM Tris-Cl, pH 8.5. Strong absorbance at 280 nm, resulting in a low A_{260}/A_{280} ratio, indicates the presence of contaminants such as proteins.

We can use conversion factor 50 to convert O.D. to concentration in μg/ml as DNA at a conc. of 50 μg/ml has an absorbance of 1 at 260 nm.

(ii) Agarose gel electrophoresis

Electrophoresis of DNA samples on 0.8% agarose gel with DNA markers and their staining with ethidium bromide and examination under UV should reveal a large band migrating close to the origin.

Problems During Extraction (DNA isolation and storage)

Problem

1. DNA degradation due to old tissue; senescence (in vivo)
2. Improper storage of tissues (slow freezing, freeze-thaw)
3. Breakage during isolation
4. Breakage after isolation (freeze-thaw, nuclease or bacterial contamination, particularly in dialysis)

Solution

1. Use fresh tissue.
2. Store properly.
3. Extract gently.
4. Be certain that the dialysis tubing has been treated appropriately, store DNA clean and frozen. Test storage solution (TE) for nuclease activity. When the sample is degraded, use frequent-cutting enzymes to reduce the average size of fragments compared, thereby minimizing degradation effects. If the sample has been contaminated, clean by (1) velocity gradient centrifugation, (2) Phenol-chloroform extraction, or (3) commercially available wash solutions.

15.3 Isolation of RNA (Brawerman 1974)

Ribonucleic acid (RNA) occurs as ribonucleoprotein particles in intact cells. The total RNA includes three classes – ribosomal, messenger, and transfer RNA. There are a number of methods described for the extraction of total RNA and the method selected largely depends upon the source material and the experimental purpose for which RNA is extracted. Among the available methods used with plant materials, the phenol-chloroform method is the common method used to recover the total RNA intact.

Principle

The ribonucleoprotein complex is dissociated by SDS into RNA and protein, deproteinized by phenol and the free RNA left in aqueous solution is precipitated in the cold after adding alcohol.

Requirements

- Magnetic stirrer
- Bench top centrifuge
- Cold room (4°C)
- Phenol (freshly redistilled)
- Extraction buffer (pH 9.0)

Tris-HCl (0.1 M) – 1.21 g
NaCl (0.075 M) – 0.44 g
EDTA Na_2 (0.005 M) – 0.19 g
Water to – 100 ml

- Ethanol
- SDS 10% (w/v in water)
- Ether

Procedure

All operations need to be conduced at 0–4°C.

1. Freeze 0.5–5 g of the material in a mortar and pestle with liquid N_2, grind to a fine powder, then to a paste and extract using 10 volumes of extraction buffer.
2. Centrifuge the homogenate at 2,000 × *g* for 3 in.
3. Transfer the supernatant to a volumetric flask and stir with 0.1 volume of 10% SDS for 2–3 min.
4. Add an equal volume of buffered phenol (freshly redistilled phenol saturated overnight with 100 mM Tris-HCl pH 8.5).
5. Partition the content by centrifuging at 2,000 × *g* for 5 min and collect the upper aqueous phase into a separate flask.
6. Shake the lower and interphase again with an equal volume of extraction buffer for 5 min and centrifuge.
7. Combine the aqueous phase with the first one (step 5) and stir with an equal volume of buffered-phenol for 5 min.
8. Repeat the extraction and centrifugation steps at least 5 times or until the interphase shows no proteins.
9. Finally, collect the upper aqueous phase containing RNA, dissolve in it about 250 mg NaCl, add two volumes of cold ethanol (96%), and leave the flask overnight at −20°C for RNA precipitation.
10. Collect RNA by centrifugation at 2,000 × *g* for 10 min. Wash the pellet (RNA) with 70% ethanol, ethanol, ethanol:ether (1:1 v/v), and finally with either. Dry the pellet gently *in vacuo* for a few minutes.
11. Dissolve the RNA completely in elution buffer for further analysis by vortexing.
12. Dilute 20 μl aliquot to 2 ml with buffer and read the absorbance using 1 cm light path cuvette at 260 nm in a spectrophotometer. One A_{260} unit is assumed equivalent to 40 μg RNA/ml. Otherwise, the RNA content is estimated colorimetrically (see estimation of RNA).

Precautions

- All the glassware and solutions should be sterile. Any contaminating RNase is inactivated by rinsing the glassware with 1% diethyl pyrocarbonate solution.
- Phenol: Chloroform (1:1) mixture is an effective deproteinizing agent that retains the poly (A) tail in mRNA intact.
- Any contaminating DNA will appear just like cotton wool during ethanol addition.

- Use freshly distilled colourless phenol. It can be stored frozen in small aliquots in brown-coloured bottles. Wear gloves while handling phenol.
- A variety of extraction medium, chelating agents, deproteinizing agents, etc. are used for the extraction of RNA.
- The denatured proteins gather at the interphase after low-speed centrifugation which should be discarded.
- The volume of aqueous phase will be drastically reduced if the phenol is not fully saturated.
- Deproteinization using phenol leads to the loss of poly (A)-tail of mRNA considerably.

15.4 Quantitative Estimation of DNA (Burton 1956)

The quantitative estimation of DNA could be carried out by number of methods; one of the method is described below.

Principle

Under extreme acid conditions, DNA is initially depurinated quantitatively followed by the dehydration of sugar to ω-hydroxylevulinylaldehyde. This aldehyde condenses, in acidic medium, with diphenylamine to produce a deep-blue coloured condensation products with absorption maximum at 595 nm.

Reagents

- DNA standard (0.5 mg/ml).
- Saline citrate solution (0.15 M NaCl, 0.015 M Na_3 citrate).
- Diphenylamine reagent: Mix 5 g fresh or recrystallized diphenylamine, 500 ml glacial acetic acid, and 13.75 ml conc. H_2SO_4. Stable for 6 months at 2°C; warm to room temperature and swirl to remix before use.

Procedure

1. Prepare separate marked tubes containing 1, 2, and 3 ml aliquots of the isolated DNA dissolved in standard saline citrate and similar aliquots of a 0.5-mg DNA/ml standard.
2. Make all sample tubes, and a separate blank, up to 3 ml with H_2O.
3. Add 6 ml of diphenylamine reagent to each tube, and after mixing, heat the tubes in a boiling water bath for 10 min. Cool the tubes.
4. Read the absorbance of blue solution at 600 nm against the blank.
5. Construct a standard graph A_{600} (ordinate) vs. quantity of DNA (abscissa) and then calculate the concentration of DNA dissolved in the saline citrate solution.

   This method is commonly applied for samples of 50–500 μg DNA.

15.5 Quantitative Estimation of RNA (Ashwell 1957)

Methods suitable for pentose determination are used for measurement of RNA, which include reactions with orcinol, phloroglucinol, aniline, etc.

Principle

The method depends on conversion of ribose (pentose) in the presence of hot acid to furfural, which then reacts with orcinol to yield a green colour. The colour formed largely depends on the concentration of HCl, ferric chloride, orcinol, the time of heating at 100°C, etc. up to certain maxima.

Materials

- Standard RNA.
- Sample RNA solutions.
- Orcinol acid reagent: Add 2 ml of a 10% solution (w/v) of $FeCl_3.6\ H_2O$ to 400 ml of conc. HCl.
- 6% Orcinol: Dissolve 6 g orcinol in 100 ml 95% ethanol. It is stable for 1 month. Refrigerate in a brown bottle until use.
- Colorimeter.

Procedure

1. Prepare a standard RNA (50 μg RNA/ml) solution in ice-chilled 10 mM Tris-acetate, 1 mM EDTA buffer (pH 7.2), or any other suitable buffer by dissolving RNA completely.
2. Dissolve the isolate RNA in the above buffer solution to an approximate concentration of 50 μg/ml.
3. Prepare a series of tubes containing 0.5, 1.0, 1.5, and 3.0 ml of isolated RNA, 0.5, 1.0, 1.5, and 3.0 ml of 50 μg standard RNA/ml.
4. Make up each tube to 3.0 ml with water. In addition, set a blank containing 3.0 ml of water.
5. Add 6 ml of orcinol acid reagent to each tube.
6. Add 0.4 ml of 6.0% alcoholic orcinol to each tube. Shake the tubes to mix the contents, and then heat all tubes in a boiling water bath for 20 min.
7. Cool the tubes, and read the absorbance at 660 nm against the blank.
8. Draw a standard curve using A_{660} and the concentration of standard RNA. Calculate the amount in the isolated RNA solution using the graph.

   - The yield and purity of RNA preparation can be assessed by measuring the absorbance of ultraviolet light by a solution of nucleic acid. A pure RNA solution should give a 260 nm:280 nm of 2; 1 U of A_{260} measured in 1 cm light path length is equivalent to 40 μg/ml.

15.6 Southern Blot Analysis of Plant DNA

This procedure is used for hybridizing labelled-DNA probes to DNA fragments separated on an agarose gel and then blotted on nitrocellulose (or nylon) membrane. Southern blot analysis involves the following steps.

1. Restriction (digestion) of DNA
   Plant DNA is restricted using restriction endonucleases. The restriction endonucleases used are type II enzymes which recognize either 4, 5, 6, or 8 bp long sequences and generate either 3′ protruding ends, 5′ protruding ends, or blunt-ended fragments. The next step is the separation of restricted fragments using agarose gel electrophoresis.
2. Agarose gel electrophoresis
   Agarose gel electrophoresis is used to separate DNA molecules according to their size. The negatively charged DNA molecules migrate in an electrical field from negatively charged cathode to positively charged anode, with smaller molecules migrating faster than the bigger ones through the pores of the matrix. The pores of the matrix can be changed by altering the agarose concentration. After electrophoresis, DNA fragments are visualized by staining with ethidium bromide which fluoresces in UV.
3. DNA transfer to nylon membranes
   DNA hybridization requires the breakage of hydrogen bonds between the complementary strands of DNA, which can be achieved either by high temperature treatment or treatment with denaturing agents like formamide. These treatments cannot be done on agarose gels. Hence, DNA is transferred from the gel onto a synthetic membrane. Nylon coated with nitrocellulose is used which combines the physical strength of nylon membrane with high resolution of nitrocellulose membrane. The banding pattern is preserved during the transfer. Since only single-stranded DNA binds to nitrocellulose, the DNA has to be denatured before transfer. The weak interaction between the DNA and the membrane achieved during transfer is further modified into a covalent body by heating to 80°C.

Solutions and Reagents

(a) Restriction of DNA

1. Mix in a sterile 1.5 μl Eppendorf tube

   DNA – 5 μg
   10× buffer – 5.0 μl
   Restriction enzyme – 2.5 μl (10 units/μl)
   Sterile water to – 50 μl

2. Incubate the mixture at 37°C for 5 h overnight.
3. Add 5 μl of 10× dye to the sample, centrifuge for 5 s in an Eppendorf centrifuge tube, and load on an agarose gel.

(b) Agarose gel electrophoresis

1. 10× TBE buffer

Tris base – 108 g
Boric acid – 55 g
0.5 M EDTA (pH 8.0) – 20 ml
Dissolve in DDW and make up to 1 l. Autoclave before use.

2. Gel-loading buffer (10× dye)

Glycerol – 5 m
10× TBE – 1 ml
Bromophenol blue saturated – 1 ml
Xylene cyanol (10%) – 1 ml
Add DDW and make up to – 10 ml
Mix well. Divide into 1 ml aliquots, autoclave, and store at –20°C.

(c) DNA transfer to nylon membranes

1. 0.25 M HCl

Conc. HCl – 21.5 ml
H_2O – 978.5 ml

2. Denaturing solution (1 l)
(1.5 M NaCl and 0.5 M NaOH)

NaCl – 87.66 g
NaOH – 40.0 g
Add DDW – to 1 l

3. Neutralizing solution

1 M Tris-Cl (pH 8.0) + 1.5 M NaCl
Tris base – 121.12 g
NaCl – 87.66 g
Adjust pH to 8.0 with HCl and make up the volume to 1 l with DDW and autoclave before use.

4. 20× SSC
(3 M NaCl + 0.3 M Sodium citrate, pH 7.0)

NaCl – 175.32 g
Na citrate – 88.213 g
DDW – to 1 l
Adjust pH to 7.0 with citric acid and make up the volume to 1 l with DDW and autoclave before use.

Protocol

A. Agarose Gel Electrophoresis (0.8%)

1. Weigh 0.8 g agarose and put in a 250-ml conical flask. Add 100 ml of 1× TBE buffer and gently boil the solution in a microwave oven with occasional mixing until all agarose particles are completely dissolved. Allow it to cool to 50°C. Add ~10 µl of ethidium bromide (10 mg/ml) and mix well. Prepare the gel mould and keep the comb in position. Pour the cooled gel solution into the gel mould and allow the gel to set for 20 min.
2. Fill the horizontal electrophoresis chamber with 1× TBE. Remove the comb from the gel and place gel with the tray in the electrophoresis chamber.
3. Load the digested DNA sample carefully into the wells. In one well, load a standard marker (Lambda DNA restricted with *Hind*III)
4. Run the gel at 20 mA overnight.
5. Stain the gel in 100 ml sterile distilled water containing 1 µg/ml ethidium bromide for 10 min. Briefly destain with sterile water (ethidium bromide may be added directly to the gel solution prior to pouring or to the running buffer).
6. Visualize the DNA bands on a UV transilluminator and place a ruler next to the gel to be able to determine the fragment sizes later on. Take a picture for records.

Problems and Solutions During Digestion Process

Problem: Endonuclease or exonuclease contamination
Solution: Titrate enzymes properly; switch to cleaner enzymes.
Problem: Partial digestion
Solution: Use more enzyme; two-step digestion.

Electrophoretic Artefacts

Problem: Retardation due to excess DNA
Solution: Use less DNA; use purified or semipurified organellar DNA
Problem: Unclear bands
Solution: Reduced buffering capacity of running buffer (make new buffer)
Problem: Missing small bands
Solution: Reduce electrophoresis time or use a combination of agarose and polyacrylamide gels.
Problem: Missing large bands
Solution: Too much BSA in digests; could also be due to DNA degradation.
Problem: Non-specific background (i.e. flecking) in gels loaded with end-labelled samples (particularly agarose gels)
Solution: Use higher grade agarose, making certain that it is completely dissolved; make sure plates and apparatus are clean; rinse gels before drying down.

B. Transfer of DNA to Nylon Membrane (Reed and Mann 1985; Ausubel et al. 1987)

1. After staining and photography of the gel, dip and shake it in 200 ml of 0.25 M HCl for 10–15 min at room temperature in a glass baking dish until the bromophenol blue barely turns yellow (Acid depurinates the DNA, breaking large fragments into smaller pieces for more efficient transfer).
2. Decant the HCl and rinse with distilled H_2O for 1 min and denature the DNA by soaking the gel in several volumes of 1.5 M NaCl and 0.5 M NaOH for 1 h at room temperature with constant shaking.
3. Decant the denaturing solution, rinse the gel with distilled H_2O for 1 min and neutralize the gel by soaking in several volumes of 1 M Tris-Cl (pH 8.0) and 1.5 M NaCl for 1 h at room temperature with constant shaking.
4. Wrap a piece of Whatman 3MM paper around a piece of glass plate and place it inside a large baking dish. Fill the dish with 20× SSC almost to the top (20× SSC is 3 M NaCl + 0.3 M sodium citrate, pH 7.0) and smooth all air bubbles in the 3 MM paper with a glass rod.
5. Invert the gel so that its original underside is now upper most and place it on the wet 3 MM paper. Remove air bubbles, if any, between the gel and the paper.
6. Cut a piece of nylon membrane about 1–2 mm larger than the gel. Dip it in sterile water (>20 min) and then float the membrane in 20× SSC for 5–10 min.
7. Place the wet nylon membrane on top of the gel and remove the air bubbles that are trapped between the gel and the membrane (Air bubbles trapped between the gel and membrane cause uneven or incomplete transfer).
8. Wet two pieces of Whatman 3 MM paper, cut to exactly the same size as the gel in 20× SSC, and place it on top of the nylon membrane followed by 6 cm stack of dry paper towels. Put a glass plate on top of the stack and weigh it down with 500 g weight.

   To prevent short circuiting of fluid between the paper towels and 3 MM paper under the gel, surround the gel with a watertight border of Saran wrap (Do not touch nylon membrane with naked hands. Always wear gloves and use forceps while handling nylon membranes. Ethidium bromide is carcinogenic).
9. Allow transfer of DNA to proceed for 6 h to overnight. Small fragments of DNA (>1 kb) takes 15 h or more.
10. Disassemble the blot in reverse order, and using a soft pencil, clearly label the slots on the membrane. Also mark the filter to define its orientation relative to the gel.
11. Soak the membrane in 10× SSC at room temperature for 5–10 min.
12. Place the filter on 3 MM paper to air-dry it.
13. Bake the membrane at 80°C for 1 h in a vacuum oven.
14. Soak the membrane in warm 2× SSC, seal it in plastic bag, and store in the dark in refrigerator until use.

C. Preparation of Probe by Oligolabelling (Using Hexalabel DNA Labelling Kit)

This method relies on the priming of the polymerase reaction on the template DNA with random hexanucleotide primers. The complementary strand is synthesized from the 3′ end of the primer with the help of the large fragment of DNA polymerase I, Exonuclease Minus (Klenow Fragment, exo$^-$) in the presence of labelled deoxyribonucleoside triphosphates.

Protocol

1. Add the following components into 1.5 ml microcentrifuge tube

   DNA template (100 ng) – 10 μl
   Hexanucleotide in 5× reaction buffer – 10 μl
   Deionized water up to – 40 μl
   Vortex the tube and spin down in a microcentrifuge for 3–5 s. Incubate the tube in a boiling water bath for 5–10 min and cool it on ice. Spin down quickly.

2. Based on labelled triphosphate (dATP or dCTP), use Mix A or Mix C, respectively.
3. Add the following components in the same tube:

   Mix A (or Mix C) – 3 μl
   [α-^{32}P]-dATP [α-^{32}P]-dCTP**
   (1.85 M Bq = 5 μCi)
   Klenow fragment, exo$^-$ (5 U) – 1 μl
   Shake the tube and spin down in a microcentrifuge for 3–5 s. Incubate for 10 min at 37°C.

4. Add 4 μl of dNTP and incubate for 5 min at 37°C.
5. Stop the reaction by the addition of 1 μl 0.5 M EDTA, pH 8.0
6. The labelled DNA is used directly for hybridization or stored at −20°C. Removal of the unincorporated label is not necessary for most applications, since the levels of its utilization are usually high. If required, the unincorporated dNTP can be removed by chromatography on Sephadex™ G-50 or by selective precipitation of DNA with ethanol in the presence of ammonium acetate.

Removal of Unincorporated Label (Optional)

1. Add 1 volume of 4 M ammonium (pH 4.5) to the labelled DNA, mix and vortex.
2. Add 2 volumes of ethanol, mix and chill in ice for 15 min.
3. Heat at 37°C for 2 min to re-dissolve free deoxyribonucleotides precipitated with occasional mixing.
4. Centrifuge at 12,000 × *g* for 15 min and carefully aspirate the supernatant.
5. Wash the pellet once in 90% ethanol and dry the pellet.
6. Re-dissolve the labelled DNA in TE buffer and use the probe for hybridisation.

Determination of Per Cent Incorporation (DE 81 Filter Binding Assay)

1. Dilute 1 μl of the labelling reaction with 0.2 M EDTA (1:100). Spot 3 μl (in duplicate) of the diluted sample on Whatman DE 81 circular (2.3 cm) filters.

2. Dry the filters under a hot lamp. Keep on filter aside for the determination of total cpm in the sample.
3. Wash the filter in 50 ml of 0.5 M sodium phosphate, pH 6.8 twice for 5 min to remove unincorporated label.
4. Dry the washed filter under a hot lamp.
5. Add scintillation fluid to each filter and count in a liquid scintillation counter.

D. Hybridisation

Solutions

1. 20× SSC: 3 M NaCl, 0.3 M Na citrate, pH 7.0
2. Non-homologous DNA (2 mg/ml Salmon sperm DNA). Prepare by dissolving in water overnight, then sonicate for 1 h. Store at 4°C.
3. 100× Denhardts solution: 2% (w/v) Ficoll (400,000 MW), 2% (w/v) polyvinyl pyrrolidone (360,000 MW), 2% (w/v) bovine serum albumin. Store at 4°C.
4. 10% sodium dodecyl suphate solution (SDS).
5. 1 M Na phosphate, pH 6.5.
6. Formamide deionized using a mixed bed ion exchange resin such as Bio-Rad AG 50001 – X8.

1. *Pre-hybridisation*

   (i) Prepare hybridisation solution which contains 5× Denhardts, 5× SSC, 50 mM phosphate pH 6.5, 0.1% SDS, 250 μg/ml non-homologous DNA, and 50% formamide. Incubate the solution for 10 min at 95–100°C to denature DNA, then cool on ice.
   (ii) Keep the nylon membrane in a plastic bag and pour hybridization solution (4 ml/100 cm^2) and seal the bag. Seal the bag inside a second bag containing damp paper towel.
   (iii) Immerse the bags in 42°C waterbath for 1–2 h.

2. *Hybridisation*

   1. Open the bags containing the membrane and remove the hybridisation solution.
   2. To the membrane in the bag add fresh 2 ml/cm^2 hybridisation solution to which probes have been added after denaturing by heating at 95–100°C for 10 min.
   3. Seal the bag and massage in order that the solution is distributed well.
   4. Immerse the bags in the 42°C waterbath for 8–10 h or for the duration of the hybridisation.

3. *Washing the membrane and autoradiography*

   1. After completion of the hybridisation, remove the membrane from the bag and wash briefly in 2× SSC and 0.1% SDS.

2. Keep the membrane in a fresh bag and pour 250 ml/100 cm^2, 2× SSC + 0.1% SDS buffer. Seal the bag and agitate vigorously for 5 min at room temperature.
3. Repeat Step 2 three times.
4. Add 250/100 cm^2 a solution containing 0.1× SSC + 0.1% SDS and agitate vigorously for 15 min at 50°C.
5. Repeat step 4 twice.
6. Remove the membrane from the bag and keep it in between Saran wrap. Remove any remaining liquid by pressing with tissue paper and expose to X-ray film.
7. Develop the film after appropriate number of days using X-ray developer and fixer.

Problems and Solutions

Transfer Hybridization: Poor Transfers

Problem: Bubbles and spots of no transfer

Remedy: Treat filter carefully – do not touch with bare hands; roll out bubbles in setting up blot.

Problem: Bottom filter weaker than top

Remedy: Avoid excess weight in blotting; use bottom filters (which may be weaker due to less DNA transferred) with high-copy number of probes; hybrids with total mtDNA or cpDNA to assess bad spots.

Problem: Double images

Remedy: Don't slide gel or filters around while setting up the blot.

Problem: Large fragments weak and poorly transferred.

Remedy: Use acid depurination in transfer protocol.

Transfer Hybridisation: Hybridisation Problems

Problem: Non-specific background

Remedy: Don't let filters dry in wrapping, washing, and exposing; improper pre-hybridisation (especially if large, bubble-shaped blotches), strip and repeat; use larger volume of pre-hybridisation solution.

Problem: Hybridisation to contaminating vector DNA

Remedy: Use isolated inserts; avoid vector contamination of DNAs.

Problem: Inability to strip completely previously hybridized probe

Remedy: Let filters decay; use low-copy number and more divergent probes first, high-copy number, and conserved probes last; don't let filters dry after probing.

Problem: Weak bands on autoradiogram (new filter)

Remedy: Be sure transfer was complete (stain gel with EB following transfer); be sure probe is labelled to high specific activity and in sufficient concentration.

General Precautions

- Keep the gel and nitrocellulose filter grease-free. Wear disposable gloves and use forceps whenever handling the filter.
- There are a number of ways by which blotting is set. Make sure in any case that the buffer flows through the gel and filter. The upper layer of papers should not touch the bottom layers.
- Nitrocellulose filter should not be dried in a hot air oven as the filter may explode.
- The dried filter may be further analysed by hybridization (Southern hybridisation) using a radioactively labelled DNA as a probe to identify a particular gene.
- Transfer of RNA from gel to filter is termed as "Northern blotting" and that of protein as "Western blotting". The latter procedure is described elsewhere in the book.

15.7 Isolation of Plasmids (Maniatis et al. 1982)

Plasmids are extrachromosal, self-replicating double-standard, circular DNA molecules found in most prokaryotes. These molecules carry genetic information for a variety of special functions such as resistance to antibiotics, nitrogen fixation, ability to utilize novel substrates, etc. The plasmids can be transferred from one cell to another and therefore function as vectors or carriers in genetic engineering techniques. A number of plasmids used in genetic engineering have a relaxed mode of replication. This means that the plasmid replicates independently of chromosomal control and accumulates up to one third of the cellular DNA content when cell protein synthesis is inhibited by a drug. Thus, milligram quantities of plasmid DNA may be isolated from a single litre of cells. Plasmid DNA is required in substantially pure form for cloning. The procedure for culturing and harvesting of bacterial cells, lysing of cells, and extraction of plasmid is described below.

Principle

The bacterial cells are grown to stationary phase, harvested, and gently lysed by weakening the cell walls with lysozyme treatment followed by use of the SDS. As a result, the cells release their high MW DNA which is removed by high-speed centrifugation leaving the plasmid DNA in the cleared lysate. This fraction is deproteinized and nucleic acids are then precipitated by ethanol. Purification of the plasmid is performed by equilibrium density centrifugation in cesium chloride.

Requirement

- Bacterial Strain carrying the Plasmid (e.g. *E. coli* J A 221 carrying pBR 328).

- LB Broth

  Yeast extract – 5 g
  NaCl – 10 g
  Tryptone – 10 g
  Water – 1 l

- TE + Sucrose (pH 8.0)

  0.05 M Tris – 0.61 g
  25% (w/v) sucrose – 25 g
  Water to – 100 ml

- Lysozyme Solution: 5 mg/ml in 0.25 M Tris-HCl (pH 8.0)
- Phenol-chloroform Mixture: 1:1 (v/v)
- Saline Sodium Citrate (SSC) solution:

  0.15 M NaCl – 0.88 g
  0.015 M sodium citrate – 0.44 g
  Water – 100 ml
  Dilute it 10 times to get 0.1 SSC

- Ethidium Bromide 5 mg/ml in 0.1 SSC solution
- 0.25 M EDTA solution
- TES Buffer (pH 8.0)

  30 mM Tris – 0.36 g
  5 mM EDTA Na^{2+} – 0.19 g
  50 mM NaCl – 0.28 g
  Water – 100 ml

Other requirements: High-speed refrigerated centrifuge, ultracentrifuge, polycarbonate ultracentrifuge tubes, UV Lamp (long wavelength), and Pasteur pipettes

Procedure

Step 1. Harvesting of cells

1. Grow the bacterial strain in 250 ml LB broth + antibiotic (ampicillin) at 37°C with shaking (vigorous aeration) to stationary phase.
2. Harvest the cells by centrifugation at 5,000 rpm in a refrigeration centrifuge for 10 min at 4°C.
3. Wash the cells by re-suspending in the TES buffer and centrifuging at 6,000 rpm for 10 min. Repeat the washing step.
4. Re-suspend the cells in a small volume of TE + Sucrose buffer. The cells can also be stored frozen at this stage, if required. Make up the volume of suspension to 3.75 ml by adding TE + Sucrose buffer.

Step 2. Cells lysing and isolation of DNA

5. Transfer the cell suspension to a pre-cooled 100 ml flask.
6. Add 0.75 ml of lysozyme solution followed by 1.25 ml of 0.25 M EDTA (pH 8.0) solution and shake the contents on ice for 10 min.
7. Add 0.75 ml of 20% SDS (final concentration 2%) and ensure uniform mixing.
8. Incubate without shaking at 37°C in a water bath until the suspension clears (cell lysis; 10–60 min). Cool on ice.
9. Centrifuge the lysate in thick-walled polycarbonate tubes in an ultracentrifuge at 40,000 rpm for 1 h at 20°C. Use 0.25 Tris pH 8.0 for balancing the centrifuge tubes, if necessary. This will clear the lysate and the supernatant will contain most of the plasmids with RNA and proteins as contaminants. High-molecular-weight chromosomal DNA is removed in the pellet.
10. Carefully decant the supernatant into a measuring cylinder, note its volume and transfer into a 100-ml flask. Add 0.1 volume supernatant 2.0 M Tris base (pH unadjusted).
11. Add an equal volume of phenol-chloroform. Shake thoroughly at room temperature for 4 min.
12. Centrifuge the emulsion in a bench centrifuge at 5,000 rpm for 10 min to separate the aqueous and organic phases.
13. Transfer the upper aqueous phase to fresh flask using a Pasteur pipette taking care not to disturb the protein precipitate at the interface. Repeat steps 11 and 12.
14. Carefully remove the aqueous phase and note its volume. Add 0.25 times the volume of 4.5 M potassium acetate to give a 0.9-M solution to ensure quantitative precipitation of DNA.
15. Add two volumes of chilled ethanol and place in freezer for 60 min to allow complete precipitation of DNA.
16. Centrifuge the contents at 10,000 rpm for 10 min at 0°C to pellet the tubes on paper towels. Dry gently in a vacuum desiccator or using a stream of nitrogen gas.
17. Dissolve the precipitate in 0.4 ml 0.1 SSC and withdraw 20 μl for testing by electrophoresis; then make up the remaining solution to 3.6 ml with 0.1 SSC.

Step 3. Purification by cesium chloride centrifugation

18. Dissolve 3.9 g CsCl in the preparation completely. Then add 0.4 ml of ethidium bromide.
19. Load the sample into ultracentrifuge tubes to within a few mm of the top and balance the tubes in pairs.
20. Centrifuge at 140,000 × *g* for 40 h at 20°C in a swing-out rotor.
21. After centrifugation, view the tubes under long-wave UV light. The DNA-ethidium bromide complex fluoresces and two defined bands could be seen near the middle of the tube. The more intense lower band consists of supercoiled, circular plasmids and the top band consists of linear plasmids and fragments of nuclear DNA.

22. The plasmid band can be recovered by a number of methods. First draw-off the upper part of the gradient and the DNA band using a Pasteur pipette. Then suck the plasmid band into a sterile syringe fitted with a wide-bore needle carefully. Alternatively, a long needle fitted to a syringe is carefully drawn into the syringe.
23. Remove the ethidium bromide from the plasmic fraction by extracting thrice with two volumes each of isopropyl alcohol. Cesium chloride and ethidium bromide are removed by dialysis for 16 h against several changes of 0.1 × SSC or any other suitable buffer for future analysis.
24. Following dialysis, transfer the plasmid solution to sterile tubes. Measure the absorbance at 260 and 280 nm. The A_{260} should be nearly twofold of A_{280} for a good preparation. Calculate the concentration of plasmid DNA using the relationship A_{260} of 1.0 = 50 μg/ml of DNA. The preparation can be stored frozen for several weeks. If a more concentrated preparation is required, concentrate by precipitation with ethanol (step 14–16).

- Avoid any contamination and dispose properly. Ethidium bromide and ultraviolet light are harmful; wear gloves and safely glasses, respectively, while using them.
- After treatment with SDS, the bacterial suspension should become highly viscous and gel-like indicating successful lysis. Treat the suspension as gently as possible to avoid damaging unwanted high MW DNA released from the cells.
- Dissolve the DNA precipitate, which may be even invisible in the centrifuge tubes, very gently to avoid shearing of DNA molecules.
- For a good resolution in CsCl gradient, ultracentrifugation should preferably by done in rotors that take short, wide tubes.
- The dialysis tubing used in the procedure should be pre-treated by boiling in 10 mM EDTA for 15 min followed by two 15 min treatments in boiling distilled water.
- At appropriate stages, aliquots may be withdrawn and analysed on agarose minigels to follow the course of purification of plasmid DNA.

15.8 Rapid Isolation of Plasmid DNA (Marko et al. 1982)

In some experiments, isolation of small amounts of partially purified plasmid DNA from a large number of bacterial clones in a relatively short period of time is necessary, whereas the methods for large-scale isolation and purification of plasmid DNA are elaborate, time-consuming, and require complicated steps. At these stages, "miniscreen" procedure is required, which does not require a large input of cells and can be carried out in small Eppendorf tubes.

Principle

The isolation procedure is based upon the release of soluble high MW DNA from disrupted cell wall and membranes, dissociation of nucleoprotein complexes by denaturation and proteolysis, and the separation of DNA from other macromolecules.

Reagents

- *Solution A* (pH 8.0)

  25 mM Tris-HCl – 30.3 mg
  50 mM glucose – 90 mg
  Lysozyme – 100 mg (add fresh before use)
  Water – 10 ml

- *Solution B*

  0.2 M NaOH – 0.8 g
  1.0% SDS – 1.0 g
  Water – 100 ml

- *Solution C*

  3 M sodium acetate (pH 4.8) – 24.6 g
  Water – 100 ml
  Adjust pH with glacial acetic acid and then make up the volume

- *Solution D*

  50 mM Tris-HCl (pH 8) – 0.60 g
  100 mM sodium acetate – 8.20 g
  Water – 100 ml

Procedure

1. Transfer 1 ml of an overnight cell culture (*E.coli* JA 221 carrying plasmid pBR 328 or any other strain) into an Eppendorf tube. Sediment the cells by centrifuging briefly in the microfuge. Drain off excess liquid.
2. Re-suspend the cells in 100 μl of solution A. Incubate on ice for 30 min.
3. Add 200 μl of freshly prepared solution B. Vortex briefly and keep on ice for 5 min. The cells will lyse immediately and the solution will become viscous.
4. Add 150 μl of solution C. Vortex briefly and keep on ice for 60 min. The bulk of chromosomal DNA and cell material will precipitate into a white viscous clump. Remove this by centrifuging in the microfuge for 5 min.
5. Transfer 500 μl of the cleared lysate to a clean Eppendorf tube. Add 1 ml of ethanol, mix thoroughly, and cool in a freezer (−70°C) to precipitate DNA. Sediment the DNA by spinning for 5 min.

6. Discard the supernatant from the DNA pellet and dissolve the precipitate in 100 μl of solution D. Add two volumes of ethanol, mix, and store in the freezer for 10 min. Pellet the DNA by centrifuging for 2 min.
7. Repeat step 6 twice more. Remove residual ethanol from the final DNA pellet by drying under vacuum. Dissolve the DNA in a suitable buffer for further analysis. Test the plasmid DNA preparation by agarose electrophoresis.

15.9 Miniprep of Plasmid DNA (Sambrook et al. 1989)

The plasmid miniprep method is useful for preparing partially purified plasmid DNA in small quantities from a number of transformants. This method provides enough partially purified plasmid DNA for a rapid analytical restriction enzyme analysis of plasmids. The most common method used for isolation of plasmid DNA is the alkaline lysis method developed by Birnboim and Doly. In this method, bacteria are lysed by treatment with a solution of sodium dodecyl sulphate (SDS) and NaOH (SDS denatures proteins and NaOH denatures chromosomal DNA and plasmid). The mixture on neutralization with sodium acetate and/or potassium acetate allows the plasmid DNA to re-anneal rapidly. The chromosomal DNA and bacterial proteins and SDS together form a precipitate and are removed by centrifugation. Plasmid DNA is then recovered from the supernatant by ethanol precipitation.

Solutions/Reaction Mixtures

1. Solution I

   20% glucose – 2.25 ml
   0.5 M EDTA (pH 8.0) – 1.00 ml
   1 M Tris (pH 8.0) – 1.25 ml
   Sterile DDW to – 45.50 ml
   Store at 4°C, add 2 mg/ml lysozyme immediately before use.

2. Solution II

   10 N NaOH – 0.4 ml
   20% SDS – 1.0 ml
   Sterile H_2O – 18.6 ml
   Do not autoclave.

3. Solution III

   3M sodium acetate, pH 4.8
   Sodium acetate – 40.82 ml
   DDW to – 100 ml
   Adjust pH to 4.8 with glacial acetic acid and make volume to 100 ml. Autoclave before use.

4. RNase A
   10 mg/ml in 10 mM Tris-Cl (pH 7.5) containing 15 mM NaCl. Heat to 100°C for 15 min and cool slowly to room temperature. Store in small aliquots at −20°C.
5. TE buffer

   10 mM Tris-HCl (pH 8.0)
   1 mM EDTA.

6. Saturated phenol
7. Chloroform:isoamyl alcohol mixture (24:1)

Protocol

1. Inoculate 5 ml of LB medium containing the appropriate antibiotic with a single bacterial colony carrying the plasmid. Incubate at 37°C overnight with vigorous shaking.
2. Pour 1.5 ml of the culture into a microfuge tube. Centrifuge for 1 min in a microfuge.
3. Remove the medium by aspiration leaving the bacterial pellet as dry as possible.
4. Re-suspend the pellet in 100 μl of ice-cold solution I. Vortex thoroughly. No lumps should be seen.
5. Incubate in ice for 5 min.
6. Add 20 μl of freshly prepared solution II. Mix gently by inverting the tube. The contents at this stage become clear and viscous. Do not vortex. Leave on ice for 5 min.
7. Add 150 μl of ice-cold solution III. Mix by inverting the tube slowly several times. Keep on ice for 5 min.
8. Centrifuge for 5 min at 1,400 × *g* at 4°C.
9. Transfer the supernatant carefully to a fresh tube.
10. Add an equal volume of phenol: chloroform-isoamyl alcohol mixture (1:1). Mix by vortexing, centrifuge for 2 min in a microfuge at 12,000 rpm. Collect supernatant in fresh tube.
11. Add two volumes of ethanol at room temperature. Mix by vortexing. Allow it to stand at room temperature for 2 min.
12. Centrifuge for 5 min in a microfuge at 12,000 rpm.
13. Discard the supernatant completely by keeping the tube inverted on paper towels.
14. Add 1 ml of 70% ethanol. Vortex briefly and re-centrifuge.
15. Remove ethanol and dry the pellet.
16. Dissolve the pellet in 50 μl TE and analyse an aliquot on agarose gel.

RNase Treatment of Plasmid

To 2 μl of the DNA from the above step, add 2 μl of RNase A (10 mg/ml) and incubate at 37°C for 30 min. Dilute DNA to 50 μl with TE and add an equal volume of phenol-chloroform-isoamyl alcohol mixture. Mix well and centrifuge in microfuge at full speed for 2–5 min. Collect the aqueous phase and repeat phenol-chloroform extraction until the interface is clear. To the aqueous phase, add 1/10 vol. of 3 M sodium acetate, pH 4.8 and 2 Vol. of 100% ethanol. Mix well and keep at −70°C for 15 min. Centrifuge to collect the DNA, wash with 70% ethanol, dry the pellet, and dissolve it in 20 μl TE.

Isolation of DNA Fragments Using Low-Melting Agarose Gel

1. Restrict 5 μg of the recombinant plasmid DNA with desired restriction enzyme in an Eppendorf tube in a total volume of 20 μl for 60 min at 37°C.
2. Prepare 1% low gelling agarose in 1 × TBE, load the sample, and do electrophoresis to separate the fragments.
3. Stain the gel using ethidium bromide and cut out the target band with a clean scalpel.
4. Melt the gel slice at 65°C and add enough TE buffer to decrease agarose percentage to 0.4% or less.
5. Add an equal volume of buffered phenol, mix vigorously for 5–10 min, and centrifuge at 10,000 rpm for 10 min at room temperature.
6. Collect the aqueous phase and keep it aside. Re-extract the phenol and interphase with an equal volume of TE buffer. Centrifuge and collect the aqueous phase and combine with the earlier one.
7. Add 2 volumes of cold ethanol and leave at −70°C for 15 min.
8. Centrifuge and collect the pellet. Wash the pellet with 70% ethanol. The final pellet is air-dried and dissolved in TE buffer.

15.10 Cultivation of Lambda (λ) Phage

Phage λ is used as a vector for cDNA and genomic libraries construction. Unlike bacterial cells, viruses have to be multiplied in living hosts. Bacteriophages are prepared by allowing them to infect and multiply on bacterial cells.

Principle

Bacteriophage λ is cultured in E. coli cell suspension. The phage is allowed to harbour and multiply till the complete lysis of bacterial cells. Then the bacterial cells are killed with chloroform and the phage particles are separated by centrifugation.

Materials

1. Stock λ phage (EMBL-3)
2. Suitable *E. coli* strain
3. Chloroform
4. Polyethylene glycol (PEG 6000)
5. Pancreatic DNase and RNase
6. Sodium chloride

7. Cesium chloride
8. Luria Broth

   Bacto-tryptone – 10 g
   Yeast extract – 5 g
   Sodium chloride – 5 g
   D-glucose – 1 g
   Water – (sterile) – 1 l

9. SM Buffer

   Sodium chloride – 5 g
   Magnesium sulphate – 1.2 g
   1 M Tris (pH 8.0) – 50 ml
   Gelatin – 0.1 g
   Water – 1 l
   Sterilize and use.

Procedure

1. Inoculate 1 ml of overnight grown culture of suitable *E.coli* strain to 100 ml LB containing 10 mM $MgCl_2$ and 0.2% maltose in triplicate. Grow for 2–3 h till the OD 600 reaches 0.4.
2. With the assumption that 1 $OD_{600} = 8 \times 10_8$ cells/ml, calculate the cell concentration.
3. Centrifuge the cell suspension in autoclaved tubes at 4,000 × *g* for 10 min at room temperature.
4. Discard the supernatant and re-suspend the bacterial cells in 2 ml of SM.
5. Add bacteriophage from the stock to a concentration of $5 \times 10_8$ phage/ml and mix rapidly.
6. Incubate at 37°C for 20 min. Shake intermittently.
7. To 125 ml of LB containing 10 mM $MgSO_4$ and 0.2% maltose, add 1 ml of infected cells (step 4) in 500 ml flask. Incubate at 37°C with very vigorous shaking till it lysed. Disappearance of silkiness is the indication of complete lysis. It takes 7–9 h for complete lysis.
8. To check complete lysis, take 1 ml of the culture from the flask and add few drops of chloroform. If it becomes clear, proceed with the next step. If not, continue the incubation till complete lysis is achieved.
9. Add 2.5 ml chloroform to each flask and continue vigorous shaking for another 30 min at room temperature.
10. Collect the supernatant and pool for virus particles.
11. Bring the culture to room temperature and add pancreatic DNase and RNase, both to a final concentration of 1 μg/ml. Incubate for 30 min at room temperature.
12. Now, add NaCl to a final concentration of 1 M (29.2 g/500 ml of culture). Dissolve by swirling. Let it stand for 1 h on ice.

13. Remove white silky mass of bacterial cells by centrifugation at 11,000 × *g* for 10 min at 4°C. Pool the supernatant and repeat this step.
14. To the supernatant in a flask, add solid polyethylene glycol (PEG 6000) in small quantities with slow shaking to a final concentration of 10% *w/v* (i.e. 50 g/500 ml).
15. Cool in ice water and keep it at 4°C overnight.
16. Centrifuge at 11,000 × *g* for 10 min at 4°C and recover the precipitated phage particles.
17. Re-suspend the bacteriophage pellet by gentle shaking in 7 ml SM buffer (i.e. 7 ml/500 ml of original supernatant).
18. To this suspension, add equal volume (7 ml) of chloroform and mix thoroughly. Centrifuge at 1,600 × *g* for 10 min at 4°C. Save the aqueous phase containing bacteriophage.
19. Measure the volume of this phage suspension and add 0.5 g/ml of solid cesium chloride. Mix gently. After completely dissolving cesium chloride, carefully layer this suspension onto cesium chloride step gradients that are prepared in cellulose nitrate centrifuge tubes (refer note below for the preparation of CsCl step gradient).
20. Centrifuge using SW 27 rotor at 25,000 rpm for 3 h at 4°C. A bluish band of bacteriophage particles will be visible at the interface between the 1.45 and 1.50 g/ml layers.
21. Collect the band of bacteriophage particles using siliconized Pasteur pipette. Store at 4°C in tightly capped tube.

Prepare cesium chloride solutions of 5 M and 3 M separately in distilled water. Transfer10 ml of 5 M solution to the bottom of cellulose nitrate centrifuge tube (SW 27). Carefully lay on top of it 10 ml of 3 M CsCl. Load the phage suspension over 3 M layer. Cellulose nitrate tube should be filled full. Otherwise, the tube will collapse during centrifugation.

15.11 Extraction Lambda DNA

After harvesting of purified phage particles, DNA from purified phage could be separated and used in certain recombinant DNA experiments. Usually, bacteriophage DNA is required for many experiments in molecular biology.

Principle

Proteinase K is used to digest the viral protein coat and then DNA is precipitated with acetate and alcohol.

Materials

1. EDTA (0.2 M) sodium salt.
2. Pronase or proteinase K.

3. SDS (20% solution).
4. Phenol dialysis bag.
5. Dialysis buffer.

   50 mM Tris-HCl (pH 8.0)
   10 mM NaCl
   10 mM $MgCl_2$

6. Chloroform.
7. 3 M Sodium acetate (pH 5.5).
8. Absolute alcohol.
9. 70% ethanol.
10. TE buffer.

    10 mM Tris-HCl (pH 7.8)
    10 mM EDTA.

11. Phenol (freshly redistilled phenol equilibrated with TE Buffer): Mix equal volumes of phenol and TE buffer, make it stand for a few hours, and drain off excess aqueous upper phase.

Procedure

1. Transfer the purified phage particles into a dialysis bag and dialysis against a 1,000-fold volume of dialysis buffer for 1 h.
2. Change the dialysis buffer and dialysis for another 1 h.
3. Now, carefully transfer the content of the dialysis bag into a centrifuge tube.
4. Add EDTA to give a final concentration of 20 mM.
5. Add proteinase K to a final concentration of 50 g/ml or protease to 0.5 mg/ml.
6. Add SDS to a final concentration of 0.5%. Mix thoroughly by inverting the tube several times.
7. If proteinase K is used, incubate for 1 h at 65°C (for pronase 1h at 37°C).
8. Add an equal volume of phenol equilibrated with TE buffer. Mix by inverting the tubes several times. Centrifuge at 1,600 × *g* for 5 min at room temperature. Transfer the aqueous phase to a clean tube.
9. Extract the aqueous phase again with 50:50 mixture of equilibrated phenol and chloroform.
10. Extract the aqueous phase again with an equal volume of chloroform.
11. Add one-tenth volume of 3 M sodium acetate, equal volume of absolute alcohol, mix thoroughly, and make it stand at −20°C for a few hours. Centrifuge and wash the pellet with 70% ethanol. Re-dissolve the DNA in a small volume TE buffer.
12. Determine the DNA concentration spectrophotometrically. Store the DNA solution in small aliquots at −20°C.

15.12 Restriction (Digestion) of DNA (Maniatis et al. 1982)

Restriction enzymes are found in bacteria and in vivo are involved in recognition and destruction of foreign DNA The restriction endonucleases cut at the interior part of DNA. Invading phage DNA, for instance, will be restricted by such enzymes. These enzymes recognize a particular sequence of bases. Type II restriction enzymes cut the DNA within the recognized sequences. Each enzyme has its own characteristic recognition sequence and it may be 4–7 bases long with dyad symmetry. The bacteria protect their own DNA from the restriction endonucleases by modification process. With the commercially available restriction enzymes, it is now possible to construct the physical map of genes after digestion of the DNA with different restriction enzymes and separation of DNA fragments on agarose gels.

Reagents

- Plant DNA
- Lambda DNA
- Plasmid DNA
- Restriction buffer (10×)
- Restriction enzyme

The following restriction buffer (10×) can be used with most restriction endonucleases.

100 mM Tris-HCl pH 7.4
100 mM $MgCl_2$
100 mM NaCl, 10 mM DTT

- Stopping mix

  20% Sucrose
  10% Ficoll
  10 mM EDTA Na_2
  1% bromophenol blue
  1% xylene cyanol FF

- Sterile double-distilled water
- Microfuge tubes (0.5 ml)

Procedure

1. Label microfuge tubes 1–10 and arrange them.
2. Bring all the reactants on ice box. DNA samples stored frozen are thawed quickly and brought on ice. Each reaction tube should contain about 1 μg DNA. Restriction enzymes should be diluted before use to a concentration of 1 U/μl. (One unit is defined as the enzyme activity which completely digests 1 μg of DNA in 1 h at 37°C)

Table 15.2 Reaction mixture

	Tube											
Contents (µL)	1	2	3	4	5	6	7	8	9	10	11	12
λ DNA	2	2	2	2								
Plasmid DNA					2	2	2					
Plant DNA								2	2	2		
Restriction	2	2	2	2	2	2	2	2	2	2	2	2
Sterile water	16	14	14	12	16	14	14	16	14	14		
EcoRI	–	2	–	2	–	2	–	2	2	–		
Hind III	–	–	2	2	–	–	2	–	–	–		
Total volume (µl)	20	20	20	20	20	20	20	20	20	20	20	20
	20 µl in each tube											

3. Prepare the following reaction mixes by carefully pipetting into the bottom of microfuge tubes (Table 15.2).
4. Mix contents of tube carefully. Centrifuge all the tubes in a microfuge (Eppendorf) for 2 s. to settle the contents at the bottom of tubes.
5. Incubate at 37°C in a water bath for 60 min or longer if necessary.
6. Terminate the digestion by adding 5 µl of stopping mix to each tube or by heating at 65°C for 10 min.
7. Mix and spin down as before.
8. Load equal volume (15/20 µl) of each digest in agarose minigel wells and run.

   - The enzyme activity is considerably reduced even with proper storage after about 6 months.
   - Siliconize the microfuge tubes so that no solution sticks to the side of the tubes. Prepare a 5% solution of dichlorodimethyl silane in chloroform, fill in the tubes, and leave for a few minutes. Rinse the tubes with water thoroughly. Glassware can also be similarly siliconized and baked at 180°C overnight.
   - Use a clean pipette tip (or microcapillary) for each transfer in order to avoid cross-contamination between solutions.
   - Use the restriction buffer (10×) supplied along with the restriction enzymes for satisfactory results wherever possible. Each restriction enzyme requires specific ionic strength for optimum and precise cut. For most frequently used enzymes, the restriction buffer (10×) composition is given below:

     EcoRI and – 100 mM Tris-HCl pH 7.9, 60 mM $MgCl_2$
     Sal I – 1.5 M NaCl, 60 mM 2 MSH, 1 mg/ml BSA (Fraction V)
     Hind III – 200 mM Tris-HCl pH 7.5, 70 mM $MgCl_2$,
     500 mM NaCl, 70 mM 2-mercaptoethanol
     Pst I – 200 mM Tris-HCl pH 7.5, 100 mM $MgCl_2$,
     500 mM $(NH_4)_2HO_4$

   - The reaction volumes are usually less than 50 µl and contain about 1 µg of DNA (More if a large number of fragments are expected to be generated).

- The restriction enzymes are usually supplied with glycerol as stabilizing agent. Note that if the concentration of glycerol exceeds 10% in the digestion mixture, the specificity of most enzymes is altered.

15.13 DNA Electrophoresis in Agarose Gel (Maniatis et al. 1982)

Electrophoresis of DNA or RNA in gels is a rapid and relatively inexpensive method by which DNA can be checked for size, intactness, homogeneity, and purity. Importantly, this technique is core of molecular biology techniques. Separation of DNA can be achieved in polyacrylamide or most preferably in agarose gels. Agarose vertical gels give reproducible results and can be completed more rapidly than many horizontal systems. Horizontal agarose gel electrophoresis has the advantage that much lower concentration of agarose can be utilized than in a vertical gels system allowing the separation of large DNA fragments, and also gel handling is easier.

An effective combination of standard vertical and horizontal slab gel is the "submarine minigel". These small (10 × 6 × 0.5 cm) agarose gel slabs are run horizontally under approximately 2–5 mm buffer and have the advantage of being easier to prepare and faster-running than the conventional gels.

Principle

Agarose forms a gel by hydrogen bonding and the gel pore size depends on the agarose concentration. The DNA molecules are separated by electrophoresis on the basis of their size, shape, and the magnitude of net charge on the molecules. Ethidium bromide dye intercalates between the bases of RNA and DNA and fluorescences orange when irradiated with UV light. Low concentration agarose gels with large pore permit fractionation of high MW molecules and vice versa.

Materials

- Tris-Borate Buffer (10×; pH 8.2).

  0.9 M Tris-HCl – 113.0 g
  0.025 M EDTA Na_2 – 9.3 g
  0.9 M boric acid – 55 g
  Double-distilled water – 1 l

- *Agarose*: 1.0 (w/v) in single-strength Tris-borate (gel-running) buffer. Autoclave to dissolve the agarose; then maintain at 50°C in a stoppered flask until used.

- *Gel-loading Solution*:
  Sucrose 30%, bromophenol blue 0.25%, and Xylene cyanole FF 0.25% (all w/v) in single-strength, Tris-borate buffer.
- DNA Preparations (Plant DNA, plasmid DNA, recombinant DNA, etc).
- Standard DNA for Size Determination
  (λ DNA cut with restriction endonucleases Eco RI and Hind III)
- Gel Casting Plate, gel tank, power supply, etc.
- UV transilluminator and gel documentation system.

Procedure

1. The plate is placed in a suitable gel casting tray set-up perfectly horizontal over a levelled plate.
2. Pour 30 ml of 1% agarose solution maintained at 50°C onto the casting plate. Immediately place a suitable well-forming comb about 1 cm from one end of the plate. The teeth of comb should not touch the glass plate. Allow the gel to set for 1 h.
3. Remove the comb from the gel, carefully. Transfer the gel along with the glass plate to the electrophoresis tank such that the wells are near the cathode. Pour single-strength Tris-borate buffer into the tank until the gel (including zinc oxide wall) is submerged.
4. Connect the electrodes to the power supply with the cathode (−ve) at the well end of the gel.
5. Load DNA samples (5–20 μl) and standard taken in gel-loading solution using a micro-syringe.
6. Turn on the power supply and run at 100 V (10–15 mA). Monitor the progress of fast-running (bromophenol blue) tracking dye during electrophoresis. Terminate the run when the tracking dye is near the end of the gel.
7. After disconnecting the power supply, transfer the gel to a staining tray containing 250 μg ethidium bromide (50 μl of 5 mg/ml solution) in 250 ml of used Tris-borate buffer. Stain the gel for about 30 min.
8. Transfer the gel onto a thick plastic sheet, place on a UV transilluminator, and view the gel under ultraviolet light (300 nm). Nucleic acids on the gel will appear orange owing to the fluorescence of bound ethidium bromide. Photograph as soon as the gel has been checked for the presence of bands. Use the photograph for further interpretation of band patterns.
9. Measure (from photograph) the distance moved by each band from the front of the loading well. Plot the distances against log molecular weight of standards (λ DNA fragments) to give a calibration curve. Deduce the size of DNA and/or restricted fragments of samples using the curve.
   - Agarose gels of different thickness, concentration (0.2–2%), and dimensions can be used depending upon the experimental requirement.
   - The gel should be poured carefully without entrapping any air bubbles. Keep the agarose solution always at 50–55°C until poured on the casting plate to avoid setting.

Table 15.3 Sizes of restriction fragments from λ DNA (49 kb) in kilobase pairs (kb)

Eco RI	Hind III	Eco RI + Hind III
21.80	23.70	21.80
7.52	9.46	5.24
5.93	6.75	5.05
5.54	4.26	4.21
4.80	2.26	3.41
3.41	1.98	1.98, 1.90, 1.71, 1.32, 0.93, 0.84, 0.58

- Minigels are run at high current for quick results. Otherwise, the gels of large size are usually run at a constant low voltage (25–40) overnight for high resolution of bands.
- Wear gloves while using ethidium bromide as it is mutagenic.
- Ethidium bromide may be added to the gel-running buffer itself. The separation of bands then can be followed even during electrophoresis using a hand-held UV lamp.
- Wear safety glasses for viewing the gel under UV light. In order to prolong the life of the transilluminator and to avoid excessive exposure to UV radiation, photograph the gel at the shortest period and use the photograph for further interpretation. Position a mini-scale and photograph to measure the relative mobilities of bands later.
- The electrophoretic mobility of bands is very much affected by the salt concentration in the sample. It is important that the salt concentration in samples loaded in various wells is (nearly) uniform.
- The sizes of restriction fragments from lambda DNA (49 kb) are given above (Table 15.3).

15.14 Recovery of DNA Fragments from Agarose Gels

Among several available methods, the recovery of DNA by electroelution is one of the common techniques in molecular experiments.

Materials

- TBE Electrophoresis Buffer
- 3 M Tris-HCl (pH 7.5)
- Dialysis Tubing

Procedure

1. After electrophoretic fractionation, visualize stained DNA bands in UV light in transilluminator. Cut out the agar containing the DNA to be recovered with a sterile scalpel.

2. Transfer the agar inside a piece of dialysis tubing. Add electrophoresis buffer so that the agarose piece is surrounded by buffer with no air bubbles. Use maximum volume of buffer in the dialysis bag.
3. Place the dialysis bag at the bottom of a gel apparatus filled with buffer in such a way that the piece of agar is in the same position with respect to the electrodes as it was in the gel.
4. Apply 100–120 mA for nearly 2 h to elute DNA out of the gel.
5. At the end, reverse the current for 30 s in order to mobilize any DNA which may be stuck to the dialysis membrane.
6. Transfer the buffer containing the electroeluted DNA through a cotton-plugged 1 ml tip into an Eppendorf tube by centrifuging for 2 min. This process will remove contaminating agarose particles.
7. Extract the supernatant with phenol, then with chloroform:isoamyl alcohol (24:1) to remove any soluble contaminants.
8. Mix the DNA solution with 3 M Tris-HCl (pH 7.5) to give a final concentration of 0.5 M. Add two volumes of isopropanol and make it stand at −20°C for 30 min.
9. Centrifuge for 10 min at 15,000 × *g*. Repeat the precipitation step once more to remove any remaining ethidium bromide.
10. Dry the DNA pellet under vacuum and re-suspend in a suitable buffer for any subsequent experiment.

- Sterilize dialysis tubing for 30 min at 120°C in solution of 5% w/v sodium carbonate, 1 mM EDTA, and keep at 4°C in sterile distilled water.
- DNA fractionation can be carried out on low-melting agarose gels and processed subsequently for ligation reaction.

15.15 Isolation of mRNA by Affinity Chromatography (Bantle et al. 1976)

mRNA (messenger RNA) accounts for only about 2% of the total RNA in plants, the rest being occupied by tRNA and rRNA. Only mRNA carries a stretch of polyadenylate at its 3′ end, which may vary in length ranging from 3 to 200 adenylate residues.

Principle

Under high ionic conditions, the poly (A) tail of mRNA forms a duplex with complementary base sequences such as poly (U) or oligo (dT), while tRNA and rRNA do not form any duplex since they lack poly (A) tail. When the ionic condition is lowered, the hydrogen bonding between poly (A) and poly (U) or oligo (dT) is broken, thus releasing the Poly (A)-containing molecules. Therefore, by affinity, chromatography mRNA can be isolated from the total RNA. The Poly

(U) or oligo (dT) immobilized (covalently attached) to sepharose or cellulose can be used as the stationary phase.

Reagents

- Oligo(dT) Cellulose or Poly(U) Sepharose
- Binding buffer (pH 7.5)

  0.01 M Tris – 0.12 g
  0.5 M NaCl – 2.92 g
  0.001 M EDTA Na_2 – 0.04 g
  10% SDS – 0.5 ml
  Water – 100 ml
  Adjust pH with HCl and add SDS solution to the buffer before use.

- Elution buffer (pH 7.5)

  0.01 M Tris – 0.12 g
  0.001 M EDTA Na_2 – 0.04 g
  10% SDS – 0.5 ml
  Water – 100 ml
  Adjust pH with HCl

- Chromatography Column (Pasteur Pipette)

Procedure

1. Suspend about 500 mg of affinity material (oligo-(dT) cellulose of Poly (U)-Sepharose) in 20 ml of sterile binding buffer.
2. Pack the column with the affinity material as usual and equilibrate it by passing through 25–30 ml of binding buffer.
3. Dissolve the isolated total RNA in the binding buffer at a concentration of approximately 1 mg per ml.
4. Apply the RNA solution to the column. Recycle the initial effluent of the column.
5. Wash the column extensively with the binding buffer until the absorbance at 260 nm reaches zero. The poly(A)-containing mRNA is retained in the column, while tRNA and rRNA pass through. Collect the high absorbance initial fractions and precipitate with ethanol for rRNA and tRNA, if needed.
6. Elute the column with low salt sterile elution buffer. Collect and pool high A_{260} fractions. These fractions contain many mRNA with slight contamination of rRNA.
7. Wash the column with elution buffer and then pre-equilibrate with the binding buffer.
8. Combine and heat the eluted fractions (step 6) at 55°C for 5 min in order to disrupt the aggregation, cool sudden, and add solid NaCl to a concentration of 0.5 M.
9. Apply this solution to pre-equilibrated column (step 7) with the binding buffer.

10. Repeat steps 5 and 6. The eluted high A_{260} fractions are pooled and 0.5 M NaCl made by adding solid substance. The RNA is precipitated, after adding two volumes of ice-cold ethanol, at −20°C for at least 20 h.
11. The poly(A)-rich RNA (mRNA) is collected, dissolved, and estimated spectrophotometrically as described under procedure for "Isolation of RNA".

- All the glassware should be sterile (baked at 120°C overnight). The solutions should be sterilized by autoclaving at 20 lb/cm_2 for 20 min.
- The dissolution of RNA in the binding buffer is slow due to high salt concentration. On the other hand, RNA is first dissolved easily in the elution buffer and then made to 0.5 M NaCl by adding solid substance. After the elution of bound mRNA, the affinity material is charged by passing through 0.1 N NaOH and stored in the binding buffer containing 0.1% sodium azide at 0–4°C.

Precautions and Requirements for Molecular Work

1. Precaution from UV light
   UV light may be divided into three wavelength groupings near UV (315–400 nm), mid-range UV (280–315 nm), and far UV (200–280 nm). Maximal sensitivity in humans is at about 280 nm. Exposure to direct or indirect mid-range or far UV can cause acute eye irritation after a latent period of 2–24 h. Because retina is not sensitive to UV, eye damage may result without the subject being aware of the exposure. Skin is also sensitive to UV which may be a cause for skin cancer. Hence, protect your eyes and skin from the effects of UV irradiation by wearing goggles with side shields, by clothing, and by limiting exposure.
2. Safety during preparation of buffer saturated phenol
   Many preparations of commercially available liquefied phenol can be used without re-distillation. Some batches that are pink or yellow and all those which are crystalline should be redistilled before use.

   Distillation of phenol should be performed under a fume hood with good aspiration. An air condenser is used to condense the phenol vapour and not water or any refrigerant. Some quantity of zinc can be added to phenol while distilling. Commercial phenol contains an impurity which raises the melting point to 182°C. When the vapour temperature reaches 181°C, it begins to collect. 8-hydroxy quinoline is added to distilled phenol to a final concentration of 0.1%. Hydroxy quinoline is an antioxidant, a partial inhibitor or RNase and a weak chelator of metal ions. The yellow colour also provides an easy way of identifying the phase.

   The liquefied, distilled phenol is equilibrated with Tris buffer. Add an equal volume of 1.0 M Tris-Cl buffer (pH 8.0). Mix well and allow to separate the phase. Remove the aqueous phase and repeat extraction until pH of the aqueous phase is >7.6. Finally, extract with 0.1 M Tris-Cl (pH 8.0) containing 0.2% β-mercaptoethanol. The phenol is then stored in aliquots at 4°C under equilibration buffer.

Phenol is corrosive to skin, initially producing a white softened area, followed by severe burns. It is rapidly absorbed through the skin, and because of the local anaesthetic properties of phenol, skin burns may not be felt until there has been serious damage. Safety glasses and gloves should be worn while working with phenol. If any phenol is spilled on the skin, flush off immediately with large quantity of water. Do not use ethanol. If eyes are contaminated, wash them with running water for about 15 min and call for medical help.

3. Preparation of chloroform:isoamyl alcohol (24:1)
   A mixture of chloroform and isoamyl alcohol (24:1, v/v) is generally used to deproteinize nucleic acid preparations. Chloroform denatures the proteins and isoamyl alcohol helps to reduce foaming during extraction as well as helps to separate the aqueous and organic phases. They do not need any treatment before use.
4. Precautions while using radioisotopes
   (a) Appropriate clothing
   Whenever working at the lab bench, it is good safety practice to wear a lab coat for protection. Radiation safety badges gloves and protective eye wears should be used. it is better to wear two pair of gloves at once when using radioactivity. When outer pair gets contaminated, it can be discarded.
   (b) Protection of work area as well as workers
   Lab benches and the bases of shields, etc. should be covered with disposable absorbent paper sheets.
   (c) Use of appropriate designated equipment
   It is advisable to designate certain equipment like microfuge, shakers, water baths, pipettes, etc. for the exclusive use with radioactive samples.
   (d) Disposal of radioactive waste
   All the radioactive waste should be disposed at the designated place. They can be stored behind shields to decay prior to disposal.
   (e) Proper labelling
   It is always necessary to label the radioactive material with yellow hazard tape printed with international symbol for radioactivity which is commercially available in a variety of widths.
   (f) Monitoring of radioactivity
   Portable radiation detection monitors are essential equipment for every laboratory using radioactivity. Use a monitor to detect any spillage or contamination at work place or body parts especially fingers, equipment, etc.

^{32}P-labelled nucleotide is often used to label probes for Northern or Southern blotting and is about 50–100 μCi per reaction. However, handling even those small amounts can result in an unacceptable level of exposure if proper shielding is not employed (The β particles emitted by ^{32}P have an energies of 1.71 MeV – 6.1 m range). With no intervening shielding, the dose rate 1 cm away from 1 mCi ^{32}P is 200,000 m rads/h, the local dose rate to basal cells resulting from a skin contamination of 1 μCi/cm^2 is 92,000 m rad/h (Shlein 1987). Such a skin

Table 15.4 Spectrophotometric conversions for nucleic acids

$A_{260} = 1$ (1 cm, detection path)	Concentration (μg/ml water)
Double-stranded DNA	50
Single-stranded DNA	33
Single-stranded RNA	40
Oligonucleotide	20–30

Table 15.5 Antibiotic solutions

	Stock solution concentration	Storage (°C)	Working concentration (dilution)
Ampicillin (sodium salt)	50 mg/ml in H_2O	−20	100 μg/ml (1:500)
Chloramphenicol	34 mg/ml in ethanol	−20	170 μg/ml (1:200)
Kanamycin	10 mg/ml in H_2O	−20	50 μg/ml (1:200)
Streptomycin	10 mg/ml in H_2O	−20	50 μg/ml (1:200)
Tetracycline-Cl	5 mg/ml in ethanol	−20	50 μg/ml (1:100)

contamination could be easily got through careless handling, because the concentration of stock solution labelled nucleotide is usually 10 μCi/μl.

Plexi glass (3/4 in. thick) screen should be used besides lab coat, protective glass, gloves, etc. while using ^{32}P . When water bath is used for conducting specific hybridization experiments, etc., the water surrounding the tubes or hybridization bags will effectively stop β-radiation, but a shielding of a flat plexi glass should be kept over the tubes where there is no water. When hybridization is done in bags, care should be taken to monitor the apparatus used for sealing. All the waste material to be discarded should be kept properly behind the shield (Tables 15.4 and 15.5).

Molecular-Weight Conversions for Nucleic Acids

MW of a double-stranded DNA molecular (sodium salt)
= (number of base pairs) × (662 daltons/base pair).

MW of a single-stranded DNA molecular (sodium salt)
= (number of bases) × (331 daltons/base).

MW of a single-stranded RNA molecule (sodium salt)
= (number of bases) × (343 daltons/base).

MW of a DNA oligonucleotide (sodium salt, pH = 7):

$$MW = (N_A \times 335.2) + (N_C \times 311.2) + (N_G \times 351.2) + (N_T \times 326.2) + P.$$

Where N_X = number of residues of the respective nucleotide within the oligonucleotide (the MW listed for each nucleotide is the MW of that nucleotide, with associated sodium, incorporated in the oligonucleotide) (Tables 15.6–15.9)

For dephosphorylated oligonucleotides: $P = -84.0$
For phosphorylated oligonucleotides: $P = -40.0$

Table 15.6 Molar conversions for nucleic acids

1 μg	Pmol	Molecules
20 b oligonucleotide	1.52	9.1×10^{13}
1,000 bp DNA	1.52	9.1×10^{11}
pUC 19 DNA (2,686 bp)	0.57	3.4×10^{11}
pBR 322 DNA (4,363 bp)	0.35	2.1×10^{11}
Lambda DNA (48,502 bp)	0.03	1.8×10^{10}
Average mRNA (1930 bp)	1.67	1.0×10^{12}
1 pmol	**μg**	
20 b oligonucleotide	0.006	
1,000 bp DNA	0.66	
pUC 19 DNA (2,686 bp)	1.77	
pBR 322 DNA (4,363 bp)	2.88	
Lambda DNA (48,502 bp)	32.01	
Average mRNA (1,930 bp)	0.6	

Table 15.7 Molar conversions for protein

100 pmol	μg
10,000 Da protein	1
30,000 Da protein	3
100,000 Da protein	10

Protein/DNA conversions

1 kb of DNA encodes 333 amino acids = 3.7 × 104 Da

Protein	DNA
1,000 Da	270 bp
30,000 Da	810 bp
100,000 Da	2.7 kb

Table 15.8 Lambda DNA markers for gel electrophoresis (bp)

λ *Hind*III	λ *Hind*III-*Eco*RI	λ *Eco*RI	ΦX 174-*Hae*III
23,130	21,226	21,226	1,353
9,416	5,148	7,421	1,078
6,557	4,973	5,804	872
4,361	4,268	5,643	603
2,322	3,530	4,878	310
2,027	2.027	3,530	281
564	1,904		271
125	1,584		234
	1,375		194
	947		118
	831		72
	564		
	125		

Table 15.9 Sizes and molecular weights of various genomic DNAs

	Base pairs per haploid DNA	Molecular weight genome (daltons)
pBR 322	4,364	2.8×10^6
SV40	5,243	3.4×10^6
ΦX174	5,386	3.5×10^4
Adenovirus 2	35,937	2.3×10^7
Lambda	48,502	3.2×10^9
Escherichia coli	4.7×10^6	3.1×10^9
Saccharomyces cerevisiae	1.5×10^7	9.8×10^9
Dictyostelium discodieum	5.4×10^7	3.5×10^{10}
Aradibopsis thaliana	7.0×10^7	4.6×10^{10}
Caenorhabditis elegans	8.0×10^7	5.2×10^{10}
Drosophila melanogaster	1.4×10^8	9.1×10^{10}
Gallus domesticus (chicken)	1.2×10^9	7.8×10^{11}
Mus musculus (mouse)	2.7×10^9	1.8×10^{12}
Rottus norvegicus (rat)	3.0×10^9	2.0×10^{12}
Xenopus laevis	3.1×10^9	2.0×10^{12}
Homo sapiens	3.3×10^9	2.1×10^{12}
Zea mays	3.9×10^9	2.5×10^{12}
Nicotiana tabacum	4.8×10^9	3.1×10^{12}

Appendix A
Glossary

Water In biochemical experiments, it is advised to use distilled water for all purposes. So, water means distilled water in the methodologies given.

Atomic weight Atomic weight of an element is the relative weight of the atom on the basis of oxygen as 16.

Molecular weight The sum of the atomic weights of all the atoms in a molecule is known as its molecular weight.

Equivalent weight Equivalent weight of a substance is the number of grams of the substance required to react with, replace or furnish one mole of H_2O^+ or OH^-.

The equivalent weight of an acid is the weight that contains one atomic weight of acidic hydrogen, i.e. the hydrogen that reacts during neutralization of acid with base.

For example, the equivalent weight of H_2SO_4 is 49. Since H_2SO_4 contains two replaceable hydrogens, equivalent weight is molecular weight/2, i.e. $98/2 = 49$.

Percent solution (w/v) One per cent solution of a substance contains 1 g of the substance in 100 mL of solvent. If *v*/*v* is given, then it means 1 mL in 100 mL of solvent.

Molar solution (M) One molar solution of a substance contains one mole or 1 g molecular weight of the substance in 1 L of solution.

For example, 1M NaOH contains 40 g sodium hydroxide in 1 L solution.

Likewise, 1 mmol solution of a substance contains 1 mg molecular weight of the substance in 1 L of solution.

For example, 1 mM NaOH contains 40 mg of sodium hydroxide in 1 L solution.

Normal solution (N) One normal solution of a substance contains one equivalent or one gram equivalent weight of the substance in 1 L of solution (i.e. molecular weight is divided by the hydrogen equivalent of the substance)

For example, 1 N H_2SO_4 contains 49 g of H_2SO_4 in 1 L solution.

R. Katoch, *Analytical Techniques in Biochemistry and Molecular Biology*,
DOI 10.1007/978-1-4419-9785-2, © Springer Science+Business Media, LLC 2011

Buffer A solution containing both a weak acid and its conjugate weak base whose pH changes only slightly on the addition of acid or alkali.

pH pH is a value taken to represent the acidity or alkalinity of an aqueous solution. It is defined as logarithm of the reciprocal of the hydrogen ion concentration of the solution i.e., $pH = \log \frac{1}{[H^+]}$

Preparation of dilute acids While preparing dilute acids, add acid slowly to water under cooling conditions.

Preparation of alkali solutions While preparing concentrated alkali solutions (e.g. 40% NaOH) dissolve the alkali in distilled water under cooling.

Expression of concentration of solutions

Molarity (M): the number of moles of solute per litre of solution (the term molar is equivalent to mol 1^{-1}).

Molality (m): the number of moles of solute per kg of solvent.

% (*w*/*v*): the weight in grams of a solute per 100 mL of solution.

% (*w*/*w*): the weight in grams of a solute per 100 g of solution.

Units of length

$1 \text{ micron} = 1\mu = 1\ \mu m = 1 \times 10^{-6}\ m = 1 \times 10^{3}\ nm = 1 \times 10^{4}\ \text{Å}$

$1\ \text{Å} = 0.1\ nm = 1 \times 10^{-4}\ \mu m = 1 \times 10^{-10}\ m$

$1\ nm = 10\ \text{Å} = 1 \times 10^{-3}\ \mu m = 1 \times 10^{-9}\ m$

Conversion units of temperature

Temp in °F = (Temp. in °C × 1.8) + 32

Temp in °C = (Temp. in °F − 32) × 5/9

Temp in K = (Temp. in °F + 459.467) × 0.5/9

Appendix B
Abbreviations

Units	Abbreviations
Weights	
Microgram	μg
Milligram	mg
Gram	g
Kilogram	kg
Ounce	Oz
Pound	lb
Metric tonne	MT
Quintal	q
Measures	
Microlitre	μL
Millilitre	mL
Centilitre	cL
Decilitre	dL
Litre	L(1)
Ounce	Oz
Gallon	gal
Concentration	
Milli equivalent	mEq
Equivalent	Eq
Micromole	μmol
Millimole	mmol
Parts per million	ppm
Parts per billion	ppb

(continued)

(continued)

Units	Abbreviations
Length	
Inch	in.
Foot	ft.
Yard	yd
Micrometer	m
Millimeter (cubic centimeter)	mm (cc)
Centimeter	cm
Decimeter	dm
Meter	m
Kilometer	km
Angstroms	Å

Appendix C
Reference Tables

Concentration of some acids and bases

Acid or base	Molecular weight	Specific gravity	Percent by weight	Normality *N*	mL/L for 1 N solution
Acetic acid	60.1	1.05	99.5	17.6	56.9
Ammonium hydroxide	35.0	0.89	28.0	15.1	66.5
Formic acid	46.0	1.20	97.0	25.5	39.2
Hydrochloric acid	36.5	1.18	36.0	11.7	85.5
Nitric acid	63.0	1.42	69.5	15.6	64.0
Perchloric acid	100.5	1.69	70.0	11.6	85.7
Phosphoric acid	98.0	1.69	85.0	44.1	22.7
Sulphuric acid	98.08	1.84	96.0	35.9	28.4
Thioglycolic acid	92.1	1.26	80.0	10.9	91.3
Pyridine	79.1	0.98	100.0	12.4	80.6
2-mercaptoethanol	78.13	1.14	100.0	14.6	68.5

pK values and pH range of some commonly used biological buffers

Name	pKa_1	pKa_2	pKa_3	Useful pH range
Acetic acid	4.75	–	–	3.6–5.6
ADA (*N*-2-acetamidoiminodiacetic acid)	6.60	–	–	6.0–7.2
Bis–Tris propane	6.80	–	–	6.3–9.5
Bicine [*N*,*N*-bis (2-hydroxyethyl) glycine]	8.30	–	–	7.6–9.0
Boric acid	9.24	–	–	7.6–9.2
Citric acid	3.08	4.75	5.40	3.6–5.6
Glycine	2.35	9.78	–	
Glycine amide hydrochloride	8.20	–	–	7.4–8.8
Glycyl glycine	3.14	8.07	–	7.5–8.9
HEPES *N*-2-hydorxyethylpiperazine-*N*′-2-ethano sulfonic acid	7.55	8.00	–	6.8–8.2
Imidazole	7.07	–	–	6.5–7.9
MES [2-(*N*-morpholino)ethanesulfonic acid]	6.15	–	–	6.5–7.9
MOPS [3-(*N*-morpholino)propanesulfonic acid]	7.20	–	–	6.5–7.9
Phosphoric acid	1.96	7.12	12.32	5.8–7.8

(continued)

(continued)

Name	pKa_1	pKa_2	pKa_3	Useful pH range
TES [*N*-tris(hydroxymethyl)methyl-2-aminoethanesulfonic acid]	7.50	–	–	6.8–8.2
TRICINE [*N*-tris(hydroxymethyl)methyl glycine]	8.15	–	–	7.4–8.8
TRIS [tris-(hydroxymethyl)aminomethane]	8.30	–	–	7.2–9.0
Succinic acid	4.19	5.57	–	3.8–6.0

Common stains for electrophoresis

Stain specificity

Stain	Glycoprotein	Protein	Nucleic acids	Polysaccharides	Comments
Acridine orange			√		RNA specific
Amido black 10B		√			
8-Anilinonaphthalene-1-sulphonic acid		√			Fluorescent stain
Commassie brilliant blue G-250	√	√			Commonly used stain
Commassie brilliant blue R-250	√	√			Sensitive, durable stain
Commassie brilliant violet R-150		√			Weaker background stain with IEF* gels compared with Commassie brilliant blue
4′,6′-Diimidino-2-phenylindol			√		Fluorescent stain
Ethidium bromide			√		Most commonly used stain
Fast green FCF		√			Quantitative determination of protein by colour intensity possible
Methylene blue			√		Specific for RNA
Methyl green			√		
Panceau S		√			Particularly suitable for staining of albumins and globulins on cellulose acetate foil
Silver stain	√	√	√		
Stains all (E.Merck)	√	√	√	√	Universal stain
Sudan black		√			Special stain for lipoproteins

Characteristics of some commonly used ion-exchangers

Type	Matrices	Functional group	Name of functional group
Weak cation exchangers	Agarose		
	Cellulose	$-COO^-$	Carboxy
	Dextran	$-CH_2COO^-$	Carboxymethyl
	Polyacrylate		
Strong cation exchangers	Cellulose	$-SO_3^-$	Sulpho
	Dextran	$-CH_2SO_3^-$	Sulphomethyl
	Polystyrene	$-CH_2CH_2CH_2SO_3^-$	Sulphopropyl
Weak anion exchanger	Agarose		
	Cellulose	$-CH_2CH_2N^+H_3^-$	Aminoethyl
	Dextan	$-CH_2CH_2N^+H(CH_2CH_3)_2$	Diethylaminoethyl
	Polystyrene	$-CH_2N^+(CH_3)_3$	Trimethylaminomethyl
Strong anion exchangers	Cellulose	$-CH_2CH_2N^+(CH_2CH_3)_3$	Triethylaminoethyl
	Polystyrene	$-CH_2N^+(CH_3)_2$ CH_2CH_2OH	Dimethyl-2-hydroxyethyl aminoethyl
		$-CH_2CH_2N^+(CH_2CH_3)_2$ CH_2CHCH_3 OH	Diethyl-2-hydroxypropyl aminoethyl

References

A.O.A.C. (1984). *Official Methods of Analysis*. Association of Official Analytical Chemists. 13th edition. (Ed. Harwitz, W.), A.O.A.C., Washington, D.C.

A.O.A.C. (1995). Modification of tannin estimation by Folin-Denis reagent (Konard, D., Christensen, I.A.E.A., 1974) A.O.A.C., Washington, D.C.

Abaza, R.H., Blake, J.T. and Fisher, E.J. (1968). Oxalate determination: Analytical problems encountered with certain plant species. *J. Assoc. Off. Anal. Chem.* 51: 963–967.

A.O.A.C. (1962). Official Methods of American Association of Cereal Chemists. A.O.A.C., Washington, D.C., pp. 14–50.

A.O.A.C. (1965). *Official Methods of Analysis*. Association of Official Analytical Chemists. 10th Edition. A.O.A.C., Washington, D.C.

A.O.A.C. (1970). *Official Methods of Analysis*. Association of Official Analytical Chemists. 11th edition. A.O.A.C., Washington, D.C., p 815.

A.O.A.C. (1995). In: Natural Toxins (49th chapt.). Official Methods of Analysis of A.O.A.C. International, 16th Edn., (Vol. II), Virginia, USA, pp. 37–49.

Arnon, R. (1970). In: *Methods in Enzymol.* vol. 19 (Eds. Perlmann and Lorand, L.) Academic Press, New York, p 228.

Ashwell, G. (1957) In: *Methods in Enzymol.* vol. 3 (Eds. Colowick, S.P. and Kaplan, N.O.) Academic Press, New York, p 87.

Atkins, A., Pate, J.S. and Shelp, B.J. (1984). Effects of short-term N_2 deficiency on N metabolism in legume nodules. *Plant Physiol.* 76: 705.

Ausubel, F.M., Brent, R., Kingston, R.E., Moore, D.D., Seidman, J.G., Smith, J.A. and Struhl, K. (1987). Current protocols in Biochemistry. Wiley, New York.

Bantle, J.A., Maxwell, I.H. and Hahn, W.E. (1976). Specificity of oligo(dT)-cellulose chromatography in the isolation of polyadenylated RNA. *Anal. Biochem.* 72: 413–427.

Bergmeyer, H.U. and Bernt, E. (1984). In: *Methods of Enzymatic Anal.* vol. 2 (Ed. Bergmeyer) Academic press, New York, p 735.

Blanchard, J.S. (1984). Buffers for enzymes. In: *Methods in Enzymol.*, Vol. 104 Part C, (Ed. Jackoby, W.B.), pp 414–417, Academic Press, New York.

Bonner, W.M. and Laskey, R.A. (1974). A Film detection method for tritium-labelled proteins and nucleic acids in polyacrylamide gels. *Eur. J. Biochem.* 46: 83.

Brawerman, G. (1974) In: *Methods in Enzymol.* vol. 30 (Eds. Moldave, K. and Grossman, L.) Academic Press, New York, p 605.

Brezezinski, W. and Mendelewski, P. (1984). Determination of total glucosinolate content in rapeseed meal with thymol reagent. *Z Pflanzenzuchtg.* 93: 177–183.

Bruening, G., Criddle, R., Preiss, J. and Rubert, F. (1970). Biochemical experiments. New York: Wiley-Interscience.

Brueske, C.H. (1980). Phenylalanine ammonia-lyase activity in tomato roots infected and resistant to the root-knot nematode, *Meloidogyne incognita*. *Physiol. Plant. Pathol.* 16: 409.

Burton, K. (1956). A study of the conditions and mechanism of the diphenylamine reaction for the colorimetric estimation of deoxyribonucleic acid. *Biochem. J.* 62: 315.

Byrant, S.D. and Lane, F.E. (1979). Indole-3-acetic acid oxidase from peas. *Plant Physiol.* 63: 696.

Chinard, E.P. (1952). Photometric estimation of proline and ornithine. *J. Biol. Chem.* 199: 91.

Clegg, K.M. (1956). The application of anthrone reagent to the estimation of starch in cereals. *J. Sci. Food. Agric.* 7: 40–44.

Colowick, S.P. and Kaplan, N.O. (1955) In: *Methods in Enzymol.* New York, Academic press.

Conn, E.E. and Stumpf, P.K. (1987). Outlines of Biochemistry, 4th ed., pp. 3–24, Wiley Eastern Ltd., New Delhi.

Dawra, R.K., Makkar, H.P.S., Singh, B. (1988). Protein binding capacity of microquantities of tannins. *Anal. Biochem.* 170: 50–53.

De, K.K. and Roy, S.C. (1984). Role of all acid phosphatase isozyme in callus tissue during cytodifferentiation. *Theor. Appl. Genet.* 68: 285.

Delma, D. (1970) In: *Methods in Ezymol.* Vol 17, Part A (Eds. Tabor, H. and Tabor, C.W.), Academic Press, New York, p 850.

Dong, M.W. and Gant, J.R. (1989). Advances in fast reversed-phase chromatography of proteins. *BioChromatography*, 4 (1): 19–34.

Douce, R., Christenssen, E.L. and Bonner, W.D. (1972). Preparation of intact plant mitochondria. *Biochem. Biophys. Acta* 175: 148.

Duncombe, W.G. (1963). The colorimetric micro-determination of long chain fatty acids. *Biochem. J.* 88: 7–10.

Eggum, B.O. (1973). A study of certain factors influencing protein utilization in rats and pigs. *Beretning fra forsogstaborats rict, Kobenhvnv*, pp. 135–147.

Esterbaner, H., Schwarzl, E. and Hayn, M. (1977). A rapid assay for catechol oxidase and laccase using 2-nitro-5-thio benzoic acid. *Anal. Biochem.* 77: 486.

FAO Nutritional Studies No. 24 (1970) Amino Acid Content of Foods and Biological Data on Proteins. FAO, Rome.

Francis, K. and Gnanam, A. (1979). Phosphoenolpyruvate carboxylase assay on polyacrylamide gels. *Experientia* 35: 23.

Gabb, M.H. and Latchem, W.E. (1967). Handbook of Laboratory Solutions, Andre Deusch Ltd., London.

Gnanam, A. and Francis, K. (1976). Distribution of photosynthetic enzymes in the leaf cells types of *Sorghum vulgare*. *Plant Biochem. J.* 3:11.

Goering, H.D. and Van Soest, P.J. (1975) Forage Fibre Analysis. US Dept of Agriculture Research Service, Washington, DC.

Goering, H.K. and Van Soest, P.J. (1970). Forage Fibre Analysis Agricultural Handbook No. 379. USDA, Washington, D.C., pp 8–12.

Gomori, G. (1955). Preparation of buffers for use in enzyme studies. In: *Methods in Enzymol.*, Vol. 1 (Eds. Colowick, P.S. and Kaplan, N.O.), pp 138–146, Academic Press, New York.

Green, A.A. and Hughes W.L. (1955) Methods in Enzymology. Vol. 1 (Ed. Colowick, S.P. and Kaplan, N.O.) Academic Press New York, pp 67–90

Hageman, R.H. and Reed, A.J. (1980). In: *Methods in Enzymol.* Vol. 69 Part C (Ed. San Pietro, A.) Academic Press, New York, p 270.

Hinton, D.M. and Pressey, R. (1974). Cellulase activity in peaches during ripening. *J. Food Sci.* 39: 783.

Hobson, G.E. (1964). Polygalacturonase in normal and abnormal tomato fruit. *Biochem. J.* 92: 304.

Honold, G.R., Farkas, G.L. and Stahmann, M.A. (1966). A qualitative investigation of the dehydrogenases. *Cereal Chem.* 43: 517.

Horn, M.J., Jones, D.B. and Blum, A.E. (1946). Colorimetric determination of methionine in proteins and foods. *J. Biol. Chem.* 166: 313.

Hurn, B.A.L. and Chantler, S.M. (1980). Production of reagent antibodies. In: *Methods in Enzymol.* 70 (Eds. Van Vunakis, H. and Langone, J.J.) Academic Press, New York, p 104.

Inoue, K.H., Hangerman, A.E. (1988). Determination of gallotannin with rhodanine. *Anal. Biochem.* 169: 363–369.

Jackson, M.L. (1973). Iron, manganese, copper, zinc, molybdenum and cobalt determination. In: *Soil Chem. Anal.*, Prantice Hall of India Pvt. Ltd. New Delhi, p 388.

Jayaraman, K.S., Ramanuja, M.N., Vijayaraghavan, P.K. and Vaidyanathan, C.S. (1987). Oxidative enzyme in pearl millet. *Food Chem.* 24: 203.

Kakade, M.L., Rackie, J.J., McGhee, J.E. and Puski, G. (1974). Determination of trypsin inhibitor activity of soy products: A collaborative analysis of an improved procedure. *Cereal Chem.* 51: 376.

Kakade, M.L., Simons, N. and Liener, I.E. (1969). An evaluation of natural vs. synthetic substrates for measuring the antitryptic activity of soybean samples. *Cereal Chem.* 65 (5): 518–526.

Kartha, A.R.S. and Sethi, A.S. (1957). A cold percolation method for rapid gravimetric estimation of oil in small quantities of oil seeds. *Indian J. Agric. Sci.* 27: 211–217.

Kasturi, R. and Vasantharajan, V.N. (1976). Properties of acetylcholinesterase from *Pisum sativum*. *Phytochem.*15: 1345.

Knegt, E. and Bruinsma, J. (1973). A rapid, sensitive and accurate determination of indolyl-3-acetic acid. *Phytochemistry*. 12: 753.

Kruger, J.E. (1972). Changes in the amylases of hard red spring wheat during germination. *Cereal Chem.* 49: 379.

Kung, S.D., Chollet, R. and Marsho, T.V. (1980). In: *Methods in Enzymol.* 69 (Ed. San Pietro, A.) Academic Press, New York, p 326.

Laemmli, U.K. (1970). Cleavage of structural proteins during the assembly of the head of bacteriophage T_4. *Nature* 227: 680–685.

Landry, J. and Moureaux, T. (1970). Heterogeneite des glutelines du grain de mais. Extraction selective et composition and acids amines des trios fractions isolees. *Bull. Soc. Chi. Biol.* 52: 1021–1024.

Lowry, O.H., Rosebrough, N.J., Farr, A.L. and Randall, R.J. (1951). Protein measurement with the Folin Phenol Reagent. *J. Biol. Chem.* 193: 265.

Luck, H. (1974). In: *Methods in Enzymatic Anal. 2* (Ed. Bergmeyer) Academic Press, New York, p 885.

Makkar, H.P.S., Dawra, R.K., Singh, B. (1988). Determination of both tannin and protein in a tannin-protein complex. *J. Agric. Food Chem.* 36: 523–525.

Malik, C.P. and Singh, M.B. (1980) In: *Plant Enzymol. Histoenzymol.* Kalyani Publishers, New Delhi, p 53.

Maniatis, T., Fritsch, E.F. and Sambrook, J. (1982). Molecular Cloning-A Laboratory Manual. Cold Spring Harbour Laboratory, New York.

Marcus, A., Efron, D. and Weeks, D.P. (1974). The wheat embryo cell-free system. *Methods Enzymol.* 30: 749.

Marko, M.A., Chipperfield, R. and Birnboim, H.C. (1982). A procedure for the large scale isolation of highly purified plasmid DNA using alkaline extraction and binding to glass powder. *Anal. Biochem.* 121: 382.

Mc Carthy, T.E. and Paoille, M.M. (1959). A rapid determination of methionine in crude protein. *Biochem. Biophys. Res. Com.* 1: 29–31.

McGhee, J.E., Kirk, L.D. and Mustakas, G.G. 1965. Total glucosinolates method for *Brassica* species. *J. Assoc. Off. Anal. Chem.* 42: 889–891.

Mertz, E.T., Jambunathan, R. and Misra, P.S. (1975). In: *Protein Quality,* Stn. Bull No. 70, Agric. Experiment Stn. Purdue University, USA, p 14.

Michael Eskin, M.A., Hochn, E. and Frenkel, C. (1978). A simple and rapid quantitative method for total phenols. *J. Agric. Food Chem.* 26: 973.

Misra, B.K. and Gupta, R.K. (1995). Protocols for evaluation of wheat quality. Technical Bulletin No. 3, Directorate of Wheat Resarch, Karnal (Haryana).

Misra, P.S., Mertz, E.T. and Glover, D.V. (1975). Studies on corn protein. Free amino acid content of opaque-2 and double mutants. *Cereal Chem.* 52: 844.

Moore, S. and Stein, W.H. (1948) In: *Methods in Enzymol. (*Eds. Colowick, S.P. and Kaplan, N.D.) Academic Press, New York 3, 468.

Morrison, W.R. and Smith, L.M. (1964). Preparation of methyl esters. *J. Lipid Res.* 5: 600–604.

Mullis, K., Faloona, F., Scharf, S., Saiki, R., Horn, G. and Erlich, H. (1986). Specific enzymatic amplification of DNA in vitro. The polymerase chain reaction. Cold Spring Harbor Symp. *Quant. Biol.* 51: 263–273.

Murmur, J. (1961). A procedure for the isolation of deoxyribonucleic acid from micro-organisms. *J. Mol. Biol.* 3: 208.

Murray, M.G. and Thompson, W.F. (1980). Rapid isolation of high molecular weight plant DNA. *Nucl. Acid Res.* 8: 4321.

Nash, D.T. and Davies, M.E. (1975). Isoenzyme changes during the growth cycle of Paul's Scarlet rose cell suspension. *Phytochemistry.* 14: 2113.

Nelson, N. (1994). A photometer adoption of the Somogyi's method for determination of glucose. *J. Biol. Chem.* 153: 375–380.

Oberbacher, M.F. and Vines, H.M. (1963). Spectrophotometric assay of ascorbic acid oxidase. *Nature* 197: 1203.

Olsnes, S., Saltvedt, E. and Alexander, P. (1974). Isolation and composition of galactose binding lectins from *Abrus precatorius* and *Ricinus communis. J. Biol. Chem.* 249: 803.

Ouchterlony, O. (1968). In: *Hand-book of Immunodiffusin and Immunoelectrophoresis.* An Arbor Science, Michigan.

Palumbi, S.R. (1996). The Polymerase Chain Reaction Nucleic Acids. In: *Molecular Systematics* (IInd Edn.) (Eds. Hillis, D., Moritz, C. and Mable, B.) Sinauer, USA.

Pateman, J.A. (1969). Regulation of synthesis of glutamate dehydrogenase and glutamine synthetase in micro-organisms. *Biochem. J* . 115: 769.

Pellet, P.L. and Young, V.R. (1980). In: *Nutritional Evaluation of Protein Foods.* Tokyo, UN University, *Food. Nutr. Bull. Suppl.* p 4.

Plummer D.T. (1971). An introduction of practical Biochemistry, pp 228. Tata McGraw-Hill, New Delhi.

Pressey, P. and Avants, J.K. (1976). Separation and characterization of endopolygalacturonase and polygalacturonases. *Phytochemistry.* 15: 1349.

Ranganna, S. (1986). Handbook of analysis and quality control for fruit and vegetable products. Tata McGraw Hill, New Delhi, pp 84–88.

Rao, S.L.N. (1978). A sensitive and specific colorimetric method for the determination of α, β-diaminopropionic acid and the *Lathyrus sativus* neurotoxin. *Anal. Biochem.* 86: 386–395.

Reddy, M.M. and Gasber, E.O. (1971). Genetic studies of variant enzyme III. Comparative electrophoretic studies of esterases and peroxidases for species, hybrids and amphidiploids in the genus *Nicotiana. Bot. Gaz.* 132: 158–166.

Reed, K.C. and Mann, D.A. (1985). Rapid transfer of DNA from agarose gels to nylon membranes. *Nucl. Acids Res.* 13: 7207–7221.

Roberts, B.E. and Paterson, B.M. (1973). Efficient translation of tobacco mosaic virus RNA and rabbit globin 9S RNA in a cell-free system from commercial wheat germ. *Proc. Natl. Acad. Sci. USA.* 70(8): 2330–2334.

Roy, D.N. and Rao, P.S. (1971). Evidence, isolation, purification and some properties of a trypsin inhibitor in *Lathyrus sativus. J. Agric. Food Chem.* 19: 257–259.

Sambrook, J., Fritsch E.F. and Maniatis, T. 1989. Molecular cloning: A laboratory Manual, 3rd volume. Cold Spring Harbour Laboratory Press, New York.

Sardar, R. and Joseph, R. (1992). Purification and properties of lipase from anaerobe propionibacterium acidi-propionicai. *JAOCS* 69(10): 974–977.

Satterlee, L.D., Marshall, H.F. and Tennyson, J.M. (1979). Measuring protein quality. *J. Am. Oil Chemists Soc.* 56: 103–109.

Saunders, R.M. and Kohler, G.O. (1972). In vitro determination of protein digestibility in wheat mill feeds for monogastric animals. *Cereal Chem.* 49: 98–103.

Segel, I.H. (1976). Biochemical Calculations, 2nd ed., Wiley, New York.

Sharma, N.D., Mehta, S.L., Patil, S.H. and Eggum, B.O. (1981). Oil and protein quality of groundnut mutants. *Qualitas Plantarum* 31: 1–4.

Shastry, B.S. and Rao, M.R. (1975). Studies on lipoxygenase from rice bran. *Cereal Chem.* 52 (5): 597–603.

Shekhar, B.P.S. and Reddy, G.M. (1982). Studies on lipoxygenase from rice. *Cereal Chem.* 33: 1160–1163.

Shlein, B. (ed). (1987). Radiation safety manual for users of radioisotopes in research and academic institutions. *Nucleon Lactern associates*, Olney, Md.

Smith, B.J., Toogood, C.I.A. and Johns, E.W. (1980). Quantitative staining of submicrogram amounts of histone and high-mobility group proteins on sodium dodecylsulphate-polyacrylamide gels. *J. Chromatog.* 200: 200–205.

Smith, H.H., Hamill, D.E., Weaver, E.A. and Thompson, K.H. (1970). Multiple molecular forms of peroxidases and esterases among *Nicotiana* species and amphiploids. *J. Hered.* 61(5): 203–212.

Smithies, O. (1955). Zone electrophoresis in starch gels: group variations in the serum proteins of normal human adults. *Biochem. J* . 61: 629.

Southern, E. (1979) In: *Methods in Enzymol.* 58 (Ed. Wee, G.) Academic Press, New York, p 152.

Spies, J.R. and Chambers, D.C. (1994). Chemical determination of tryptophan in proteins. *Anal. Chem.* 21: 1249–1252.

Stoll, V.S. and Blanchard, J.S. (1990). Buffers: Principles and Practice. In: *Methods in Enzymol.*, Vol. 182, pp 24–38 (Ed. Deutscher, M.P.), Academic Press, NewYork.

Swain, T. and Hills, W.E. (1959). The phenolic constituents of *Prumus domestica* L. The quantitative analysis of phenolic constituents. *J. Sci. Food. Agric.*10: 63.

Tempest, D.W., Meers, J.L. and Brown, C.M. (1970). Synthesis of glutamate in *Aerobacter aerogenes* by a hitherto unknown route. *Biochem. J.* 117: 405.

Tilly, J.M.A. and Terry, R. A. (1963). A two-stage technique for the in vitro digestion of forage crops. *J. Br. Grassl. Soc.* 18 : 104–111.

Towbin, H., Staehelin, T. and Gordon, J. (1979). Electrophoretic transfer of proteins from polyacrylamide gels to nitrocellulose sheets: procedure and some applications. *Proc. Natl. Acad. Sci. USA* 76(9): 4350–4354.

Tsai, C.Y., Hansel, L.W. and Nelson, O.E. (1972). A colorimetric method of screening maize seeds for lysine content. *Cereal Chem.* 49: 572–579.

Van Soest, P.J. (1963). Use of detergents in the analysis of fibrous feeds. A rapid method for the determination of fibre lignin. *J. Assoc. Off. Agric. Chem.* 46: 829–835.

Van Soest, P. J. (1967). Development of a comprehensive system of feed analysis and its application to forages. *J. Anim. Sci.* 26: 119–128.

Van Soest, P.J. and Wine, R.H. (1968). Determination of lignin and cellulose in acid- detergent fibre with permanganate. *J. Assoc. Off. Agric. Chem.* 51: 780–785.

Van soest, P.J. (1963). Use of detergents in the analysis of fibrous feeds I. Preparation of fibre residues of low nitrogen. *J. Assoc. Off. Agric. Chem.* 46 (5): 825–829.

Vega, J.M., Jacobo, C. and Manuel, L. (1980) In: *Methods in Enzymol.* (Ed. San Pietro, A.) Academic Press, New York 69, p 255.

Walker, D.A. (1980) In: *Methods in Enzymol.* 69 (*Eds* Colowick, S.P. and Kaplan, N.O.) Academic Press, New York, p 94.

Wheeler, E.L. and Ferrel, R.E. (1971). A method for phytic acid determination in wheat and wheat fractions. *Cereal Chem.* 48: 312.

Williams, P.C., Kuzina, F.D. and Hlynka, I. (1970). A rapid colorimetric procedure for estimating the amylase content of starches and flours. *Cereal Chem.* 47: 411–420.

Wim Gaastra (1984) In: *Methods in Mol. Bio.* Vol. 1 *Proteins* (Ed. Walker, J.M.) Humana, New Jersey, p 349.

Wiseman, H.G. and Jacobson, W.C. (1965). Nitrate in plant material, determination of nitrate in silages and forages. *J. Agric. Food Chem.* 13: 36–39.

Wrigley, C.W. (1968). Analytical fractionation of plant and animal protein by gel electrofocusing. *J. Chromatogr.* 36: 362.

Index